NMR of Newly Accessible Nuclei
Volume 2

CHEMICALLY AND BIOCHEMICALLY IMPORTANT ELEMENTS

NMR of Newly Accessible Nuclei
Volume 2

CHEMICALLY AND BIOCHEMICALLY IMPORTANT ELEMENTS

Edited by
PIERRE LASZLO

Institut de Chimie
Université de Liège
Liège, Belgium

1983

ACADEMIC PRESS

A Subsidiary of Harcourt Brace Jovanovich, Publishers
New York   London
Paris   San Diego   San Francisco   São Paulo   Sydney   Tokyo   Toronto

7227-0500

CHEMISTRY

ACADEMIC PRESS, INC.
111 Fifth Avenue, New York, New York 10003

United Kingdom Edition published by
ACADEMIC PRESS, INC. (LONDON) LTD.
24/28 Oval Road, London NW1 7DX

Library of Congress Cataloging in Publication Data

Main entry under title:

NMR of newly accessible nuclei.

Includes bibliographies and indexes.
Contents: v. 1. Chemical and biochemical applications
-- v. 2. Chemically and biochemically important elements.
1. Nuclear magnetic resonance spectroscopy. I. Laszlo,
Pierre. II. Title: N.M.R. of newly accessible nuclei.
QD96.N8N58 1983 543'.0877 83-4619
ISBN 0-12-437102-7 (v. 2)

PRINTED IN THE UNITED STATES OF AMERICA

83 84 85 86   9 8 7 6 5 4 3 2 1

Contents

1. Deuterium NMR

Ian C. P. Smith

2. Tritium NMR

A. L. Odell

14. Thallium NMR Spectroscopy

J. F. Hinton and K. R. Metz

15. NMR of Less Common Nuclei

P. Granger

Contributors

Numbers in parentheses indicate the pages on which the authors' contributions begin.

Ian M. Armitage (337), Department of Molecular Biophysics and Biochemistry, Yale University, New Haven, Connecticut 06510

Yvan Boulanger[1] (337), Department of Molecular Biophysics and Biochemistry, Yale University, New Haven, Connecticut 06510

Brian Coleman (197), Koninklijke/Shell-Laboratorium, (Shell Research B.V.), Amsterdam, The Netherlands

J. J. Delpuech (153), Laboratoire de Chimie Physique Organique, Université de Nancy I, F-54506 Vandoeuvre-les-Nancy, France

Christian Detellier (105), Department of Chemistry, University of Ottawa, Ottawa, Ontario K1N 9B4, Canada

Torbjörn Drakenberg (229), Physical Chemistry 2, University of Lund, S-220 07 Lund 7, Sweden

Sture Forsén (229), Physical Chemistry 2, Lund University, S-220 07 Lund, Sweden

P. Granger (386), IUT de Rouen, RMN des noyaux peu courants, F-76130 Mont-Saint Aignan, France

P. Mark Henrichs (319), Chemistry Division, Research Laboratories, Eastman Kodak Company, Rochester, New York 14650

J. F. Hinton (367), Department of Chemistry, University of Arkansas, Fayetteville, Arkansas 72701

R. Garth Kidd (50), Chemistry Department, University of Western Ontario, London, Ontario N6A 5B7, Canada

Jean-Pierre Kintzinger (79), Institut de Chimie, Universite Louis Pasteur, Strasbourg, France

---

[1] Present address: Institut de génie biomédical, Université de Montréal, Montréal, Québec H3C 3T8, Canada.

Pierre Laszlo (254), Institut de Chimie, Université de Liège, Sart-Tilman, par 4000 Liège, Belgium

B. E. Mann (301), Department of Chemistry, University of Sheffield, Sheffield S3 7HF, England

H. C. E. McFarlane (275), Department of Chemistry, City of London Polytechnic, London EC3N 2EY, England

W. McFarlane (275), Department of Chemistry, City of London Polytechnic, London EC3N 2EY, England

K. R. Metz[2] (367), Department of Chemistry, University of Arkansas, Fayetteville, Arkansas 72701

Klaus J. Neurohr[3] (229), Physical Chemistry, The Lund Institute of Technology, S-220 07 Lund 7, Sweden

A. L. Odell (27), Urey Radiochemistry Laboratory, Department of Chemistry, The University of Auckland, Auckland, New Zealand

Ian C. P. Smith (1), Division of Biological Sciences, National Research Council, Ottawa, Ontario K1A 0R6, Canada

[2] Present address: Department of Radiology, The Milton S. Hershey Medical Center, The Pennsylvania State University, Hershey, Pennsylvania 17033.

[3] Present address: Department of Molecular Biophysics and Biochemistry, Yale University, New Haven, Connecticut 06511.

General Preface

Books must follow sciences, and not sciences books.
Francis Bacon

In heeding this recommendation from the author of "The Advancement of Learning," we capitalize upon the spectacular progress of nuclear magnetic resonance (NMR) in recent years. Not only have NMR methods reached beyond chemistry and biochemistry to fertilize biomedical research, but more recently even radiologists and clinicians have started using NMR imaging techniques to diagnose and to monitor treatment. Chemistry and biochemistry have also benefited from recent advances, which have been numerous, impressive, and very useful. The introduction of Fourier transform and pulse excitation techniques has had all the features of a genuine mutation. As a consequence we have witnessed a steady and impressive gain in spectrometer sensitivity, together with a delightful improvement in ease of operation: one can now "dial" nearly any nucleus and communicate with it in a manner of minutes, in contrast to the changes of probe configuration that required hours in the 1960s and 1970s.

Beyond such quantitative changes, there are also qualitative changes, especially the availability of literally dozens of sophisticated and extremely useful pulse sequences (whose acronyms would require separate booklets to define and explain their workings). As a consequence, NMR spectroscopy has acquired new dimensions. Two-dimensional NMR is now (almost) routine, and some ancient prohibitions have fallen by the wayside. High-resolution NMR of solids, multiple quantum NMR, and many other very sophisticated modes of spectral acquisition have been perfected through the pioneering work of the likes of Richard Ernst, Ray Freeman, and John Waugh, to whom I wish to pay homage here.

This work, in two volumes, focuses on the "newer" nuclei, those that only a few years ago were not readily accessible with continuous wave excitation. For this reason, they were then nicknamed "exotic nuclei" or "other nuclei," to set them apart from the more familiar nuclei: ^{1}H, ^{13}C, ^{15}N, ^{19}F, and ^{31}P. Even though this distinction between the familiar and the exotic is already outdated and on the wane, we have set it as our goal to provide a state-of-the-art report on the newer nuclei. In so doing, we avoid duplicating the excellent existing monographs for the familiar nuclei.

Another pitfall we wished to avoid was an attempt at presenting an all-embracing work on the chemical applications of NMR. It was possible for the Emsley, Feeney, and Sutcliffe volumes to achieve such comprehensiveness in the 1960s, but today this task would require thousands of pages. Hence, we have elected to emphasize methods of study ("Methods are the habits of the mind and the savings of the memory"—Rivarol). There are two ways to present these methods: the problem-oriented approach and the technique-oriented approach. Accordingly, our first volume is devoted to general principles of magnetic resonance relative to the new nuclei and to special applications selected for their importance and timeliness. The second volume is a systematic survey, consisting of concise, albeit rather comprehensive treatments, of some of the most important nuclei and families of nuclei in the periodic table. Such a two-pronged treatment should help provide in-depth coverage. This is a novel approach, commending itself, we feel, by its simplicity and appropriateness. In this manner the reader will be able to find information more quickly and effectively.

A central feature of the work is worthy of special comment. The book lays special emphasis on the exploitation of relaxation processes, both as a new dimension of NMR that has come to the fore in the 1960s and 1970s and as a source of all-important parameters for studying the thermodynamics and kinetics of binding. Thus the "three dimensions" of a spectral line: frequency, scalar couplings, and relaxation rates are treated not as discrete entities but as inseparable elements in a single informational continuum.

We hope that these two volumes will serve as handbooks to be found in every NMR laboratory. But their natural readership, besides the NMR experts who we hope will find these volumes a useful summary, will consist of chemists and biochemists wishing to get initiated into the magnetic resonance methods at the graduate level. Indeed, Volume 1 might serve as the text for a one-semester introductory course in NMR aimed at first-year graduate students, with Volume 2 as the companion reference book. Lecturers in the chemistry and biochemistry departments of schools of medicine and pharmacy will find the books appropriate for such courses as self-contained texts. The two-volume format of presentation will also be an aid in the "self-teaching" of the new NMR.

Preface to Volume 2

This second volume of "NMR of Newly Accessible Nuclei" gathers introductory overviews and research reviews about some very important nuclides, written by some of the world's experts, who have contributed considerably both toward elucidating the characteristic properties of these nuclei and toward using them as probes in the collection of chemical information. These individual chapters update the material presented just a few years ago in the excellent book edited by Harris and Mann, "NMR and the Periodic Table." Indeed, in the meantime there have been exciting breakthroughs and a host of novel applications. Deuterium NMR has become *the* method for accurate determination of local microdynamics and for probing membranes (I. C. P. Smith). Tritons compete with deuterons as isotopic labels to elucidate details of reaction mechanisms, and tritium NMR has come of age, to supplement ^{1}H and ^{2}H NMR (A. Odell).

Use of other nuclei is only starting, and their full potential has yet to be grasped: I would put oxygen-17 (J. P. Kintzinger) and silver-109 (M. Henrichs) in this category. In particular, because oxygen is such a frequent component in molecules, oxygen NMR will continue to develop considerably. After the first applications, confined mostly to organic molecules, ^{17}O NMR is being used more and more in biochemical problems and for elucidating quite complex inorganic and organometallic structures.

Rhodium-103 is already a classic and has given beautiful solutions to problems of structure in organometallic chemistry (B. E. Mann). Boron-11 is a parallel case, for it is a nucleus that has been studied by NMR spectroscopists since the beginning, with unique value in the structural characterization of the polyhedral boranes, for example (R. G. Kidd).

Aluminum-27 and silicon-29 share a common future, because they can be fruitfully applied to structural determination of very many minerals, such as clays and zeolites, which are increasingly used as reaction catalysts in the chemical industry (J. J. Delpuech; B. Coleman).

Cobalt-59 is a very special nucleus, a record-holder, with its huge chemical shift range, its large temperature coefficients, and its sensitivity to minor perturbations of local symmetry. Some of these properties can be exploited as valuable assets by the design of appropriate cobalt(III) complexes. Cobalt-59 NMR also offers, as it were, a compendium in physical chemistry nicely illustrating a variety of textbook concepts from ligand field theory, for instance (P. Laszlo).

Cadmium-113 and thallium-205 NMR are also important, particulalry because both ions, Cd(II) and Tl(I) are very useful replacement probes in biomolecules for Ca(II) and K(I), both of which suffer from relatively poor magnetic properties (I. M. Armitage and Y. Boulanger; J. F. Hinton and K. R. Metz).

Increasingly, such nuclei as ^{25}Mg and ^{43}Ca are being studied directly, despite the technical drawbacks. Their importance derives from their high abundance on the crust of this planet, as well as their literally vital significance for biology. Alkali metals also conform to such a description. The current intensive use and studies of phase-transfer catalysis, ionic transport through membranes, and anionic activation make the details of alkali metal interactions with crown ethers and cryptands important. They can be obtained from the NMR spectra of the corresponding nuclei (T. Drakenberg, K. Neurohr, and S. Forsén; C. Detellier).

Selenium-77 and tellurium-125 NMR reflect the recent importance of these elements in synthetic organic chemistry, biochemistry, and pharmacology, and even to some extent in toxicology and cancerology (H. C. E. and W. McFarlane). Pierre Granger has put his enormous experience behind the writing of the final chapter, summarizing properties of other less familiar nuclei.

Each of my co-authors has produced his promised chapters. Let me add my thanks to these colleagues for having written lucid, authoritative, concise reviews of fields they have done so much to render fertile.

Contents of Volume 1

1 Deuterium NMR

Ian C. P. Smith

Division of Biological Sciences
National Research Council
Ottawa, Ontario, Canada

I. Fundamental Properties of Deuterium

For many years deuterium was a "much maligned magneton"; it was continually cited as being of little use to chemists because of its unfavorable fundamental properties. Later this was shown to be quite incorrect in a landmark article by Diehl and Leipert (1964). Progress was in part restricted by the limited availability of instrumentation capable of ^{2}H observation. The final blow came when ^{2}H was chosen for the heteronuclear field frequency lock on the next generation of spectrometers—apparently the instrument manufacturers had all read the same books. Despite these unfavorable circumstances, ^{2}H NMR grew rapidly during the 1970s with applications in chemistry, physics, and biology (Mantsch *et al.*, 1977). Several reviews cover a variety of other applications (Brevard and Kintzinger, 1978; Jarrell and Smith, 1983a,b; Smith and Mantsch, 1982).

TABLE I

Properties of ^{2}H

Spin	1
Quadrupole moment (m²)	2.73×10^{-31}
Frequency, 4.7 T (MHz)	30.7
Sensitivity referred to ^{13}C[a]	0.6
Natural abundance (%)	0.15
Receptivity referred to ^{13}C[a]	8.21×10^{-3}

[a] Value for ^{13}C set equal to 1.

The nuclear properties of ^{2}H are summarized in Table I. The spin of 1 implies a quadrupole moment, albeit relatively small (second smallest after ^{6}Li). This was one of the first criticisms of the nucleus—that resonances would be severely broadened by quadrupole relaxation, thus lowering spectral resolution. Although it is true that the resonances of ^{2}H are broader than those of ^{1}H or ^{3}H, this is rarely a limitation in the study of small molecules (molecular weight <500) (Fig. 1). The dominance of ^{2}H relaxation by the relatively simple quadrupolar mechanism facilitates the analysis of T_1 values and provides a means for monitoring the binding of small molecules to large ones (see the following).

The magnetogyric ratio of ^{2}H is considerably smaller than that of ^{1}H or ^{3}H. This means that its detection sensitivity is much lower (Table I) and that the separation between resonances of different chemical shift is less in frequency units (but not in parts per million). These are both disadvantages from the chemist's point of view, although the former has become of little impediment with modern high-sensitivity spectrometers and the latter becomes less of a problem at high fields (in a 400-MHz spectrometer for protons, deuterium resonates at 61.4 MHz—the proton frequency of the 1960s). Deuterium NMR has now become so popular that a comprehensive review of its applications, such as my colleagues and I attempted in 1977 (Mantsch et al., 1977), is now impossible. I shall therefore only stress the special features of ^{2}H NMR, citing several examples.

II. Chemical Shifts

In parts per million the chemical shifts of ^{1}H, ^{2}H, and ^{3}H are essentially identical. Differences exist only in the separations of the resonances, in frequency units, which decrease in the order of the magnetogyric ratios: ^{3}H > ^{1}H > ^{2}H. Figure 1 demonstrates the separation of resonances that

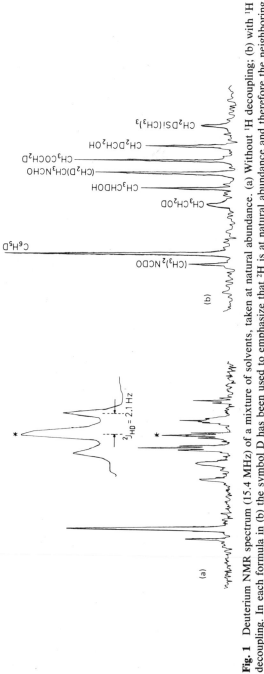

Fig. 1 Deuterium NMR spectrum (15.4 MHz) of a mixture of solvents, taken at natural abundance. (a) Without ^{1}H decoupling; (b) with ^{1}H decoupling. In each formula in (b) the symbol D has been used to emphasize that ^{2}H is at natural abundance and therefore the neighboring hydrogens are 99.985% ^{1}H. (After Smith and Mantsch, 1982. Copyright 1982 American Chemical Society.)

Labels in (b): $(CH_3)_2NCDO$; C_6H_5D; CH_3CH_2OD; CH_3CHDOH; $(CH_2D)(CH_3)NCHO$; CH_3COCH_2D; CH_2DCH_2OH; $CH_2DSi(CH_3)_3$

$^2J_{HD} = 2.1$ Hz

can be accomplished on a low-field (23-kG) spectrometer. Note that these spectra were taken at a natural abundance of deuterium. Discrimination between peaks is aided by proton decoupling. The resulting simple spectrum corresponds, in the positions of the resonances (in parts per million), to a *proton NMR spectrum with complete proton decoupling,* which can aid greatly in the analysis of complex proton NMR spectra where both chemical shifts and homonuclear coupling constants are required.

Figure 2 demonstrates the utility of ^{2}H NMR in the assignment of ^{1}H-NMR spectra. The ^{1}H pyrrolidine resonances of nicotine yield a very complex manifold which defies analysis (top spectrum). The ^{2}H spectra yield unambiguous values for the chemical shifts, especially for the various endo and exo hydrogens, leading to a total analysis of the ^{1}H-NMR spectrum and a conformational model (Pitner *et al.,* 1978).

A widespread application of ^{2}H NMR is in the study of reaction mechanisms, both chemical (Burger *et al.,* 1980; Casadevall and Metzger, 1970; De Puy *et al.,* 1974; Edwards *et al.,* 1981; Habich *et al.,* 1965; Johnson *et al.,* 1975; Montgomery *et al.,* 1967) and biochemical (Abell and Staunton, 1982; Ackland *et al.,* 1982; Battersby *et al.,* 1982; Cane and Nachbar, 1980; Kakinuma *et al.,* 1981; Kitching *et al.,* 1981; Mantsch *et al.,* 1977; Richards and Spenser, 1978; Sato *et al.,* 1976). As an example let us consider a recent study by Wigle *et al.,* (1982) on the biosynthesis of nicotine (**II**) from (R)-[1-^{2}H]putrescine (**I**). The reaction proceeds by

I   II

methylation of either of the two amino groups of putrescine (**I**), oxidative deamination, and coupling with a pyridinyl moiety. The question of interest was whether oxidation of (**I**) proceeded with stereospecific loss of the pro-R hydrogen atom, which would yield (**II**) labeled only at the 5′-pro-R position, or of the pro-S hydrogen, which would yield II labeled in both the 2′- and 5′-pro-R positions. Figure 3 shows the 61.4-MHz ^{2}H-NMR spectrum of the product, which clearly contains two resonances of equal intensity, corresponding to stereospecific loss of the pro-S proton of the —CH_2—NH_2 group of putrescine on conversion to nicotine.

In cases where ^{2}H resonances occur too close together for adequate discrimination, paramagnetic shift reagents have been used with considerable success (Cheng *et al.,* 1977; Nordlander and Haky, 1980). An extra advantage with ^{2}H is its lower magnetogyric ratio, which results in a

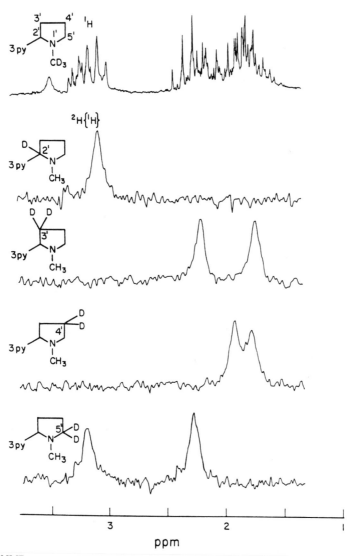

Fig. 2 NMR spectra of the pyrrolidine hydrogens of nicotine (in CDCl₃–CFCl₃, 2 : 1). (Top spectrum) ¹H at 100 MHz; (lower spectra) ²H at 15.4 MHz (with proton decoupling) of various deuterated isotopmers. (After Pitner *et al.*, 1978. Copyright 1978 American Chemical Society.)

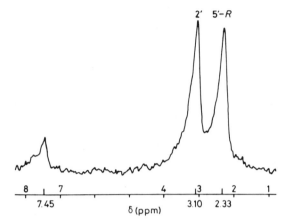

Fig. 3 Deuterium NMR spectrum (61.42 MHz) of nicotine (in CHCl₃) isolated from tobacco plants to which (R)-[1-²H]putrescine had been administered. The resonance at 7.45 ppm is due to ²H at natural abundance in CHCl₃. (After Wigle *et al.*, 1982.)

substantially smaller contribution to the NMR line width from nucleus–electron interactions. This advantage was pointed out recently by Wheeler *et al.* (1982) in a study on liganded Cr(III). Figure 4 compares the ²H- and ¹H-NMR spectra of malonate and 2,5-diazahexanedionate ligands

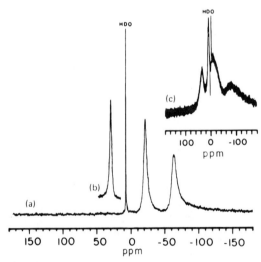

Fig. 4 Deuterium NMR spectra (31 MHz). (a) [Cr(edda-α-d_4)(mal)]⁻; (b) [Cr(edda)(mal-d_2)]⁻; (c) ¹H-NMR spectrum (200 MHz) [Cr(edda)(mal)]⁻. Abbreviations used: edda, Diazahexanedionate; mal, malonate. (After Wheeler *et al.*, 1982. Copyright 1982 American Chemical Society.)

of Cr(III). The superior resolution in the ^{2}H case is striking. Similar studies on a variety of Cr(III) complexes led to conclusions regarding their symmetry and conformations that could not have been obtained from the corresponding ^{1}H-NMR spectra.

III. Spin–Spin Coupling Constants

Deuterium interacts with other nuclei in a manner totally analogous to ^{1}H, but in all cases the magnitude of the interaction is scaled down because of the lower magnetogyric ratio of ^{2}H. For example, ^{1}H—^{2}H coupling constants range up to 3 Hz instead of 20 Hz, and ^{2}H—^{2}H coupling constants are only 2.3% of those of their ^{1}H—^{1}H analogs and therefore are usually lost in the line widths. Some excellent examples are seen in the spectrum on the left side of Fig. 1. The geminal coupling of 2.1 Hz in the methyl group of acetone is clearly visible, whereas the vicinal coupling between the methyl and methylene groups of ethanol is barely visible on the methyl resonance and lost in the width of the methylene resonance. Note that in these examples coupling between nuclei that are magnetically equivalent in the ^{1}H-NMR experiment (such as the methyl protons of acetone) can be observed in the ^{2}H-NMR spectra of partially deuterated groups or at natural abundance; in this case the ^{2}H experiment yields information that is unavailable in principle from ^{1}H-NMR spectra.

Although coupling constants to ^{2}H are not used for diagnostic purposes as often as they are for ^{1}H NMR, they do have utility. An example is the study by Edwards *et al.* (1981) on the sites of incorporation of ^{2}H in nortricyclanone.

IV. Quadrupole Coupling Constants

As mentioned previously, deuterium has a relatively small quadrupole moment. Its manifestations in the ^{2}H-NMR spectra depend on the product of the quadrupole moment and the electrical field gradient at the ^{2}H nucleus—the quadrupole coupling constant e^2qQ/h, where e is the charge on the electron, q the second derivative of the electric potential at the nucleus, and Q the quadrupole moment. Thus, depending on q, the full or a negligible effect of the quadrupole moment may be manifest. This depends largely on the symmetry of the environment of the deuteron; the higher the symmetry, the less the effect of the quadrupole moment. Several tables of quadrupole coupling constants for ^{2}H are available (Brevard and Kintzinger, 1978; Mantsch *et al.*, 1977; Millett and Dailey, 1972;

TABLE II

Representative Quadrupole Coupling
Constants of Deuterium

Compound[a]	Quadrupole coupling constant (kHz)	Reference
NH_2D (g)	290.6	Kukolich (1969)
ND_3 (s)	217	Ripmeester (1976)
HDO (g)	307.9	Verhoeven et al. (1969)
D_2O (s)	213.2	Waldstein et al. (1964)
SiD_4 (s)	95	Lähteenmäki et al. (1967)
CH_3D (g)	191.48	Wofsy et al. (1970)
C_2D_6 (s)	168	Burnett and Muller (1971)
C_6H_{12} (s)	173.7	Barnes and Bloom (1972)
$CDCl_3$ (s)	166.9	Ragle et al. (1975)
C_2H_3D (l)	175.3	Kowalewski et al. (1976)
C_6D_6 (l)	183	Millett and Dailey (1972)
C_2HD (l)	198	Millett and Dailey (1972)
$C_6H_5C_2D$ (l)	227	Jackman et al. (1974)

[a] Symbols g, s, and l refer to studies done in gaseous, solid, and liquid (or liquid crystalline) solvents, respectively.

Rinné and Depireux, 1974). Some representative values are given in Table II. They vary from 170 kHz for deuterium bound to aliphatic carbon to over 200 kHz in alkynes. Another parameter, the asymmetry parameter, expresses the deviation from axial symmetry of the quadrupole interaction. For ^2H it is usually quite low. Unsymmetric motions in solids and semisolids can have effects on ^2H-NMR spectra similar to those due to finite asymmetry parameters (Spiess, 1978).

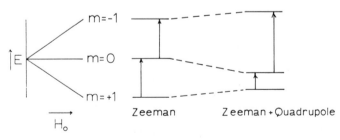

Fig. 5 The Zeeman nuclear energy levels of an $I = 1$ nucleus with and without the influence of the quadrupole coupling constant. The two allowed single-quantum transitions are shown.

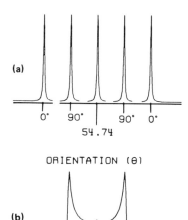

Fig. 6 (a) Deuterium NMR quadrupole doublets due to different angles between the principal axis of the quadrupole coupling tensor and the applied magnetic field. (b) Powder spectrum due to the sum of all possible quadrupole doublets. Note that the peaks in (b) correspond to the 90° doublet in (a), and the shoulders to the 0° doublet. The spectra are for the case of an axially symmetric electric field gradient ($\eta = 0$). (After Jarrell and Smith, 1983b.)

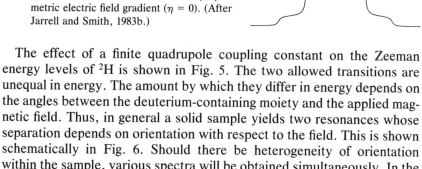

The effect of a finite quadrupole coupling constant on the Zeeman energy levels of ^2H is shown in Fig. 5. The two allowed transitions are unequal in energy. The amount by which they differ in energy depends on the angles between the deuterium-containing moiety and the applied magnetic field. Thus, in general a solid sample yields two resonances whose separation depends on orientation with respect to the field. This is shown schematically in Fig. 6. Should there be heterogeneity of orientation within the sample, various spectra will be obtained simultaneously. In the case of a perfectly random distribution of orientations, all possible spectra will be superimposed to yield a powder pattern like that shown at the bottom of Fig. 6, where well-defined turning points exist at frequencies corresponding to angles of 0 and 90° between the principal axis of the quadrupolar splitting tensor and the applied magnetic field.

Suppose we allow the molecules within the powdered solid to move isotropically with increasing frequency. When the motion is slow relative to the maximum difference in quadrupole splitting, it has little effect on the observed spectrum. However, as the rate approaches this value, the motion becomes increasingly effective at averaging the differences in quadrupole splitting due to different angles. This is demonstrated in Fig. 7. In the limit of rapid motion we obtain a single resonance whose width decreases with increasing rotational rate. This is the usual situation for small molecules dissolved in nonviscous solvents. The principal effect of the quadrupole coupling constant under these circumstances is to dominate the spin relaxation processes.

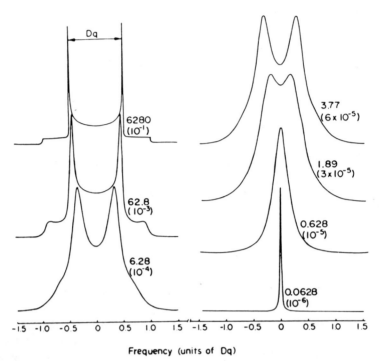

Fig. 7 The influence of rate of rotational motion on the powder spectrum of a partially ordered solid (residual quadrupole splitting of 10 kHz). The correlation times for overall motion (in seconds per radian) are shown in parentheses, whereas the product $2\pi Dq\tau$, where Dq is the residual quadrupole splitting and τ is the correlation time, are shown without parentheses. (After Stockton *et al.*, 1976. Copyright 1976 American Chemical Society.)

V. Relaxation Times

The magnitude of the quadrupolar contribution to relaxation of ^2H is usually sufficiently large that all other mechanisms may be ignored. On the other hand, it is sufficiently small that relaxation times rarely are ridiculously extreme. The spin–lattice relaxation times T_1 for small molecules vary over the range 1.5 s (benzene-d_6, Glasel, 1969) to 0.017 s (N-acetylglucosamine-3-d_1, Neurohr *et al.*, 1980), corresponding to intrinsic widths for Lorentzian resonances ($1/\pi T_2$, where $T_1 = T_2$) of 0.21 and 19 Hz, respectively. Spin–lattice relaxation times may be measured by the usual techniques, with the advantage that the relatively (compared to ^1H and ^{13}C) short T_1 values permit short recycle times in the pulse sequence. A simple relationship between the T_1 values of ^{13}C and ^2H affords an estimate of quadrupole coupling constants (Jackman *et al.*, 1974; Mantsch *et al.*, 1977; Saitô *et al.*, 1973).

The quadrupolar relaxation mechanism for ^2H relaxation results in a steep response of T_1 and T_2 to an increase in molecular weight. Thus small molecules bound to very large ones experience significant decreases in relaxation times. Use has been made of this in studying the molecular dynamics of nucleotides bound to enzymes (Zens *et al.*, 1976) and of sugars bound to lectins (Neurohr *et al.*, 1980).

Intramolecular motions may be studied via the ^2H relaxation times of different regions of a molecule in a manner similar to that employed for ^{13}C and ^1H. A variety of applications have been discussed elsewhere (Jarrell and Smith, 1983a; Mantsch *et al.*, 1974, 1977; Smith and Mantsch, 1982; Wamsler *et al.*, 1978). We shall return to their utility in Section VI,D.

VI. Powder Spectra

A. Experimental Problems

The basis of the ^2H-NMR spectra obtained from powdered samples was dealt with in Section IV. The spectra are usually very broad, as much as 1.5 times the quadrupole coupling constant. For aliphatic C—D bonds, this means finite spectral intensity over a range of 250 kHz. Furthermore, because of the very slow motions usually encountered in solids, T_2 is very short and T_1 very long. Such samples present great difficulties for high-resolution spectrometers.

The first problem is exciting nuclei equally all across the wide powder spectra. Even for high-power instruments with 90° pulse times of 1–3 μs, this condition is often not satisfied, and corrections must be made to obtain the true spectral shape (Bloom *et al.*, 1980). A second type of distortion arises from mechanical and electrical ringing in the receiver system due to the high-power pulse. In high-resolution spectrometers a delay between the excitation pulse and the acquisition of the free induction decay (FID) is used to minimize this distortion—however, for samples with small T_2 values such a delay causes further distortion and minimizes the signal. First-order phase corrections usually cannot compensate for these distortions. The quadrupolar echo technique suggested by Davis *et al.* (1976) has largely circumvented these problems. Figure 8a shows the effect of finite receiver dead time on a hypothetical powder pattern, and the spectrum of the same system in response to the sequence $90°_x$-τ-$90°_y$-τ-echo. For the spin-1 system the second 90° pulse refocuses the transverse magnetization to form an echo with a maximum

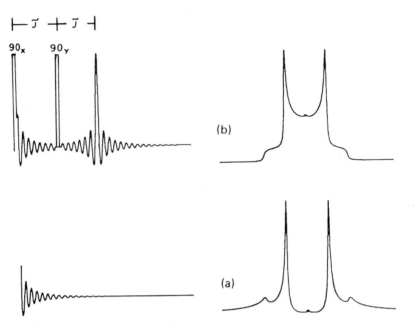

Fig. 8 Simulated ^{2}H-NMR FID and Fourier-transformed spectrum for the powder spectrum of an $I = 1$ system. (a) After a simple 90° pulse and a 30-μsec delay before acquisition; (b) after a quadrupole echo sequence $90°_x$-τ-$90°_y$-τ (Jarrell and Smith, 1983b).

at $t = 2\tau$. Usually the second delay is made somewhat shorter than the first, so that the peak of the echo can be correctly determined. This obviates the need for any first-order phase correction. The use of quadrature detection, with appropriate phase cycling of the pulses, yields spectra with minimal distortions. More detailed discussions of these acquisition techniques are given by Griffin (1981) and Davis (1983).

B. Solids

Deuterium NMR of solids is the most difficult application of the technique, but ironically was one of the first. Most of the early studies were described in detail in the review by Mantsch *et al.* (1977). These studies usually had one of two aims—to determine the quadrupole coupling constant (and asymmetry parameter, if significant) or to follow molecular motion in solids.

An example of the latter is shown in Fig. 9, the ^{2}H-NMR spectra of two solid sugars labeled in different methyl groups. If there were no motion taking place on the time scale of the quadrupole coupling constant, a

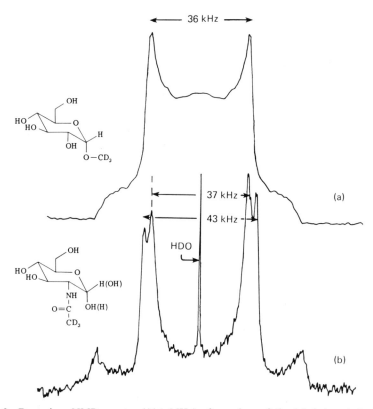

Fig. 9 Deuterium NMR spectra (46.1 MHz) of powders of the labeled carbohydrates shown. Spectra were acquired using the quadrupole echo. (From K. Neurohr, H. H. Mantsch, and I. C. P. Smith, unpublished.)

spectrum of width 250 kHz, with a quadrupole splitting of 125 kHz, would be expected. On the other hand, if rapid motion about the C_3 axis is allowed, a splitting of only 42 kHz should result. Rapid rotation about the next bond as well would reduce the splitting to 14 kHz. We see that the glycosidic methyl group in α-methylglucoside yields a splitting of 36 kHz, indicating a slight amount of motion other than simple rotation about the C_3 axis. More interestingly perhaps, the amplitude of motional averaging of the N-acetyl methyl group in N-acetylglucosamine appears to vary from the α to the β anomer. One has the splitting expected for simple C_3 motion, whereas the other has a slightly greater freedom of motion.

Figure 10 shows the spectra of solid perdeuterated nonadecane at two different temperatures (Taylor *et al.*, 1983). For the lower temperature (upper spectra), most positions give the classic doublet patterns with large

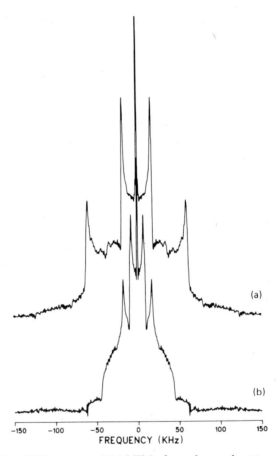

Fig. 10 Deuterium NMR spectrum (46.1 MHz) of nonadecane-d_{44} at temperatures below (a) and above (b) that of the solid–solid phase transition. (Unpublished data of M. G. Taylor, E. C. Kelusky, I. C. P. Smith, H. L. Casal, and D. G. Cameron; discussed in Taylor *et al.*, 1983.)

quadrupole splittings (~120 kHz) expected for immobile solids, except the methyl group which yields a splitting of 35 kHz as a result of C_3 motion. In strong contrast, the spectra at higher temperatures contain extra shoulders which are usually indicative of axial asymmetry in the quadrupole splitting tensor. However, we know that for deuterium attached to tetrahedral carbons the intrinsic asymmetry parameter is very close to zero. The observed patterns must therefore be due to motionally induced asymmetry, such as has been seen in polymers and lipid dispersions (Huang *et al.*, 1980; Spiess, 1978). The data were interpreted in terms of restricted rapid torsional oscillations whose amplitudes increased with distance from the center of the chain.

Another recent ^{2}H study on solid–solid phase transitions was reported by Wasylishen *et al.* (1979) for specifically deuterated camphor.

C. Liquid Crystals

The simple basis of the ^{2}H powder pattern, quadrupole interaction only at the site of deuteration, has made ^{2}H NMR popular for the study of liquid crystals, whose properties lie intermediate between those of solids and liquids. Both the liquid crystals, and guest molecules within them, have been studied (Emsley and Lindon, 1975; Mantsch *et al.*, 1977). As the methodology and types of conclusions are similar to those used for membranes (see the following), I shall merely mention several applications in the liquid crystal field (Boden *et al.*, 1979; Deloche and Charvolin, 1980; Dong *et al.*, 1981; Emsley *et al.*, 1979; Fujiwara and Reeves, 1980; Hansen and Jacobsen, 1980; Moseley *et al.*, 1982; Samuelski and Luz, 1980).

D. Membranes

Biological membranes contain amphiphilic lipids whose hydrophobic regions are usually long hydrocarbon chains—straight, branched, saturated, or unsaturated. The lipids are thought to be organized into bilayers, with the hydrophobic regions of each monolayer in contact and with the hydrophilic regions directed toward water on the outside and inside of the structure. Figure 11 gives a view of the molecules in part of such a bilayer. For a long time it was thought that the hydrophobic chains were in a "fluid," semiordered state and that the degree of fluidity determined many of the functional properties of the membrane. A technique for measuring these properties without perturbation, and with a reasonably straightforward interpretation, has been sought for many years.

The study of molecular ordering and mobility in model and biological membranes has been revolutionized by the use of ^{2}H NMR. This did not come about easily, because many deuterated compounds had to be synthesized and incorporated into membranes, and spectrometers modified to obtain the rather wide spectra. A variety of reviews on the subject are available (Davis, 1983; Griffin, 1981; Jacobs and Oldfield, 1981; Seelig, 1977; Seelig and Seelig, 1980; Smith, 1981, 1983; Smith and Jarrell, 1983).

One of the most useful parameters readily available from the ^{2}H-NMR spectra of biological membranes is the bond order parameter S_{CD}. Figure

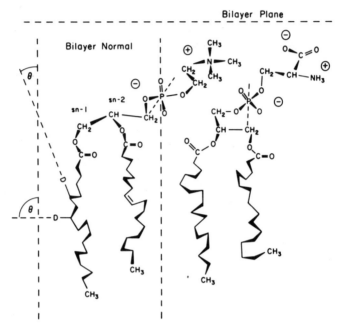

Fig. 11 Representation of lipid molecules in a bilayer such as is thought to exist in biological membranes. Two different classes of lipid are shown; from left to right they are phosphatidylcholine and phosphatidylserine. The angles θ are between particular C—D bonds and the axis of ordering. Note that this varies with conformation.

11 demonstrates some of the many conformations available to the hydrophobic chains in a bilayer, and several values of the angle θ between a C—D bond and the normal to the bilayer plane are indicated. Rapid interconversion among the various possible conformations is very likely. The order parameter measures the time and ensemble average of $(3 \cos^2 \theta - 1)/2$ and thus gives an indication of the average orientation of this region of the molecule. Conversion to the molecular order parameter S_{mol} accounts for the different intrinsic geometries of chemical groups and yields a useful scale on which 0 is total chaos and 1 is perfect order. When molecular motion within the membrane is axially symmetric, $Dq = \frac{3}{4}(e^2qQ/h)S_{CD}$, where Dq is the quadrupole splitting in the powder pattern and e^2qQ/h the quadrupole coupling constant for ^{2}H in that particular environment (usually 170 kHz).

A wide variety of model membrane systems have been studied with ^{2}H NMR, providing useful comparisons for studies on biological membranes. Most of the results for the model systems are summarized in the review articles mentioned previously. The biological membrane that has been studied in greatest detail is the plasma membrane of *Acholeplasma laidlawii;* results from this work will be used to illustrate the method.

For many years it has been known that unsaturated fatty acids fluidize lipids (butter versus margarine) and membranes. However, usually a definition of the term "fluidity" is not given. Since this property obviously involves both lipid packing and lipid motion, ^{2}H NMR is an ideal way to approach the problem.

Figure 12 shows the ^{2}H-NMR spectra of *A. laidlawii* membranes enriched in an unsaturated fatty acid, oleic acid, which is labeled with deuterium at seven different positions (Rance *et al.*, 1980). There is evidently a large dependence of quadrupole splitting, hence order parameter, on position. These data, plus the corresponding spin–lattice relaxation times, are plotted as a function of position in Fig. 13. Note in Fig. 12 the complexity of the spectrum when the oleic acid is labeled at position 2. This is actually a complex spectrum, containing as many as six quadrupole splittings (see the following). However, the principal sources of its complexity are the inequivalence of the two fatty acyl chains at the sn-1 and sn-2 positions and the spatial inequivalence of the two deuterons of the chain at the

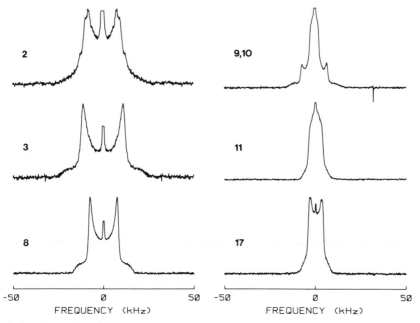

Fig. 12 Deuterium NMR spectra (41.3 MHz) of oleoyl chains, deuterated at the position indicated, incorporated into the membranes of *A. laidlawii*. The spectra were taken at 25°C using the quadrupole echo. (After Rance *et al.*, 1980.)

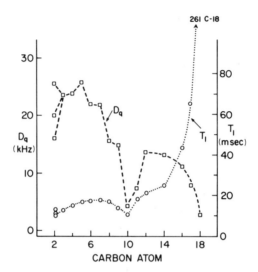

Fig. 13 Dependence of the residual quadrupole splitting (Dq, proportional to the segmental order parameter S_{CD}) and the spin lattice relaxation time (T_1, proportional to the mobility of the segment) on position of ^2H incorporation within the oleate-enriched chains of *A. laidlawii* membranes at 25°C. The T_1 value for C-18 is off-scale at 261 msec (Dq values from Rance *et al.*, 1980; T_1 values (41.3 MHz) are those of M. Rance, K. R. Jeffrey, A. P. Tulloch, K. W. Butler and I. C. P. Smith, unpublished).

sn-2 position (note the three quadrupole splittings indicated for position 2 in Fig. 13). The other striking feature in Fig. 13 is the dip in the profile in the proximity of the cis double bond. Such a dip was not seen with saturated chains (Jarrell *et al.*, 1982; Stockton *et al.*, 1977) but has been reported in the model membrane system 1-palmitoyl-2-oleoylphosphatidylcholine (Seelig and Waespe-Šarčević, 1978). It is due mainly to the geometry of the cis double bond—when corrections are made for this, the unsaturated chains appear just as ordered as their unsaturated counterparts. The T_1 values in Fig. 13 show a decrease in the region of the double bond, indicative of reduced mobility as expected, and a large increase near the end of the chain. The decrease from carbons 8 to 12 is different from the behavior seen in saturated chains, and overall the T_1 values for the unsaturated chains are lower than those for their saturated counterparts when compared at corresponding temperatures relative to that of their liquid-crystal-to-gel transition. Thus the unsaturated membranes have less mobile chains. The increased "fluidity" in unsaturated systems must be mainly due to a decrease in the temperature at which the gel phase "melts" as a result of perturbation of the properties of the gel phase rather than of the liquid crystalline phase.

The preceding argument employing the spin–lattice relaxation times of ^2H in membranes implied a simple, direct relationship between $1/T_1$ and the correlation time for effectively isotropic rotational diffusion. Motions in membranes are clearly more complex than this. Various attempts have been made to formulate models involving the degree of molecular order as well as the correlation time (Brown, 1979, 1982; Brown and Davis, 1981; Brown et al., 1979; Pace and Chan, 1982). A large amount of data taken at widely different magnetic field strengths will be required to determine which model is most appropriate. That is why for the moment I prefer to report T_1 values rather than correlation times—the T_1 values should be time-invariant. However, use of the model of Brown and Davis (1981) to extract approximate correlation times for the various positions on the oleoyl chains leads to conclusions identical to those formulated in the preceding paragraph.

As mentioned previously, at sufficiently low temperatures membrane lipids are known to undergo a solid–solid phase transition from a state of partial order and relatively high mobility to one of high order and low mobility, the so-called gel phase. This transition can usually be followed on a scanning calorimeter (McElhaney, 1982). Earlier NMR studies detected this event by the disappearance of the spectrum, because of the problems of short T_2 and large spectral width referred to earlier. Figure 14 shows the temperature dependence of the ^2H-NMR spectra of A. laidlawii membranes enriched in oleic acid-12,12-d_2 (Rance et al., 1980). The vertical expansions of the spectra show the appearance at 0°C of a component of width ~60 kHz, which grows in relative intensity with decreasing temperature. Note the coexistence of the spectra due to the two types of phases over the range of the calorimetrically defined transition (−5 to −22°C). The gel phase spectra of width 60 kHz are those expected for highly ordered methylene segments undergoing only motion about the principal axis of a strongly transoid chain segment, as seen earlier for saturated chains (Smith et al., 1979). The onset of components of width ~120 kHz at lower temperatures indicates the slowing down of all motion available to this segment. This is demonstrated very dramatically by the increase in the second moment, $M_2 = \int_0^\infty (\omega - \omega_0)^2 F(\omega - \omega_0) \, d\omega / \int_0^\infty F(\omega - \omega_0) \, d\omega$, where $F(\omega - \omega_0)$ is the spectral line shape, at low temperatures (Fig. 15). However, even at −52°C M_2 does not reach the rigid limit maximum of 1.28×10^{11} s^{-2}. The various moments of the ^2H-NMR spectra, and combinations thereof, have been found to be very useful in analyzing the phase composition (Jarrell et al., 1981), homogeneity (Paddy et al., 1981), and positional dependence of the order parameter of multiply deuterated chains (Davis et al., 1980; Jarrell et al., 1982) in biological membranes.

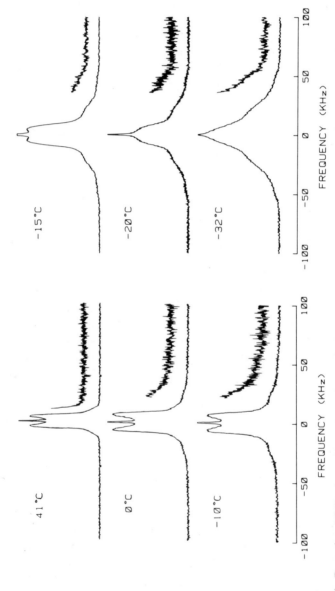

Fig. 14 Deuterium NMR spectra (41.3 MHz) of *A. laidlawii* membranes, labeled at the 12 position of the oleoyl chains, at the temperatures indicated. The vertical expansions, to show broad components, are ×4. (After Rance *et al.*, 1980.)

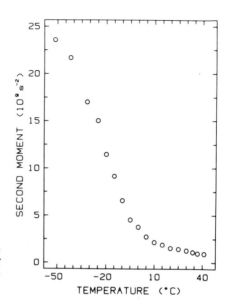

Fig. 15 Temperature dependence of the second moments of the ^{2}H-NMR spectra of the system shown in Fig. 14. (After Rance *et al.*, 1980.)

In discussing Fig. 12, several quadrupole splittings were apparent when the chains were labeled at position 2. Analysis is rather difficult because of the superposition of several powder patterns. A technique developed by Bloom and co-workers (1981) has simplified this analysis greatly. An iterative procedure analyzes the spectra for contributions due to particular angles. Figure 16 shows the powder spectra and their "de-Pake-ed" counterparts (90° angles) resulting from this procedure (Rance *et al.*, 1983). The upper right spectrum apparently contains four quadrupole splittings; in fact, it contains at least five splittings, as determined by simulation. In this membrane system it is mainly the chains at the sn-2 position that are labeled; therefore only two quadrupole splittings can be expected. The extra splittings are due to some incorporation at the sn-1 position, plus a dependence of the quadrupole splitting on the nature of the lipid head group. Note that the isolated glycolipids yield only the major quadrupole splittings due to inequivalence of the two deuterons at position 2.

Finally, by analogy with the many studies with ^{2}H NMR on guest molecules in liquid crystals, it is possible that membrane-active compounds can also be so visualized. We have had some success with model membrane systems interacting with anesthetics (Boulanger *et al.*, 1980, 1981) and with an antiarthritic drug (Smith, 1983). Figure 17 shows the ^{2}H-NMR spectra of the local anesthetic tetracaine, labeled with two NCD$_3$ groups, interacting with egg lecithin. At this pH value the partition coeffi-

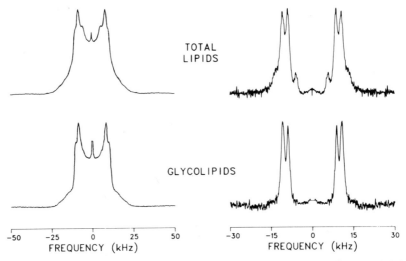

TOTAL
LIPIDS

GLYCOLIPIDS

FREQUENCY (kHz)   FREQUENCY (kHz)

Fig. 16 Deuterium NMR spectra (30.7 MHz) of aqueous dispersions of the total lipids (upper) and the purified glycolipids (lower) from *A. laidlawii* membranes enriched in [2,2-^{2}H$_2$]dihydrosterculic acid. The spectra to the left are the usual powder type; those to the right are the corresponding "de-Pake-ed" (to 90° spectra) versions. (After Rance *et al.,* 1983.)

cient for the anesthetic in the lipid–water system is 22. The relative amounts of anesthetic in each phase can be altered by varying the mass of each phase present. The percentages to the right of each spectrum represent the amount of tetracaine in the lipid phase, as calculated from the partition constant. The spectra contain two components: The broad component with a measurable quadrupole splitting is due to membrane-bound anesthetic in slow exchange with that in water; the narrower component in the center of the spectra, whose width decreases with increasing volumes of water, is due to weakly bound anesthetic in rapid exchange with that in water. Based on these and other data a detailed molecular model for anesthetic–lipid interaction was constructed (Boulanger *et al.,* 1981). Similar studies have been reported recently for cholesterol, deuterated at several positions (Taylor *et al.,* 1981, 1982). It was possible to locate the axis of motional averaging within the cholesterol molecule, as well as the degree of ordering and the mobility of various different segments.

Proteins represent a more complex system for study with ^{2}H NMR. They often exist in complex aggregates, such as with nucleic acids in nucleosomes and with lipids in membranes. The labeling of individual residues is rather difficult, and analysis of the spectra is more challenging than for membranes. Considerable information on the nature of motions

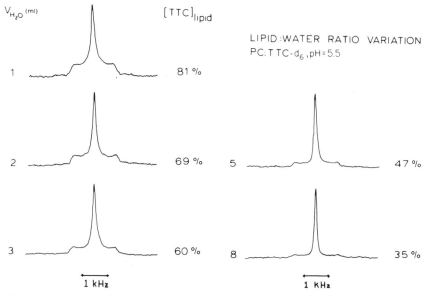

Fig. 17 Deuterium NMR spectra (15.4 MHz) of the local anesthetic tetracaine, labeled with ^{2}H at the two N-methyl groups, in the presence of 260 mmol egg phosphatidylcholine and various volumes of water (deuterium-depleted), pH 5.5, 29°C. The quantity [TTC]$_{lipid}$ indicates the relative amount of anesthetic, as calculated from the partition coefficient, bound to the lipid. (After Boulanger *et al.*, 1980.)

of amino acyl residues in collagen (Batchelder *et al.*, 1982; Jelinski *et al.*, 1980), bacteriophage coat protein (Gall *et al.*, 1982), heme proteins (Rothgeb and Oldfield, 1981), and bacteriorhodopsin in a bacterial membrane (Kinsey *et al.*, 1981) has been obtained.

VII. Conclusion

Deuterium NMR has become a routine tool in the arsenal of techniques used by the chemist. More recently it has yielded the answers to biological questions that have eluded unambiguous solution by other approaches. With the widespread interest in advanced methods such as double-quantum excitation (Eckman *et al.*, 1980a), magic angle spinning (Eckman *et al.*, 1980a, b), cross-polarization (Müller *et al.*, 1980), two-dimensional sequences, and spatial imaging of objects (Cox and Styles, 1980), even more innovative use of ^{2}H NMR will probably be seen in the future.

References

Abell, C., and Staunton, J. (1981). *J. Chem. Soc. Chem. Comm.*, pp. 856–858.

Ackland, M. J., Hanson, J. R., Ratcliffe, A. H., and Sadler, I. (1982). *J. Chem. Soc. Chem. Comm.* pp. 165–166.

Barnes, R. G., and Bloom, J. W. (1972). *J. Chem. Phys.* **57**, 3082–3086.

Batchelder, L. S., Sullivan, C. E., Jelinski, L. W., and Torchia, D. A. (1982). *Proc. Nat. Acad. Sci. U.S.A.* **79**, 386–389.

Battersby, A. R., Edington, C., Fookes, C. J. R., and Hook, J. H. (1982). *J. Chem. Soc. Chem. Comm.* pp. 181–182.

Bloom, M., Davis, J. H., and Valic, M. I. (1980). *Can. J. Phys.* **58**, 1510–1517.

Bloom, M., Davis, J. H., and MacKay, A. L. (1981). *Chem. Phys. Lett.* **80**, 198–202.

Boden, N., Bushby, R. J., and Clark, L. D. (1979). *Mol. Phys.* **38**, 1683–1686.

Boulanger, Y., Schreier, S., Leitch, L. C., and Smith, I. C. P. (1980). *Can. J. Biochem.* **58**, 986–995.

Boulanger, Y., Schreier, S., and Smith, I. C. P. (1981). Biochemistry **20**, 6824–6830.

Brevard, C., and Kintzinger, J. P. (1978). "NMR and the Periodic Table" (R. K. Harris and B. E. Mann, eds.), pp. 107–128, Academic Press, New York.

Brown, M. F. (1979). *J. Magn. Reson.* **35**, 203–215.

Brown, M. F. (1982). *J. Chem. Phys.* **77**, 1576–1599.

Brown, M. F., and Davis, J. H. (1981). *Chem. Phys. Lett.* **79**, 431–435.

Brown, M. F., Seelig, J., and Häberlen, U. (1979). *J. Chem. Phys.* **70**, 5045–5053.

Burger, U., Sonney, J.-M., and Vogel, P. (1980). *Helv. Chim. Acta* **63**, 1006–1015.

Burnett, L. J., and Muller, B. H. (1971). *J. Chem. Phys.* **55**, 5829–5831.

Cane, D. E., and Nachbar, R. B. (1980). *Tetrahedron Lett.*, pp. 437–440.

Casadevall, E., and Metzger, P. (1970). *Tetrahedron Lett.*, pp. 4199–4202.

Cheng, A. K., Stothers, J. B., and Tan, C. T. (1977). *Can. J. Chem.* **55**, 447–453.

Cox, S. J., and Styles, P. (1980). *J. Magn. Reson.* **40**, 209–212.

Davis, J. H. (1983). *Biochim. Biophys. Acta* **737**, 117–171.

Davis, J. H., Jeffrey, K. R., Bloom, M., Valic, M. I., and Higgs, T. P. (1976). *Chem. Phys. Lett.* **42**, 390–394.

Davis, J. H., Bloom, M., Butler, K. W., and Smith, I. C. P. (1980). *Biochim. Biophys. Acta* **597**, 477–491.

Deloche, B., and Charvolin, J. (1980). *J. Phys. Lett.* **41**, 39–42.

DePuy, C. H., Andrist, A. H., and Fünfschilling, P. C. (1974). *J. Am. Chem. Soc.* **96**, 948–950.

Diehl, P., and Leipert, Th. (1964). *Helv. Chim. Acta* **47**, 545–557.

Dong, R. Y., Lewis, J., Tomchuk, E., Wade, C. G., and Bock, E. (1981). *J. Chem. Phys.* **74**, 633–636.

Eckman, R., Müller, L., and Pines, A. (1980a). *Chem. Phys. Lett.* **74**, 376–378.

Eckman, R., Alla, M., and Pines, A. (1980b). *J. Magn. Reson.* **41**, 440–446.

Edwards, O. E., Dixon, J., Elder, J. W., Kolt, R. J., and Lesage, M. (1981). *Can. J. Chem.* **59**, 2096–2115.

Emsley, J. W., and Lindon, J. C. (1975). "N.M.R. Spectroscopy using Liquid Crystal Solvents." Pergamon, London.

Emsley, J. W., Khoo, S. K., and Luckhurst, G. R. (1979). *Mol. Phys.* **37**, 959–972.

Fujiwara, F. Y., and Reeves, L. W. (1980). *Can. J. Chem.* **58**, 1550–1557.

Gall, C. M., Cross, T. A., DiVerdi, J. A., and Opella, S. J. (1982). *Proc. Nat. Acad. Sci. USA* **79**, 101–105.

Glasel, J. A. (1969). *J. Am. Chem. Soc.* **91**, 4569–4571.

Griffin, R. G. (1981). *Methods Enzymol.* **72**, 108–174.

Habich, A., Barner, R., von Philipsborn, W., and Schmid, H. (1965). *Helv. Chim. Acta* **48**, 1301–1316.

Hansen, J., and Jacobsen, J. P. (1980). *J. Magn. Reson.* **41**, 381–388.

Huang, T. H., Skarjune, R. P., Wittebort, R. J., Griffin, R. G., and Oldfield, E. (1980). *J. Am. Chem. Soc.* **102**, 7377–7379.

Jackman, L. M., Greenberg, E. S., Szeverenyi, N. M., and Schnorr, G. K. (1974). *J. Chem. Soc. Chem. Commun.* pp. 141–142.

Jacobs, R. E., and Oldfield, E. (1981). *Prog. Nucl. Magn. Reson. Spectrosc.* **14**, 113–136.

Jarrell, H. C., and Smith, I. C. P. (1983a). "The Multinuclear Approach to NMR Spectroscopy" (J. B. Lambert and F. G. Riddell, eds.). NATO ASI Series **C103**, Chap. 7, pp. 133–149, D. Reidel, Dordrecht.

Jarrell, H. C., and Smith, I. C. P. (1983b). "The Multinuclear Approach to NMR Spectroscopy" (J. B. Lambert and F. G. Riddell, eds.). NATO ASI Series **C103**, Chap. 8, pp. 151–168, D. Reidel, Dordrecht.

Jarrell, H. C., Byrd, R. A., and Smith, I. C. P. (1981). *Biophys. J.* **34**, 451–463.

Jarrell, H. C. *et al.* (1982). *Biochim. Biophys. Acta* **688**, 622–636.

Jelinski, L. W., Sullivan, C. E., Batchelder, L. S., and Torchia, D. A. (1980). *Biophys. J.* **10**, 515–529.

Johnson, A. L., Stothers, J. B., and Tan, C. T. (1975). *Can. J. Chem.* **53**, 212–223.

Kakinuma, K., Ogawa, Y., Sasaki, T., Seto, T., and Otake, N. (1981). *J. Amer. Chem. Soc.* **103**, 5614–5616.

Kinsey, R. A., Kintanar, A., and Oldfield, E. (1981). *J. Biol. Chem.* **256**, 9028–9036.

Kitching, W., Atkins, A. R., Wickham, G., and Alberts, V. (1981). *J. Org. Chem.* **46**, 563–570.

Kowalewski, J., Lindblom, T., Vestin, R., and Drakenberg, T. (1976). *Mol. Phys.* **31**, 1669–1676.

Kukolich, S. G. (1969). *J. Chem. Phys.* **49**, 5523–5525.

Lähteenmäki, U., Niemelä, L., and Pyykkö, P. (1967). *Phys. Lett.* **A25**, 460–461.

McElhaney, R. N. (1982). *Chem. Phys. Lipids* **30**, 229–259.

Mantsch, H. H., Saitô, H., Leitch, L. C., and Smith, I. C. P. (1974). *J. Am. Chem. Soc.* **96**, 256–258.

Mantsch, H. H., Saitô, H., and Smith, I. C. P. (1977). Prog. Nucl. Magn. Reson. Spectros. **11**, 211–272.

Millett, F. S., and Dailey, B. P. (1972). *J. Chem. Phys.* **56**, 3249–3256.

Montgomery, L. K., Clouse, A. O., Crelier, A. M., and Applegate, L. E. (1967). *J. Am. Chem. Soc.* **89**, 3453-3457.

Moseley, M. E., Poupko, R., and Luz, Z. (1982). *J. Magn. Reson.* **48**, 354–360.

Müller, L., Eckman, R., and Pines, A. (1980). *Chem. Phys. Lett.* **76**, 149–154.

Neurohr, K., Lacelle, N., Mantsch, H. H., and Smith, I. C. P. (1980). *Biophys. J.* **32**, 931–938.

Nordlander, J. E., and Haky, J. E. (1980). *J. Org. Chem.* **45**, 4780–4782.

Pace, R. J., and Chan, S. I. (1982). *J. Chem. Phys.* **76**, 4228–4240.

Paddy, M. R., Dahlquist, F. W., Davis, J. H., and Bloom, M. (1981). *Biochemistry* **20**, 3152–3162.

Pitner, T. P., Edwards, W. B., Bassfield, R. L., and Whidby, J. F. (1978). *J. Am. Chem. Soc.* **100**, 246–251.

Ragle, J. L., Mokarram, M., Presz, D., and Minott, G. (1975). *J. Magn. Reson.* **20**, 195–213.

Rance, M., Jeffrey, K. R., Tulloch, A. P., Butler, K. W., and Smith, I. C. P. (1980). *Biochim. Biophys. Acta* **600**, 245–262.

Rance, M., Smith, I. C. P., and Jarrell, H. C. (1983). *Chem. Phys. Lipids* **32**, 57–71.

Richards, J. C., and Spenser, I. D. (1978). *J. Am. Chem. Soc.* **100**, 7402–7404.
Rinné, M., and Depireux, J. (1974). "Advances in Nuclear Quadrupole Resonance" (J. A. S. Smith, ed.), Vol. 1, pp. 357–374. Heyden, London.
Ripmeester, J. A. (1976). *Can. J. Chem.* **54**, 3677–3684.
Rothgeb, T. M., and Oldfield, E. (1981). *J. Biol. Chem.* **256**, 1432–1446.
Saitô, H., Mantsch, H. H., and Smith, I. C. P. (1973). *J. Am. Chem. Soc.* **95**, 8453–8455.
Samuelski, E. T., and Luz, Z. (1980). *J. Chem. Phys.* **73**, 142–147.
Sato, Y., Oda, T., and Saitô, H. (1976). *Tetrahedron Lett.*, pp. 2695–2698.
Seelig, J. (1977). *Q. Rev. Biophys.* **10**, 353–418.
Seelig, J., and Seelig, A. (1980). *Q. Rev. Biophys.* **13**, 19–61.
Seelig, J., and Waespe-Šarčević, N. (1978). Biochemistry **17**, 3310–3315.
Smith, I. C. P. (1983). "Biomembranes" (C. E. Manson and M. Kates, eds.), Vol. 12 (in press).
Smith, I. C. P. (1981). *Bull. Mag. Res.* **3**, 120–133.
Smith, I. C. P., and Jarrell, H. C. (1983). Acc. Chem. Res. (in press).
Smith, I. C. P., and Mantsch, H. H. (1982). *ACS Symp. Ser.* **191**, 97–117.
Smith, I. C. P., Butler, K. W., Tulloch, A. P., Davis, J. H., and Bloom, M. (1979). *FEBS Lett.* **100**, 57–61.
Spiess, H. W. (1978). "NMR Basic Principles and Progress" (P. Diehl, E. Fluck, and R. Kosfeld, eds.), pp. 55–214. Springer Verlag, Berlin.
Stockton, G. W., Polnaszek, C. F., Tulloch, A. P., Hasan, F., and Smith, I. C. P. (1976). *Biochemistry* **15**, 954–966.
Stockton, G. W. *et al.* (1977). *Nature* **269**, 267–268.
Taylor, M. G., Akiyama, T., and Smith, I. C. P. (1981). *Chem. Phys. Lipids* **29**, 327–339.
Taylor, M. G., Akiyama, T., Saitô, H., and Smith, I. C. P. (1982). *Chem. Phys. Lipids* **31**, 359–379.
Taylor, M. G., Kelusky, E. C., Smith, I. C. P., Casal, H. L., and Cameron, D. G. (1983). *J. Chem. Phys.* **78**, 5108–5112.
Verhoeven, J., Dynamus, A., and Bluyssen, H. (1969). *J. Chem. Phys.* **50**, 3330–3338.
Waldstein, P., Rabideau, S. W., and Jackson, J. A. (1964). *J. Chem. Phys.* **41**, 3407–3411.
Wamsler, T., Nielsen, J. T., Pedersen, E. J., and Schaumburg, K. (1978). *J. Magn. Reson.* **31**, 177–186.
Wasylishen, R. E., Pettitt, B. A., and Lewis, J. S. (1979). *Chem. Phys. Lett.* **67**, 459–462.
Wheeler, W. D., Kaizaki, S., and Legg, J. I. (1982). *Inorg. Chem.* **21**, 3250–3252.
Wigle, I. D., Mestichelli, L. J. J., and Spenser, I. D. (1982). *J. Chem. Soc. Chem. Commun.* pp. 662–664.
Wofsy, S. C., Muenter, J. S., and Klemperer, W. (1970). *J. Chem. Phys.* **53**, 4005–4014.
Zens, A. P. *et al.* (1976). *J. Am. Chem. Soc.* **98**, 3760–3764.

2 Tritium NMR

A. L. Odell

Urey Radiochemistry Laboratory
Department of Chemistry,
University of Auckland,
Auckland, New Zealand

I. Historical Introduction

In 1947 F. Block *et al.* announced that the triton possesses a spin of $\frac{1}{2}$, because a sample of water containing about 80% of its hydrogen in the form ^3H gave *nuclear induction* signals for ^1H and ^3H of strengths proportional to their abundance. This observation eliminates the other possible value of $S = \frac{3}{2}$, as this would lead to a fivefold increase in signal intensity for the triton. The magnetogyric ratio for ^3H was reported to be 1.067 ± 0.001 times that for ^1H. Simultaneously, Anderson and Novick (1947) reported the ratio of nuclear g values to be 1.06666 ± 0.0001 after measurements on "one drop" of water in which both ^1H and ^3H resonances were observed simultaneously. Neither group gives quite enough details to estimate the activity of the tritium used in these pioneering experiments, but it was clearly in the multicurie range, and the resulting radiological hazards must have been considerable.

The first report of a high-resolution ^3H-NMR spectrum was presented by Tiers *et al.* (1964) who studied ethylbenzene labeled in the side chain. They hydrogenated phenylacetylene over palladium charcoal using pure $^3H_2$ gas followed by an excess of $^1H_2$. Their sample contained ~ 10 Ci in 0.3 ml, representing ~ 1 atom% ^3H. These authors reported two first-order multiplets, the higher field one, due to —CH_2T, being a triplet of triplets ($J_{TCH} = 13.8 \pm 0.1$ Hz, $J_{TCCH} = 8.11 \pm 0.06$ Hz) and a weaker downfield doublet of quartets due to —CHT—.

Integration of peak areas indicated 1.96 for the —CH_2T/—CHT ratio, a result not consistent with the simple addition of two atoms of ^3H nor with exchange of the ethylbenzene product but which does indicate isotopic randomization at the styrene stage of reduction, a result of considerable mechanistic significance. Tiers writes, "... the possibility of using the *non radiochemical* NMR properties of tritium in chemical studies has been demonstrated.... The synthesis of organic compounds having high levels of tritiation, especially at specific sites, is thus feasible." He cautions, however, that the hazards of handling curie quantities of tritium in a fragile NMR tube are considerable and are made greater by radiolytic decomposition of the sample (1 ml of H_2 per day from 10 Ci), leading to dangerous gas pressures.

II. Nuclear Properties of the Triton

The nuclear characteristics of tritium and other isotopes of hydrogen are summarized in Table I with those of ^{13}C for comparison. Two points are noteworthy:

(a) ^3H is the most sensitive NMR nucleus known, ~ 1.21 times more sensitive than ^1H.

TABLE I

Nuclear Properties of the Isotopes of Hydrogen
(with ^{13}C Data for Comparison)

Property	^{1}H	^{2}H	^{3}H	^{13}C
Nuclear spin	$\frac{1}{2}$	1	$\frac{1}{2}$	$\frac{1}{2}$
Natural abundance (%)	99.984	0.0156	$<10^{-16}$	1.11
Magnetic moment (nuclear magnetons)	2.7927	0.8574	2.9788	0.7022
Magnetogyric ratio, γ, (10^7 rad T^{-1} s^{-1})	26.7510	4.1064	28.5336	6.7263
Resonance frequency (MHz at 1.41 T)	60.0	9.2	64.0	15.07
Half-life (yr)	∞	∞	13	∞
Relative sensitivity to NMR, detection at natural abundance	5682	0.008	$\simeq 0$	1.00
Relative sensitivity at 100% isotopic abundance	62.89	0.604	76.1	1.00

(b) ^{3}H resonates at a higher frequency than any other nucleus, ~6.7% above the ^{1}H resonance frequency.

III. Modern High-Resolution Tritium NMR Spectroscopy

In 1971 J. A. Elvidge of the University of Surrey and E. A. Evans of the Radiochemical Centre, Amersham, England, and their co-workers published the first description (Elvidge *et al.*, 1971) of routine measurements of tritium NMR spectra. They used a Perkin-Elmer R-10 spectrometer operating at 1.41 T, at which field the triton resonates at 64.0 MHz compared to the proton at 60.0 MHz. This group has continued to spearhead the development of ^{3}H NMR up to the present time. Motivation for this work stems largely from the use of tritium as a radiotracer in molecules of biological interest and the increasing need to know not only the level of labeling in any preparation but also the position(s) of the triton(s) in the molecule. Spectra of about six compounds each labeled in a variety of positions were reported, establishing the similarity of chemical shifts to those of protons in the same chemical environment and the ability to locate tritons and to determine whether labeling was confined to one site. Spectra were all of the spin-coupled ^{1}H—^{3}H type, allowing geminal and vicinal coupling constants to be determined, e.g., $J_{HT} = 15.5$ Hz for $CH_2TCOONa$.

This paper also reports a study on the self-radiolysis of [5-^{3}H]uridine (31 Ci mmol^{-1} at position 5). After 6 days, when the sample had received a dose of 18 Mrads, new ^{3}H-NMR signals were seen which were ascribed to hydrates and glycols. The paper foresees important developments in resolution and sensitivity when more sophisticated spectrometers become available.

Commercial development of the pulse Fourier transform (FT) NMR spectrometer in the 1970s, primarily to provide the increased sensitivity needed by ^{13}C spectroscopists, provided the spur for the next phase of development of tritium NMR. Elvidge and co-workers (1974a) described the use of a Bruker WH90 pulse spectrometer and reported greatly in-

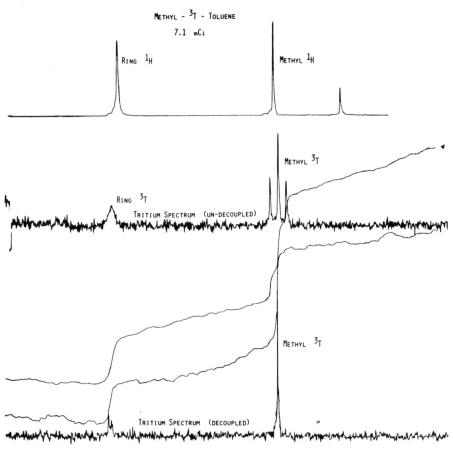

Fig. 1 Nuclear magnetic resonance spectra of toluene at 60 MHz. (Reproduced by permission of J.E.O.L.)

creased sensitivity of detection, spectra with a signal/noise ratio of 3 : 1 being obtained with a water sample containing 0.5 mCi total activity in a 30-μl microcell after 1.9×10^4 pulses. The specific activity of this preparation, 0.3 mCi mmol^{-1}, may be compared to the 5 mCi mmol^{-1} of water requiring 64 scans in the scanning spectrometer described in the earlier paper (Elvidge *et al.*, 1971).

A description of the performance of a 60-MHz (^{1}H) FT spectrometer (JEOL FX-60) equipped with a 64-MHz transmitter and a probe tuned to this frequency (Odell *et al.*, 1977) shows that even at this lower field satisfactory spectra can be obtained, using a slightly larger microcell (400 μl), with samples containing $\sim$10 mCi of ^{3}H and as little as 0.7 mCi at one molecular site.

Figure 1 shows ^{3}H spectra of tritium-labeled toluene obtained with this instrument, with ^{1}H spectra for comparison. Similarities between the chemical shifts of ^{3}H and ^{1}H at particular sites are obvious, and Elvidge *et al.* (1974a) have established that triton and proton chemical shifts are identical (when expressed as δ in parts per million) within very fine limits over a wide variety of molecular environments for the triton. This means that the vast compilation of shift data on proton NMR can be applied to tritium spectra without modification.

Coupling between methyl protons and tritons is clearly seen in the spectrum in Fig. 1b, where $J_{^1H^3H(gem)} = 15.4$ Hz. Decoupling using noise-modulated ^{1}H irradiation gives not only a single sharp resonance for methyl ^{3}H but also an improved signal/noise ratio. These spectra were obtained in a sample containing 7.1 mCi of ^{3}H in a 400-μl bulb.

IV. Referencing

Early studies (Elvidge *et al.*, 1971) on ^{3}H-NMR spectra utilized tritiated water (HTO) as an internal reference standard, but water is not an ideal standard or solvent for NMR studies as its tendency to form specific associations with solute molecules gives rise to chemical shifts that are highly dependent on concentration and temperature.

Monotritiated trimethylsilane (TMS), which would otherwise make a suitable internal reference material, is rather too volatile for safe, convenient handling, and its regular use would constitute a radiochemical hazard. A number of other reference materials have been proposed, e.g., hexamethyldisiloxane (Long *et al.*, 1979) and [*ethyl*-1,2-^{3}H]triethoxysilane (Elvidge *et al.*, 1971).

The commonly accepted method in current use is the "ghost reference'' procedure proposed by Elvidge *et al.* (1974a). The conditions for resonance of any nucleus X at a constant field H_0 are given in Eq. (1):

$$\nu_X = \gamma_X H_0 (1 - \sigma_X) \tag{1}$$

where ν_X is the resonance frequency of X, γ_X the magnetogyric ratio of X, and σ_X the screening constant of X. The screening constant σ_X is mainly a function of the molecular environment, which replacement of one proton by a triton leaves virtually unchanged.

The ratio of the Larmor frequencies ω_T/ω_H, which may be determined experimentally as ν_T/ν_H, would therefore be expected to be constant with such isotopic substitution. Elvidge *et al.* (1974a) showed that $\omega_T/\omega_H = 1.06663975 \pm 3 \times 10^{-8}$ over a wide variety of molecules.

Minor variations in this ratio were later reported (Elvidge *et al.*, 1979a) and appear to correlate with the type of hybridization of the carbon atom to which the triton or proton is attached, as shown in the accompanying tabulation.

Hybridization	Value of *xyz* in $\omega_T/\omega_H = 1.066639xyz$
sp	746 ± 0.002
sp^2	782 ± 0.002
sp^3	703 ± 0.021

The constancy of the ratio ω_T/ω_H is utilized to provide a ghost reference for tritium spectra by observing the proton spectrum of the tritium-containing sample with a trace of TMS added. The frequency of the TMS resonance is then multiplied by 1.06663975, and this frequency is assigned a zero chemical shift for the tritium spectra. The reliability of this procedure has been checked (Elvidge *et al.*, 1979b; Long *et al.*, 1982) by synthesizing monotritiated TMS and measuring its resonance frequency at both 96 and 320 MHz. The agreed value for ω_T/ω_H is $1.066639737 \pm 2 \times 10^{-9}$.

V. Use of Tritium NMR in Difficult Assignments of ^{1}H Signals

Tritium is usually present in labeled compounds at very high dilution; e.g., 10 mCi ^{3}H in 1 g of toluene corresponds to 0.003% substitution of one ^{1}H atom by a ^{3}H atom. At such levels of labeling the chance of one molecule having more than one ^{3}H atom is negligible (unless some special method of synthesis is used). Consequently ^{3}H—^{3}H coupling is absent even when the label is present at several different molecular sites. Furthermore, ^{1}H—^{3}H coupling can be eliminated by decoupling methods using noise-modulated proton radiation, giving spectra that are greatly simplified compared to proton spectra where multiple ^{1}H—^{1}H spin coupling frequently causes complexity. This simplification can be utilized in making assignments of proton signals that would otherwise be difficult, as shown in the following discussion.

A. Phenanthridine

Phenanthridine has a complex ^{1}H spectrum which had not been analyzed when Elvidge et al. (1979b) prepared a generally labeled specimen ([G-^{3}H]phenanthrine) and produced a ^{1}H-decoupled ^{3}H spectrum in which a single line signal was obtained from each labeled position. Assignments were made for seven resolvable signals by comparison with those for quinoline and isoquinoline (also discussed in their paper) and with ^{1}H data for phenanthrene. Figures for relative levels of labeling at each site were also given (see Table II).

TABLE II

Position and Extent of Tritium Labeling in
Phenanthridine from Tritium NMR
in CDCl$_3$[a]

δ	Assignment	Relative incorporation per site (%)
8.43	1, 10	2
7.67	2	14
7.56	3, 8	34
8.17	4	11
9.17	6	10
7.90	7	13
7.70	9	16

[a] Elvidge (1979) reproduced by permission of J. Chem. Soc. Perkin Trans. II.

B. Pyrrole

Analysis of the complex AA'BB'X-type proton spectrum of pyrrole proved troublesome and was not finally confirmed until Abraham and Bernstein (1959) made extensive studies on both methyl- and deuterium-substituted pyrroles followed by computer simulation. Fukui *et al.* (1970) assigned the N—H resonance and its associated couplings.

Electrophilic substitution by $^3H^+$ in gamma-irradiated silica gel gave (Odell *et al.*, 1982a) labeled pyrrole which revealed a greatly simplified spectrum, as shown in Fig. 2. The higher labeling of the downfield peak confirms that this resonance arises from atoms in positions 2 and 5, whereas the upfield resonance corresponds to positions 3 and 4.

C. Benzamides

In a study on substituent effects on the NMR spectra of substituted benzamides, O'Connor *et al.* (1981) encountered difficulties involving the assignment of proton signals due to NHR groups. Such signals are broadened by the nitrogen quadrupole and, when R = alkyl, are further split by vicinal proton coupling. Furthermore, in many cases the signals are obscured by those arising from aromatic hydrogens. Tritium exchange of NH groups was carried out by adding 10 μl of ^3H_2O (5 Ci cm^{-3}) in a standard 10-mm NMR tube containing 2×10^{-4} mol of amide in dimethyl sulfoxide (DMSO) and heating in a boiling water bath for 1 hr. The ^3H spectrum was obtained by using 10^4 pulses of width 15 μs and a repetition time of 2.2 s. The doublet structure of the N—H resonance is attributed to restricted rotation about the N—C linkage (Fig. 3).

D. Toluene

Proton–proton spin coupling makes the assignment of signals due to ortho, meta, and para hydrogen atoms of toluene difficult. Measurements (Odell *et al.*, 1978) on a mixture of [*ortho*-^3H]toluene, [*meta*-^3H]toluene, and [*para*-^3H]toluene prepared by treating ortho, meta, and para Grignard reagents with HTO have shown the order of shifts for tritons to be meta > para > ortho. This order is in agreement with that obtained by Schaefer and Hutton (1979) using ^1H-NMR spectroscopy on 99% deuter-

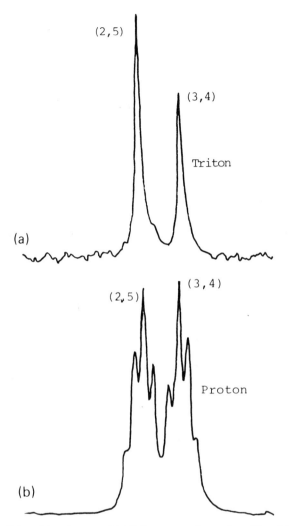

Fig. 2 Triton (a) and proton (b) NMR spectra of pyrrole at 64 and 60 MHz.

ated material at low temperatures and also agrees with an earlier detailed analysis (Williamson *et al.,* 1968) of the ¹H spectrum of toluene; but it disagrees with the order obtained in another analysis (Bovey *et al.,* 1965) of the spectrum of 25% toluene in carbon tetrachloride where the order of shifts is reported as meta > ortho > para. A study by Long *et al.* (1982), using a high-field spectrometer operating at 320 MHz, on solvent effects on the spectrum of toluene has resolved this apparent discrepancy. It has been shown that the use of carbon tetrachloride, chloroform, cyclohex-

(a)

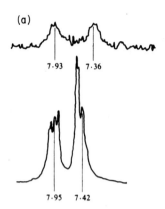

(b)

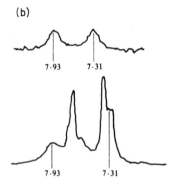

Fig. 3 Triton (above) and proton (below) NMR spectra for NH of benzamides. (a) Benzamide, (b) 3-methylbenzamide. (After O'Connor (1981). *Australian Journal of Chemistry*.)

ane, or DMSO as a solvent causes a change in the chemical shift order from para > ortho in pure toluene or toluene in hexadeuterobenzene (Odell *et al.*, 1978) to ortho > para. Discrepancies in chemical shifts of up to 0.04 ppm among different observers are ascribed about equally to solvent effects and to isotope effects on substrate and reference molecules. Tritium spectra of toluene at 64 and 320 MHz are shown in Fig. 4.

E. Fatty Acids

Proton NMR spectra of long-chain fatty acids characteristically show separate resonances for hydrogens of the terminal methyl and 2-methylene groups, whereas signals due to all other methylene groups form an unresolved band. Reagents such as Eu(III)(fod)$_3$, if used at high concentrations, can shift these methylene signals out of the band so that about six or seven peaks may be resolved (Odell *et al.*, 1982a) using a 60-MHz

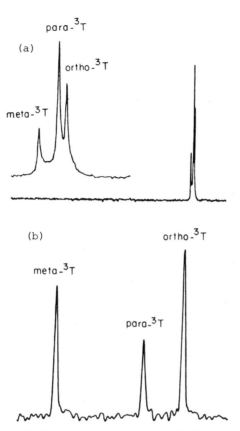

Fig. 4 Proton-decoupled ^{3}H NMR spectra of toluene at 64 MHz (a) Odell *et al.* (1978), and 320 MHz (b) Long *et al.,* (1982).

instrument, whereas use of higher frequencies, up to 400 MHz, increases the number of separate signals to seven or eight. (Spectra by Courtesy of Bruker, Spectro-Spin, Karlsruhe). Fatty acids labeled in the 2-position, however, give excellent tritium spectra (Odell *et al.,* 1980). It has therefore not been possible to locate the tritium atoms in [9,10-^{3}H]stearic acid prepared by hydrogenating oleic acid with tritium gas using tritium NMR spectroscopy, except to show the absence of labeling in position 2–7 and 18.

Oleic acid specifically labeled at the 9- and 10-positions has been prepared (Mounts *et al.,* 1971; Sgoutas and Kummerow, 1964) by hydrogenation of the methyl ester of the acetylenic analog methyl 9-octadecynoate, using tritium gas (Lindlar and Dububs, 1966; Odell *et al.,* 1982b) catalyst in the presence of quinoline. The signals in the proton-decoupled tritium spectrum due to tritons adjacent to the double bond (positions 9 and 10) are well clear of the methylene band, as are also those of the methylenes at positions 8 and 11. It has been shown that this method yields over 90% of the tritium activity in the 9- and 10-positions, with only small amounts in positions 2, 3, 8, and 11. Peak separation for 9-^{3}H and 10-^{3}H is 0.916 Hz at 64 MHz or 0.015 ppm, whereas proton–proton coupling in the proton spectrum makes this difference unobservable.

VI. Applications of Tritium NMR to Labeled Molecules of Biochemical Interest

In this section selected examples of tritium-labeled molecules of biochemical interest in which the positions of atoms have been determined by tritium NMR methods are discussed.

A. Penicillic Acid

Penicillium cyclopium incorporates [^{3}H]acetate into penicillic acid (Fig. 5), and Elvidge *et al.* (1974b) have used tritium NMR methods to show the stereospecificity of this labeling, this being the first use of tritium NMR in a biosynthetic investigation. The triplet signal for 7-CH_2T and a singlet signal for 3-T with a relative intensity ratio of 3 : 1 are easily seen. Assignment of the two 5-T signals required the use of the shift reagent Eu(III)(thd)$_3$, and this caused equal downfield shifts of the higher field 5-H signal and the C-methyl H signal, whereas the shift of the lower field 5-H was about zero. On this basis the higher field 5-H signal was assigned to H_b, since rotation about the C—C bond 1-6 causes H_b and CH_3 to pass about equally close to Eu(III) bound to 1-OH. This report illustrates well the parallel use of proton and triton spectra in assigning signals and also the use of proton–triton coupling in making assignments.

The observed pattern of labeling was presented by the author as evidence of 4,5 ring scission in the presumed orsellinic intermediate in the biosynthesis rather than the alternative 1,2 scission.

B. Testosterone

Testosterone labeled with tritium atoms at the 1- and 2-positions may conveniently be prepared by catalytic reduction with tritium gas of 17β-hydroxyandrosta-1,4-dien-3-one, but the determination, by degradative

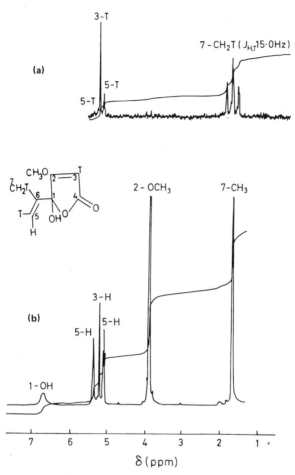

Fig. 5 Triton NMR (a) and ¹H-NMR (b) spectra of penicillic acid at 90 MHz. (Reproduced by permission of Elvidge *et al.* (1974■) and the Royal Society of Chemistry.)

methods, of the distribution of tritium in the reduction product has given variable results (see Mounts, *et al.*, 1971 and references therein). Testosterone gave (Elvidge *et al.*, 1976a), in DMSO solvent, a proton broad-band-decoupled tritium spectrum (Fig. 6) showing strong doublet signals for 1β and 2β tritons, whereas testosterone produced using a Wilkinson homogeneous catalyst gave strong doublet signals characteristic of 1α and 2α tritons. The presence of the doublet structure indicates triton–triton coupling, showing the presence of two tritons in the same molecule. The assignments were made using nondecoupled spectra, and the 1β triton is located in the equatorial conformation, whereas the 2β triton is axial.

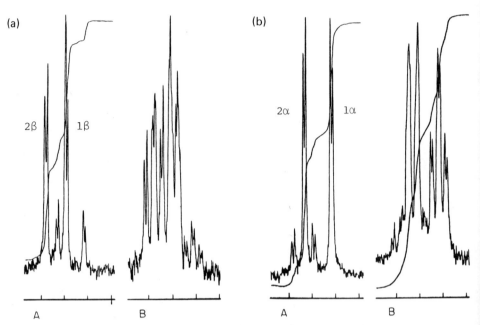

Fig. 6 (a) Triton NMR spectra of [1β, 2β-^{3}H] testosterone. (A) With ^{1}H decoupling; (B) without decoupling. Prepared using Pd—C. (b) Triton NMR spectra of [1α, 2α-^{3}H] testosterone. (A) With ^{1}H decoupling; (B) without decoupling. Prepared using (Ph$_3$P)$_3$RhCl. (Reproduced by permission of Elvidge *et al.* (1976a) and *Steroids.*)

Altman and Silberman (1977a, b) have reported on the tritium NMR spectrum of [1β,2β-^{3}H$_2$ (N)] testosterone prepared by New England Nuclear Corporation (NET-187). The spectrum shown in Fig. 7 illustrates further the power of tritium NMR to reveal details of the labeling pattern.

The spectrum indicates the presence of at least four distinct molecular species.

(a) [1β,2β-^{3}H$_2$]Testosterone (48.8%), doublets centered at 1.388 ppm (1β) and 2.078 ppm (2β), $J_{T_{1\beta}T_{2\beta}}$ = 5.67 Hz

(b) [1β-^{3}H]Testosterone (36.6%), singlet at 1.397 ppm

(c) [2β-^{3}H]Testosterone (2.3%), singlet at 2.088 ppm

(d) [1α,2α-^{3}H$_2$]Testosterone (12.3%), doublets centered at 1.16 ppm (1α) and 2.173 ppm (2α), $J_{T_{1\alpha}T_{2\alpha}}$ = 4.90 Hz.

Details of this kind concerning multiple labeling of individual molecules are not otherwise obtainable without the separation of isotopic isomers, and this has been achieved in only very favorable cases e.g., (Hunter and Odell, 1977; Bruner and Cartoni, 1965).

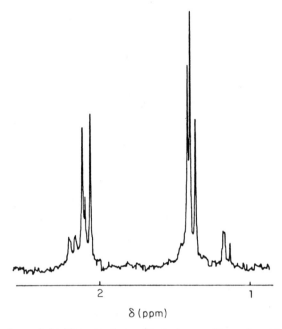

δ (ppm)

Fig. 7 Proton-decoupled tritium spectrum of testosterone. (Reproduced by permission of Altman and Silberman (1977a) and *Steroids*.)

An interesting difference of opinion arose between the Elvidge group and the Altman group over the error likely to be introduced in calculating the relative abundances of the different labeled species present in preparations of tritiated testosterone using integrated intensities of proton noise-decoupled spectra. In 1976(b), Elvidge *et al.* demonstrated the absence of nuclear Overhauser enhancements (NOEs) in proton noise-decoupled tritium spectra of polycyclic aromatic hydrocarbons by showing that the intensities of the signals obtained with and without gated decoupling were identical. They stated (Elvidge *et al.*, 1976a) that NOEs were also absent in their spectra of testosterone and attributed this to the presence of paramagnetic radicals formed by radioactive disintegration of tritons. Supporting evidence for a lack of NOE was obtained by finding an agreement between integrated signal intensities from ^{1}H-decoupled and nondecoupled ^{3}H spectra. The abundances of various species were then calculated directly from these integrated signal intensities.

Altman and Silberman (1977a, b) measured NOEs in their studies on tritiated testosterone using a gated decoupling pulse train and reported 40% NOE (compared to the maximum attainable, 47%). They comment that Elvidge's comparison of integrated signal intensities of decoupled and undecoupled spectra demonstrate only the lack of differential NOEs. They stress the importance of building up data on differential NOEs, so that tritium NMR spectroscopists can anticipate the appearance of this effect and use a NOE suppression technique with delays of up to $10 \times T_1$ seconds between pulses. Where NOE suppression is not necessary, use of the time-consuming suppression pulse train can be avoided.

Elvidge *et al.* (1977a) replied with a list of five compounds in which relative abundances of tritium at different sites had been determined (a) by degradation and counting and (b) by using proton noise-decoupled tritium NMR. Agreement was excellent. They also redetermined the NOE for freshly prepared solutions of labeled testosterone (and other molecules) and found agreement between the abundances calculated from spectra acquired with proton noise decoupling and with NOE suppression.

This interesting exchange of views has certainly alerted tritium workers to the possibility of errors in quantitation arising from NOEs, but in practice differential effects seem to be usually rather small.

VII. Tritium NMR in Studies on Reaction Mechanisms

The following brief descriptions of examples of the use of tritium NMR in studying reaction mechanisms have been chosen to illustrate the diversity of uses to which the technique has been put in this field.

A. Tieman–Reimer Reaction

When the phenoxide ion is treated with chloroform and alkali, *o*-hydroxybenzaldehyde is formed, and Hine and van der Veem (1959) have shown that diclorocarbene is the essential electrophilic reagent.

The second step involves the transfer of a proton to the side chain, and the origin of this proton was in doubt. That it indeed originated from the solvent was established by Kemp (1971) who carried out the reaction in tritiated water and showed, using a somewhat complicated degradation, that the aldehydic hydrogen was labeled.

This result has now been confirmed by Elvidge and co-workers (1974a) using tritium NMR methods. The ^{1}H-NMR spectrum showed the aldehyde singlet at δ 4.2, and the ^{3}H spectrum gave a singlet at δ 4.15 from HTO as a reference. Not only did this approach avoid the degradation procedure, but it also showed the absence of tritium at any other site.

B. Stereospecificity of Δ^1 Dehydrogenation of Testosterone

Microbial Δ^1 dehydrogenation of testosterone using *Chlindrocarpon radicicola* has been studied (Elvidge *et al.*, 1974c) using tritium NMR. [1,2-^{3}H$_2$]Testosterone containing 1α and 1β tritons in the ratio 5:1 was incubated with *C. radicicola*.

The ^{3}H spectrum showed signals from tritons on C-1 and C-2 with intensities in a ratio of 1:5, thus demonstrating 1α,2β-trans elimination.

Elimination using 2,3-dichloro-5,6-dicyanobenzaquinone was shown to be less specific, the tritium spectrum ^{3}H—^{3}H spin-coupled doublets of moderate intensity due to ^{3}H on C-1 and C-2 implying some 1β,2β-cis elimination as well as the predominant 1α,2β-trans elimination.

C. Tritium Exchange in Aromatics on a Platinum Surface

Garnett, Long, and co-workers (1979) have investigated the orientation of tritium atoms introduced into the aromatic ring or side chain of alkylbenzenes and halogenated benzenes when these substrates are sorbed on platinum in the presence of HTO. Their tritium NMR spectra obtained with proton decoupling gave well-resolved signals for ortho, meta, and para tritons which are identified by comparison (Long *et al.*, 1975) with specifically labeled compounds, and they report abundances of tritium atoms as in Table III.

TABLE III

Orientation of Tritium in Halogenated Benzenes and Alkylbenzenes Using Heterogeneous Platinum Exchange

Alkylbenzene	Tritium in Compound Per Site (%)			
	Ortho	Meta	Para	Alkyl
Fluorobenzene	8.3	29	25	—
Chlorobenzene	4.0	30	32	—
Bromobenzene	<2	33	33	—
Toluene	9.8	15	18	CH$_3$, 11
Isopropylbenzene	5.5	13	13	CH$_3$, 7; CH, 11
Cyclohexylbenzene	4.1	19	21	CH$_2$, 1; CH, 25
Triphenylmethane	2.3	9.5	9.7	CH, <1

The pattern of labeling in the alkylbenzenes indicates preferential adsorption through the ring and, further, that the molecule must be strongly tilted so as to allow preferential exchange at aromatic and allylic positions. In triphenylmethane, however, steric effects prevent sorption at the allylic position. Strong steric hindrance sheltering ortho positions and increasing with the size of the halogen atom is evident in reactions of halobenzenes. These results correct some earlier evidence (Garnett and Sollich-Baumgartner, 1961) from deuterium labeling obtained by IR methods. The pattern of results affords strong evidence of a π-dissociative process being dominant in the heterogeneous exchange reaction.

D. Exchange-Labeling Long-Chain Esters on Rhodium Black

Hexylpropionate, labeled by exchange with tritium gas on rhodium black, has been studied (Garnett et al., 1980) with tritium NMR. Signals from the terminal methyl groups are well separated, as are those from the two methylenes adjacent to the ester group. All other methylene signals are included in a broad, unresolved band. Addition of the shift reagent Eu(III)(fod)$_3$ (18 mol %) resolved all the methylene signals, and it was found that the terminal methyl groups were the most heavily labeled and that labeling decreased steadily toward the ester grouping, the activity on the methylene group next to the oxygen atom approaching zero. This evidence suggests that the two ends rather than the polar center of the ester molecule may be the favored points of attachment to the catalyst.

E. The Fate of Hydrogen in Biosynthesis

In addition to the above mechanistic studies extensive use has been made of both ^{2}H and ^{3}H isotopes in tracing the fate of hydrogen in biosynthesis, and both ^{2}H NMR and ^{3}H NMR have been extensively used. This topic has been the subject of a lengthy review (Garson and Staunton, 1979).

VIII. Recent Developments

Tritium labeling is now an accepted part of isotope technology, and new methods of labelling molecules at specific sites are appearing frequently.

The location of such labels is now regularly being determined by tritium NMR, and some recent advances are listed here.

(a) Sodium borohydride labeled with tritium of high specific activity is available, and Altman and Thomas (1980) have described a tritium NMR method of determining its tritium content. The spectrum is interesting in that splittings due to coupling to ^{11}B ($S = \frac{3}{2}$) and to ^{10}B ($S = 3$) are seen.

(b) Elvidge and co-workers (1981a) have described tritium NMR spectra of 50 neurochemicals, labeled with tritium, of importance in studies on neurotransmission.

(c) Ethylaluminium chloride (Long et al., 1975) and other Lewis acids (Elvidge, 1980; Elvidge et al., 1977b) have been shown to catalyze exchanges between aromatic hydrogens and HTO, although some activity appears in side chains.

(d) Detailed distribution of tritium in labeled methyl groups can be determined (Elvidge et al., 1981b) from high-resolution tritium NMR spectra, and relative abundances of $-CT_3$, $-CHT_2$, and $-CH_2T$ species determined utilizing primary and secondary isotope effects.

(e) The zeolite HNaY has been shown (Long et al., 1981) to catalyze exchange between HTO and aromatic ring structures with little labeling in side chains. Ortho/para ratios determined by tritium NMR appear similar to those obtained by homogeneous electrophilic substitution.

(f) The shape-selective zeolite H-ZSM-5 has been used (Odell et al., 1982c) after irradiation with gamma rays as an exchange catalyst for tritium gas–toluene exchange. Yields of [para-^3H]toluene have been shown to be lower than those for gamma-irradiated silica gel. Whether this is a "shape-selective" effect is not yet clear.

(g) Addition of T_2 or HT across a double bond using a metallic catalyst is a favored method of preparing tritium-labeled compounds, but this addition is often complicated by simultaneous exchange (Odell et al., 1982b; Tiers et al., 1964) of the ethylenic protons, giving an asymmetric distribution of tritium. Tritium NMR studies have shown (Elvidge et al., 1982) that similar asymmetric labeling patterns are usually obtained when Wilkinson's hydrogenation catalyst is used.

(h) Altman and co-workers (1978) have discussed primary isotope effects on chemical shifts in cases of strong intramolecular hydrogen bonding. They consider the shape of the potential energy function describing the hydrogen bond and distinguish three types:

    (i) Weak hydrogen bonds in which the potential energy shows two well-pronounced minima and anharmonicity is small. This gives rise to near-zero isotope effects, like those found in alcohols.

(ii) Stronger and shorter bonds in which the potential minima draw closer together and anharmonicity increases. Here the equilibrium distance for ^1H, ^2H, and ^3H differs, with that for $H > D > T$ and we expect positive isotope effects in the chemical shift i.e., they are larger for ^3H than for ^2H. The ratio of shift expected is given by

$$\Delta\delta_{^1H^3H} = 1.44 \; \Delta\delta_{^1H^2H}$$

and this is found experimentally for acetoacetic ester, acetylacetone, dibenzoylmethane, and similar compounds.

(iii) Extremely strong, short hydrogen bonds give a single minimum, and anharmonicity is thus small so that isotope effects are again small, with a small negative component because of lower vibration amplitudes of the heavier isotopes. This effect is seen in the hydrogen maleate anion.

Tritium NMR is now a firmly established technique, and its use is expanding rapidly. Two reviews (Elvidge *et al.*, 1978; 1980), more detailed than space permits this chapter to be, provide ready access to the literature on this field.

IX. The Future

At present the usefulness of tritium NMR is limited by the lack of sensitivity. Thus about 1 mCi of tritiated material is needed to obtain a satisfactory signal/noise ratio in a reasonable time, whereas fractions of 1 μCi of tritium may readily be detected by scintillation counting.

Behrendt (1976) has proposed a method of enhancing the sensitivity of tritium NMR to a point where such low activities may be utilized. The suggested method involves the following.

(a) Working at 0.01 K using helium dilution refrigeration. This implies, of course, working with frozen glasses rather than liquids

(b) Using, instead of the usual magic angle spinning (MAS) technique, the rotating field of Andrew and Eades (1962)

(c) Measuring the anisotropy of beta particles emitted from the spin-aligned tritons instead of the usual rf signals.

The author is careful not to give hard numbers for the expected sensitivity of this device (nor yet for its likely cost), but enormous improvements in sensitivity are promised.

References

Abraham, R. J., and Bernstein, H. J. (1959). *Can. J. Chem.* **37**, 1056–1065.

Altman, L. J., and Silberman, N. (1977a). *Steroids* **29**(4), 557–565.

Altman, L. J., and Silberman, N. (1977b) *Anal. Biochem.* **79**, 302.

Altman, L. J., and Thomas, L. (1980). *Anal. Chem.* **52**, 992–995.

Altman, L. J., Laungani, D., Gunnarson, G., Wennerstrom, H., and Forsen, S. (1978). *J. Am. Chem. Soc.* **100**(26), 8264–8266.

Anderson, H. L., and Novick, A. (1947). *Phys. Rev.* **71**(6) 372–373.

Andrew, E. R., Eades, R. G. (1962). *Discuss. Farady Soc.* **34**, 38–42.

Behrendt, S. (1976). *J. Radioanal. Chem.* **29**, 335–342.

Bloch, F., Graves, A. C., Packard, M., Spence, R. W. (1947). *Phys. Rev.* **71**(6), 373.

Bovey, F. A., Hood, F. P., Pier, E., and Weaver, N. E. (1965). *J. Am. Chem. Soc.* **87**, 2060–2061.

Bruner, F., and Cartoni, G. P. (1965). *J. Chromatogr.* **18**, 390–391.

Elvidge, J. A. (1980). "Isotopes: Essential Chemistry and Applications" (J. A. Elvidge and J. R. Jones, eds.), Ch. 5. *In* "Deuterium and Tritium Nuclear Magnetic Resonance Spectroscopy". The Chem. Soc. (London).

Elvidge, J. A., and Jones, J. R. (1978). "Isotopes in Organic Chemistry" (E. Buncel and C. C. Lee, eds.), Vol. 4. Elsevier Amsterdam.

Elvidge, J. A., Bloxsidge, J., Jones, J. R., Evans, E. A. (1971). *Org. Magn. Reson.* **3**, 127–138.

Elvidge, J. A. *et al.* (1974a). *J. Chem. Soc. Perkin Trans. 2,* 1635–1638.

Elvidge, J. A., Al-Rawi, J. M. A., Jaiswal, D. K., Jones, J. R., and Thomas, R. (1974b). *J. Chem. Soc. Chem. Commun.* 220–221.

Elvidge, J. A., Al Rawi, J. M. A., Thomas, R., and Wright, B. J. (1974c). *J. Chem. Soc. Chem. Commun.* 1031–1032.

Elvidge, J. A. *et al.* (1976a). *Steroids* **26**(3), 359–376.

Elvidge, J. A., Al Rawi, J. M. A., Bloxsidge, J. P., and Jones, J. R. (1976b). *J. Labelled Compd. Radiopharm.* **12**(2), 293–307.

Elvidge, J. A., Bloxsidge, J. P., Jones, J. R., and Mane, R. A. (1977a). *J. Chem. Res. Synop.* 258–259.

Elvidge, J. A., Jones, J. R., Long, M. A., and Mane, R. B. (1977b). *Tetrahedron Lett.* **49**, 4349–4350.

Elvidge, J. A., Bloxsidge, J. P., Jones, J. R., Mane, R. B., and Saljoughian, M. (1979a). *Org. Magn. Reson.* **12**(10), 574–578.

Elvidge, J. A., Jones, J. R., Mane, R. B., and Al Rawi, J. M. A. (1979b). *J. Chem. Soc. Perkin Trans. 2,* 386–388.

Elvidge, J. A., Al Rawi, J. M. A., Jones, J. R., Mane, R. B., and Saieed, M. (1980). *J. Chem. Res. Synop.* 298–299.

Elvidge, J. A. *et al.* (1981a). *J. Labelled Compd. Radiopharm.* **18**(8), 1141–1165.

Elvidge, J. A. *et al.* (1981b). *Org. Magn. Reson.* **15**(2), 214–217.

Elvidge, J. A. *et al.* (1982). *J. Chem. Res. Synop.* 82–83.

Fukui, H., Shimokawa, S., and Sohma, J. (1970). *Mol. Phys.* **18**(2), 217.

Garnett, J. L., and Sollich-Baumgartner, W. A. (1961). *Aust. J. Chem.* **14**, 441–448.

Garnett, J. L., Long, M. A., and Lukey, C. A. (1979). *J. Chem. Soc. Chem. Commun.* 634–635.

Garnett, J. L., Long, M. A., and Odell, A. L. (1980). *Chem. Aust.* **47**(6), 215–220.

Garson, M. J., and Staunton, J. (1979). *Chem. Soc. Rev.* (1979). **8**, 539–561.

Hine, J., and van der Veen, J. M. (1959). *J. Am. Chem. Soc.* **81,** 6446–6449.

Hunter, K. A., and Odell, A. L. (1977). *Int. J. Appl. Radiat. Isot.* **28,** 439–440.

Kemp, D. A. (1971). *J. Org. Chem.* **36,** 202–204.

Lindlar, H., and Dubuis, R. (1966). *Org. Synth.* **46,** 89–92.

Long, M. A., Garnett, J. L., and Vining, R. F. W. (1975). *J. Chem. Soc. Perkin Trans. 2,* 1298–1303.

Long, M. A., Garnett, J. L., and Lukey, C. A. (1979). *Org. Magn. Reson.* **12**(a), 551–552.

Long, M. A., Garnett, J. L., Williams, P. G., and Mole, T. (1981). *J. Am. Chem. Soc.* **103,** 1571–1572.

Long, M. A., Saunders, J. K., Williams, P. G., and Odell, A. L. (To be published.)

Mounts, T. L., Emken, E. A., Rohwedder, W. K., and Dutton, H. J. (1971). *Lipids* **6,** 912.

O'Connor, C. J., Martin, R. W., and Calvert, D. H. (1981). *Aust. J. Chem.* **34,** 2297–2305.

Odell, A. L., Calvert, D., Kazakevics, A., Martin, R. W. (1977). *JEOL News* **14A**(1), 5–7.

Odell, A. L., Calvert, D., Martin, R. W. (1978). *Org. Magn. Reson.* **11**(4), 213–214.

Odell, A. L., Crossley, P. C., Martin, R. W., Mawson, J. B. (1980). *J. Labelled Compd. Radiopharm.* **17**(6), 779.

Odell, A. L. Calvert, D., and Martin, R. W. (1982a). Unpublished results.

Odell, A. L., Ronaldson, K. J., Martin, R. W., and Calvert, D. J. (1982b). *Aust. J. Chem.* **35,** 1615–1620.

Odell, A. L., Chipman, N., Gill, P. (1982c). Unpublished results.

Schaefer, T., and Hutton, H. M. (1979). *Org. Magn. Reson.* **12**(11), 645–646.

Sgoutas, D. D., and Kummerow, F. A. (1964). *Biochemistry* **3,** 406.

Tiers, G. V. D., Brown, C. A., Jackson, R. A., Lahr, T. N. (1964). *J. Am. Chem. Soc.* **86,** 2526–2527.

Williamson, M. P., Kostilnik, R. J., and Castellano, S. M. (1968). *J. Chem. Phys.* **49,** 2218–2224.

3 Boron-11

R. Garth Kidd

Department of Chemistry
The University of Western Ontario
London, Canada

Present to the extent of only 3 ppm in the earth's crust, boron was among the first atoms to engage the interest of NMR spectroscopists in the 1950s. Boron has two naturally occurring magnetically active isotopes, both of them weakly quadrupolar. It is ^{11}B that is most commonly studied because of its greater abundance and higher detection sensitivity. A comparison limited to their quadrupole moments suggests that ^{11}B is also favored by slower relaxation, giving narrower lines, but such is not the case. The ^{10}B/^{11}B spin factor ratio in Eq. (6), Chapter 5, Vol. I, outweighs the Q^2 ratio, and the ^{10}B isotope gives narrower lines by a factor $\Delta\nu_{10_B}/\Delta\nu_{11_B} = 0.65$.

Although the biological activity of boron is not fully understood, it is known to be essential for green algae and higher plants. In excessive amounts it is moderately toxic to plants and slightly toxic to mammals. Boron has a high affinity for oxygen and is always found naturally as the

TABLE I
Nuclear Properties of Boron Isotopes

Nucleus	Spin	Abundance (%)	Magnetogyric ratio (10^7 rad T^{-1} s^{-1})	Quadrupole moment (10^{-28} m^2)	Receptivity referred to ^{13}C
^{10}B	3	19.6	2.87	0.074	22
^{11}B	312	80.4	8.58	0.036	754

oxide. Boron hydrides are readily oxidized and, because the heat liberated on a mass basis is roughly twice that for hydrocarbons, they received much attention as potential rocket fuels in the early days of the NASA space program. Boron-10 serves as a reactor moderator because of its high neutron capture cross section which presumably accounts for its lower natural abundance. In contrast to metallic nitrides and carbides containing ionic N^{3-} and C^{4-}, respectively, metallic borides are nonstoichiometric alloys in which metal–boron interactions are covalent in nature, and boron behaves as an electropositive metal. The ^{11}B resonance positions in metallic borides are thus expected to exhibit Knight shift behavior.

I. Boron Chemical Shifts

From its most highly shielded BI_4^- environment to the $B(CH_3)_3$ molecule at the low-field end of the scale, the boron atom exhibits a chemical shift range of 220 ppm. This is comparable to 250 ppm for ^{27}Al (Akitt, 1972), reaching 1400 ppm for ^{71}Ga (Akitt et al., 1965) and 5500 ppm for ^{205}Tl (Chan and Reeves, 1974; Koppel et al., 1976) and reproducing the characteristic pattern of similar shift ranges for the first two members of a periodic-table group followed by a dramatic increase for the heavier members. To the left, ^{9}Be has received insufficient study to establish a shift range, but to the right ^{13}C shieldings span a range of 650 ppm (Mann, 1978), followed by ^{14}N (Witanowski and Webb, 1973) with 900 ppm and ^{19}F with 800 ppm (Mann, 1978).

The external reference against which most investigators report boron chemical shifts is $Et_2O \cdot BF_3$ whose resonance lies near the middle of the range. Its ^{11}B Ξ value is 32.08394 MHz. Other reference compounds against which ^{11}B chemical shifts have been reported in the literature, together with their $\delta_{Et_2O \cdot BF_3}$ values which must for consistency be subtracted from the chemical shifts so referenced, are BF_3, δ 11.5; $B(OMe)_3$, δ 18.1; BCl_3, δ 47.0; and $B(CH_3)_3$, δ 86.2. The primary isotope effect on boron shielding is on the order of 0.1 ppm (McFarlane, 1973), and for all but the most precise comparisons, the ^{10}B and ^{11}B chemical shifts of a specific environment can be regarded as identical.

The chemical shift for the hypothetical ^1S state gaseous boron atom does not appear to have been located and, lacking this $\sigma_p = 0$ standard state for boron shielding, it is not yet possible to place the available chemical shifts on an absolute shielding scale. This standard state represents the upper shielding limit for diamagnetic boron and, because its calculated σ_d is 202 ppm relative to that of the bare boron nucleus, ^{11}B must be one of the few cases where molecules such as $B(CH_3)_3$ at the deshielded end of the range have resonance frequencies higher than that of the bare nucleus.

The nuclear shielding experienced by boron, hence the chemical shifts exhibited by its compounds, can be naturally divided into two classes of compounds, which for convenience of study are recognized and discussed separately. The two distinct shielding ranges that arise for ^{11}B are based on the local symmetry at the boron atom, with trigonal symmetry giving chemical shifts in the range δ 90 to δ −10 and tetrahedral symmetry giving shifts in the range δ 10 to δ −130. Within the 220-ppm ^{11}B shift range observed to date, the overlap between the two categories of compounds is only 20 ppm. When boron compounds containing halogen atoms, noted for causing shielding extremes, are removed from the comparison, then the trigonal coordination range is reduced to δ 90 to δ 15 and the tetrahedral coordination range to δ 10 to δ −40, indicating that only with the highly shielded halogen extremes does overlap between the two regions occur.

The distinction between the two classes of boron compounds has been drawn on symmetry grounds rather than on the basis of oxidation states. These symmetry differences beget differences in charge density at the boron atom not unlike those characterized by a change in oxidation state but formalized in a different way. Described in valence bond terms, tetrahedral coordination arises when the coordinatively unsaturated trigonal boron atom with three pairs of electrons in its sp^2-hybridized valence shell reacts as a Lewis acid with a nucleophilic donor to achieve sp^3 hybridization—and four electron pairs in the boron valence shell. Although this reaction cannot be regarded formally as reduction of the boron compound, it nonetheless represents an increase in electron density at the boron atom, which translates into chemical shifts at the low-frequency end of the total range related to reduced $\langle 1/r^3 \rangle$ and σ_p terms.

A. Theoretical Understanding of Boron Shielding

Theoretical approaches to boron shielding start with Ramsey's generalized nuclear screening equation (Ramsey, 1950) and the recognition that the chemical shifts observed arise from variations in the paramagnetic σ_p term rather than from variations in the diamagnetic σ_d term. Although the alternative of assigning a significant role to σ_d variations has been advanced (Grinter and Mason, 1970), it has not met with widespread acceptance among spectroscopists.

The σ_p portion of the shielding depends on (a) electronic excitation energy, (b) the inverse cube radius of the valence shell electrons, and (c) the orbital angular momentum of the valence shell electrons. Most boron compounds lack the low-lying electronic excited states necessary for a satisfactory correlation with the electronic excitation energy to exist. In theoretical analyses of boron shieldings, factors (b) and/or (c) appear to predominate, with factor (a) implicitly or explicitly assumed to remain constant.

Tetrahedral boron, with its high symmetry, lies closest to the hypothetical free boron atom in a ^1S state with $\sigma_p = 0$ and occupies the upper reaches of the boron shielding range. For molecules with close to tetrahedral symmetry, chemical shifts can be rationalized in terms of the ligand-produced nephelauxetic effect operating through the $\langle r^{-3} \rangle$ factor. Where boron forms four σ bonds, there are no π-bonding orbitals available in its valence shell, and only minor variations in the angular momentum factor residing in departures from strictly tetrahedral symmetry can occur.

Trigonal boron, with its p_π orbital more or less populated depending on the π-donor strength of the σ-bonded atoms, is subject to wide variations in orbital angular momentum. These variations occur around a sizable σ_p contribution to the shielding associated with the D_{3h} or lower symmetry. p-Orbital populations calculated using molecular orbital (MO) methods of varying degrees of sophistication (Hall *et al.*, 1974; Kroner and Wrackmeyer, 1976) correlate well with ^{11}B chemical shifts, confirming the essential utility of the model.

When group V, VI, and VII ligating atoms with lone pairs are σ-bonded to trigonal boron, ligand-to-boron π-bonding can occur with a resultant increase in the electron density at the boron atom. Since alkyl carbon lacks the electrons to participate in π-bonding, the boron atom in trialkylboranes is deprived of electron density through this mechanism, resulting in chemical shifts at the extreme high-frequency end of the ^{11}B range. Evidence derived mainly from bond length comparisons indicates a substantial degree of π-bonding from F, and lesser amounts from O and N. With the heavier members of groups V, VI, and VII, where the size of the orbital containing the lone pair is too great to provide optimum overlap with the boron $2p_z$ orbital, little if any π-bonding occurs, and low-frequency shifts produced by these heavier ligand atoms must be explained in terms of factors other than π-bonding.

Trigonal symmetry of this type is rare among the atoms of the periodic table, and the only analogous structures with which the shieldings of trigonal boron can be compared are the unstable carbocations. A comparison with ^{13}C shieldings has been made (Spielvogel *et al.*, 1975), and the linear correlation between ^{11}B and ^{13}C shieldings in analogous trigonal environments shows ^{13}C to be more sensitive to variations by a factor of 2.6.

B. Shielding of Trigonal Boron

Boron forms trigonal compounds with C ligating atoms from group IV, with N, P, As, and Sb ligating atoms from group V, with O, S, and Se ligating atoms from group VI, and with F, Cl, Br, and I ligating atoms from group VII. Although this variety in compound formation is noteworthy, it is not unique to boron. Other atoms, however, do not exhibit this extended variety of chemical behavior *in more than one symmetry class*. The chemistry of boron is remarkable in that not only is the variety of trigonal compounds formed extensive, but the variety of tetrahedral boron compounds formed is equally extensive. Among all magnetically active nuclei, ^{10}B and ^{11}B thus provide the best opportunity for the chemist and the NMR spectroscopist to test the effect of symmetry change on nuclear shielding, all other factors being kept as nearly constant as possible.

Chemical shifts for trigonally coordinated boron lie in the range demarcated by δ 87 for $B(CH_3)_3$ and δ 7 for BI_3 (Fig. 1). This pattern of minimum deshielding (i.e., minimum σ_p) for I and large or maximum deshielding for alkyl C is a not uncommon one, being also observed for ^{113}Cd, ^{199}Hg, ^{27}Al, and ^{205}Tl. When bonded to atoms from groups IV, V, and VI, however, alkyl C does not dominate the range limit of, for example, ^{13}C, ^{29}Si, ^{119}Sn, ^{77}Se, and ^{125}Te.

1. ^{11}B—C Environments

The extremely narrow 3-ppm range from δ 87 to δ 84 for trialkylborane chemical shifts indicates the extreme insensitivity of ^{11}B shielding to alkyl group chain length and chain branching. Alkyl group structures that reduce the local symmetry at boron below D_{3h} cause shielding variations, and ring formation by two alkyl substituents that incorporates B into a

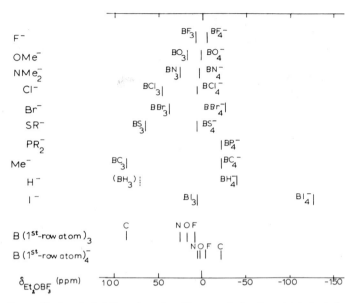

Fig. 1 Boron-11 chemical shifts selected to illustrate their trigonal-tetrahedral symmetry dependence and ligating atom dependence.

five-membered ring reduces the C—B—C bond angle below the preferred 120° and deshields B by an additional 6 ppm. *tert*-Butyl groups cause slightly higher shielding than primary and secondary carbon; this is probably an electronegativity effect rather than a steric one.

Unsaturation at the α and β positions in organoboranes introduces at the α carbon a p orbital with π symmetry that matches the symmetry of the vacant boron p_z orbital. Even limited occupancy of p_z through π-bonding increases the electron density at the boron atom, and replacement of one alkyl group in BR_3 by $—C_6H_5$, $—CH{=}CH_2$, or $—C{\equiv}CH$ causes a low-frequency shift of 7, 10, or 13 ppm, respectively.

2. ^{11}B—Si, ^{11}B—Sn, and ^{11}B—Pb Environments

Below C, the other members of group IV engage in compound formation with boron to a much more limited extent, and there are barely enough spectral data available to establish the shielding trend for these ligating atoms. Although a uniform series of compounds of the type R_2BX (X = Si, Sn, Pb) is not available, suitably substituted triammino boranes have been studied, and the chemical shifts for the compounds $(R_2N)_2CX$ (X = C, Si, Sn, Pb) are shown in Fig. 2a.

The shielding for Si substituents is identical to that for C, and as the

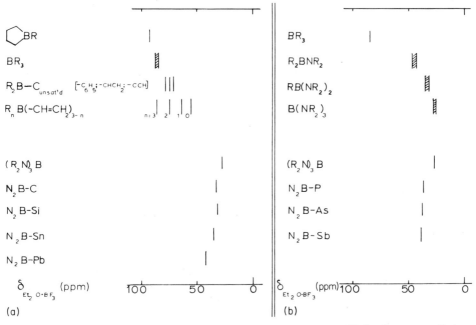

Fig. 2 (a) Chemical shift dependence of trigonal boron on group IV ligating atoms. (b) Chemical shift dependence of trigonal boron on group V ligating atoms.

ligating atoms become heavier, there is a slight trend toward lower shielding, the C—Pb difference being 10 ppm.

It is worth noting here, because of the frequency with which the pronounced shielding effect of I is described as a heavy-atom effect, that Pb is heavier than I and yet here it reflects the regular continuation of a trend toward *lower* shielding.

3. ^{11}B—N Environments

Reference to Fig. 1 shows that the shielding of trigonal boron by ligating atoms from the first row of the periodic table in compounds of the type B(first-row atom)$_3$ lies in the order C $\lll$ N $<$ O $<$ F. Because C lacks a lone pair of electrons with which to enhance the electron density at the boron atom through π-bonding, this is generally accepted as the reason for the large shielding difference between C and the other first-row atoms. Replacement of one alkyl group in BR$_3$ by an —NH$_2$, —NHR, or —NR$_2$ group increases the shielding by 40 ppm, giving a narrow chemical shift range of δ 47 to δ 43 for the R$_2$BNR$_2$ compounds shown in Fig. 2b.

Substitution of a second amino group causes a further 10-ppm shielding enhancement, with the third N substitution increasing the shielding by another 5 ppm. Thus all trigonal boron compounds of the types C_2BN, CBN_2, and BN_3 absorb in the range δ 46 to δ 27. The fact that N substitution of the first alkyl group in BR_3 causes the largest shielding increment, followed by relatively small increments for further N-substitution, suggests that the total π-bond order at the boron atom changes little after the first lone pair of electrons has been provided.

4. ^{11}B—P, ^{11}B—As, and ^{11}B—Sb Environments

Here the shielding pattern is virtually identical to that observed for the heavier members of group IV. A comparison of Fig. 2a with Fig. 2b shows in both cases a very slight shift to lower shielding with each substitution of a heavier ligating atom from the same group. Between P donor ligands and Sb donor ligands the difference is less than 5 ppm.

5. ^{11}B—O Environments

Compounds of the type R_2BO absorb in the region of δ 52 ± 2, indicating that the shielding increment achieved by replacing one alkyl group in BR_3 by O is less than the 40-ppm shielding increment achieved with one amino N. Further O-substitution to give compounds of type RBO_2 with δ 30 ± 2 and BO_3 with δ 18 ± 2 gives larger shielding increments than the corresponding N substitution, so that BO_3 compounds are more shielded than BN_3 compounds by about 10 ppm.

Again, as in the case of nitrogen, the shielding effect of the ligating oxygen atom does not depend on whether it is alkoxy oxygen or hydroxy oxygen, so that boric acids and borate esters of the same structural type absorb in the same regions. Where the two oxygens in an RBO_2 compound incorporate B into a five-member ring, one sees the familiar 5- to 6-ppm deshielding increment resulting from the strain at B. The six-member boroxine ring compounds, which are of the structural type RBO_2, absorb within the normal range for this type, presumably because the local symmetry at the boron atom retains 120° bond angles.

6. ^{11}B—S and ^{11}B—Se Environments

The pattern of ^{11}B shielding observed for the heavier ligating atoms of group VI is similar to the pattern seen for group V. The change from O to S introduces a shielding increment of δ 22 ppm for the first substitution, 15 ppm for the second, and 8 ppm for the third, seen in a comparison of the $R_nB(OR)_{1+n}$ and the $R_nB(SR)_{1+n}$ ($n = 0, 1, 2$) series of compounds shown in Fig. 3.

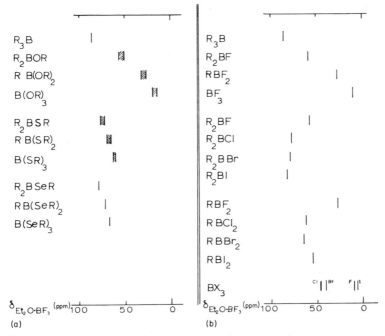

Fig. 3 (a) Chemical shift dependence of trigonal boron on group VI ligating atoms. (b) Chemical shift dependence of trigonal boron on group VII ligating atoms.

Relative to S, Se provides a further 2 ppm per Se shielding decrement consistent with the reduced Se → B π-bonding picture. It is worth noting here that, relative to δ 86 for Me_3B in which no electrons are available for π-bonding, the δ 79 for Me_2Br-SeMe represents a shielding increment that reflects very little π-bonding when one recognizes that the electronegativities of C and Se are virtually identical.

7. $^{11}B—F$ Environments

It is the B—F bond for which the strongest independent evidence in favor of π-bonding exists, and among the first-row ligating atoms it is F that causes the highest shielding of ^{11}B. The relative shielding properties of F among first-row ligating atoms are depicted at the bottom of Fig. 1. Successive substitution of F for Me in BMe_3 gives δ 59 for Me_2BF, δ 28 for $MeBF_2$, and δ 10 for BF_3 as shown in Fig. 3b, with the BF_3 value demarcating the low-frequency end of the shielding range for trigonal boron if we exclude the special case of BI_3 at δ −8 to which reference will be made later. The shielding increments relative to alkyl C of 27, 29, and

18 ppm for successive F substitution show that the full π-bonding potential of B is not satisfied by the first F atom.

8. ^{11}B—Cl, ^{11}B—Br, and ^{11}B—I Environments

In the R_2BX series of compounds shown in Fig. 3b, the pattern of a modest decrease in shielding with increasing weight of halogen mirrors precisely that already seen in groups IV, V, and VI. In particular, the iodide is *less* shielded than the bromide and chloride. Only in compounds containing more than one halogen does the extraordinary shielding ability of iodine become manifest. In R_2BI_2, boron is *more* shielded than in the analogous Br and Cl compounds, and BI_3 at δ 6 is the most highly shielded of all trigonal boron compounds.

C. Shielding of Tetrahedral Boron

At the high-field end of the boron shielding range, and spanning approximately 135 ppm, lie the chemical shifts for tetrahedrally coordinated boron. When the well-known supershielding effect of iodide is discounted by removing iodo compounds from the analysis, the chemical shifts for all other tetrahedral boron compounds converge in the extremely narrow range of 32 ppm from BCl_4^- at δ 7 to BBr_4^- at δ -25, (except for boron hydrides which are analyzed separately later). When compared with trigonal boron, which even without the inclusion of iodo compounds spans over 80 ppm, it is clear that the shielding of tetrahedral boron is much less sensitive to ligand substitution. When referenced against the $\sigma_p = 0$ standard state, its chemical shifts are also significantly smaller.

It is revealing to compare boron shieldings in BL_3 and BL_4^- compounds (L = ligand bonding through first-row ligating atoms C, N, O, and F) (Fig. 1). Conversion of the trigonal to the tetrahedral molecule by nucleophilic addition of the fourth ligand increases the boron shielding by 13 ppm in the case of F^-, by 16 ppm in the case of RO^-, and by 23 ppm in the case of R_2N^-, a modest increase consistent with the fact that in trigonal compounds all three ligands are capable of π-bonding to boron, with π-donor strengths increasing in the order N < O < F as indicated by bond length comparisons. For R_3C^- with zero π-donor strength, the shielding increase is an enormous 107 ppm, confirming the lower symmetry associated with the vacant p_z orbital as the cause of the large σ_p and downfield shifts shown by trigonal boron compounds.

An identical pattern of behavior is observed (Spielvogel and Purser, 1971) in the ^{13}C shieldings of CL_3^+ trigonal carbocations and their CL_4 analogs. Addition of the fourth RO^- moves the ^{13}C resonance upfield by

26 ppm, but the fourth R^- moves it by more than 350 ppm. π-Bonding is already recognized (Kidd, 1967) as an effective generator of σ_p and downfield shifts. It is now clear that a vacant p_π orbital is equally effective. Anyone inclined to attribute the deshielding simply to the absence of a fourth ligand should note that ^{14}N in NR_3 absorbs 37 ppm *upfield* of NR_4^+—a case where electrophilic addition of the fourth ligand does not change the orbital hybridization at the nitrogen atom.

The ligands on the left in Fig. 1 have been arranged according to increasing symmetry dependence of the ^{11}B shift, and it is clear that this dependence is inversely proportional to the π-donor ability of the ligand, not only for the first-row ligating atoms but for the others as well. The extent to which halide ions function as π donors decreases in the order F^- > Cl^- > Br^- > I^-, and the chemical shift for the hypothetical BH_3 imputed (Noth and Wrackmeyer, 1974) from the isoelectronic carbon analog is consistent with the inability of hydride ions to participate in π-bonding.

By focusing in Fig. 1 on the chemical shifts of the tetrahedral compounds only, the shielding characteristics of tetrahedral boron in an absolute sense relative to the spherically symmetric gaseous atom with $\sigma_p = 0$ can readily be perceived and summarized. In both magnitude and range, σ_p for ^{11}B in tetrahedral environments is significantly less than for ^{11}B in trigonal environments.

The Tetrahaloborate Anions

Consistent with the chemical shift patterns for other nuclei, and in large part because of the anomalous shielding properties of covalently bonded iodine, the chemical shifts for boron halides span 80% of the total ^{11}B range. Tetrahedral boron halides alone span 135 ppm, and their chemical shifts are shown in Fig. 4. They exhibit normal halogen dependence (NHD) (Kidd, 1980) wherein boron shielding increases in the order Cl^- < Br^- $\ll$ I^- and has a Cl—Br/Cl—I halide separation ratio for BX_4^- compounds of 0.25, close to the 0.3 average (Kidd, 1980) over all central atoms studied.

For each of the four binary mixed halide systems analyzed in Fig. 4, the chemical shifts for the five components of each system form a regular pattern the numerical aspects of which are discussed later in the context of pairwise additivity parameters. Particularly revealing for the four compounds $BFCl_3^-$, $BF_2Cl_2^-$, BF_3Cl^-, and BF_4^- is a comparison of the ^{11}B chemical shifts with the ^{19}F chemical shifts shown at the bottom of Fig. 4. A substituent constant picture of chemical shifts in these compounds would predict a major change in the ^{11}B shift and a minor change in the ^{19}F shift with each successive substitution of F^- for Cl^-. (A substituent constant model is one in which a given ligand makes an incremental contribu-

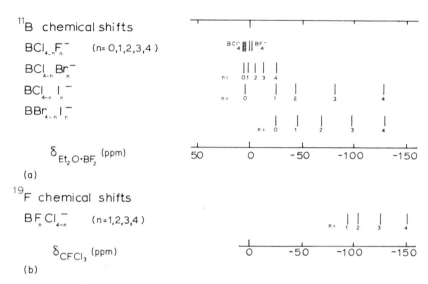

Fig. 4 (a) Boron-11 chemical shifts of tetrahaloborate anions. (b) Fluorine-19 chemical shifts of $BF_nCl_{4-n}^-$ anions.

tion to the shielding of a given central atom, the increment being independent of the other ligands present. This independence requires that the effect of substitution be felt only by the central atom and not by the other ligands.) What we observe, however, is an average shift of 2 ppm in the ^{11}B resonance and 19 ppm in the ^{19}F resonance with every substitution. Although the ^{11}B shifts are too small to assess whether or not the increments are constant, it is clear that by either a through-two-bonds or a through-space mechanism, substitution has a profound effect on ^{19}F shielding.

D. The Pairwise Additivity Model

Experience with ^{11}B as well as with the ligand dependence of metal shielding in complexes of ^{27}Al (Kidd and Truax, 1968), ^{73}Ge (Kidd and Spinney, 1973a), and ^{93}Nb (Kidd and Spinney, 1973b) indicates that ligand–ligand interactions are present to the degree that a first-order substituent constant model cannot reproduce the observed chemical shifts. A number of second-order empirical models for chemical shifts have been introduced, and the most successful of these is the pairwise additivity model (Vladimiroff and Malinowsky, 1967).

Consider the molecule represented as a polyhedron, Fig. 5; each *edge* is taken to contribute a constant shielding to the atom at the center, rather

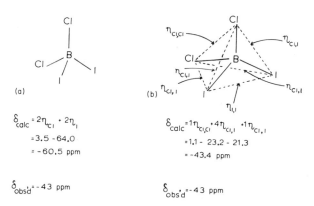

(a)

$\delta_{calc} = 2\eta_{Cl} + 2\eta_I$

$= 3.5 - 64.0$

$= -60.5$ ppm

$\delta_{obsd} = -43$ ppm

(b)

$\delta_{calc} = 1\eta_{Cl,Cl} + 4\eta_{Cl,I} + 1\eta_{Cl,I}$

$= 1.1 - 23.2 - 21.3$

$= -43.4$ ppm

$\delta_{obsd} = -43$ ppm

Fig. 5 Substituent constant (a) and pairwise additivity (b) models for boron shielding.

than each *apex* making a constant contribution as in the substituent constant model. For the most common mixed-ligand case where two different substituents are present, the chemical shifts for all molecules, including cis and trans distinctions in octahedral ones, are fully determined by the three pairwise parameters η_{xx}, η_{xy}, and η_{yy}.

The mixed tetrahaloborate anions in methylene chloride solution (Hartman and Schrobilgen, 1972) are sufficiently labile that the halogen scrambling reaction can be induced, leading to a statistical distribution of halide among the various $BX_{4-n}Y_n^-$ ($n = 0, 1, 2, 3, 4$; X, Y = F, Cl, Br, I) molecules. The chemical shifts for these binary mixed tetrahalides are shown in Fig. 4, and the pairwise additivity parameters for the halide ions and for a number of other donor ligands are given in Table II, re-referenced to $\delta_{ET_2O \cdot BF_3} = 0$ where necessary. Where halide ions are the only ligands present, ^{11}B chemical shifts calculated using the pairwise additivity model agree with those observed to within 0.5 ppm (Hartman and Schrobilgen, 1972), providing remarkable confidence in the model. In other molecules, differences of up to 3 ppm are encountered (Spielvogel and Purser, 1971).

E. Boron Hydrides and Other Boron Clusters

Most periodic-table representations of the elements are ambivalent about hydrogen and group it both with the alkali metals and with the halogens to emphasize its duality. Both the electronegativities ($B = 2.0$, $H = 2.2$) and the proton shielding range confirm the view of this fascinating class of compounds as true hydrides containing hydrogen($-I$). When assessing the shielding effects of ligating atoms on ^{11}B as a function of their

TABLE II

Pairwise Additivity Parameters η_{ij} for Tetrahedral ^{11}B Chemical Shifts Referenced to $\delta_{Et_2O \cdot BF_3} = 0$[a]

					i				
j	I⁻	H⁻	Br⁻	Me₂P⁻	F⁻	MeO⁻	Me₂N⁻	Cl⁻	MeS⁻
I⁻	−21.3	—	—	—	—	—	—	—	—
H⁻	—	−6.7	—	—	—	—	—	—	—
Br⁻	−10.9	−4.5	−3.9	—	—	—	—	—	—
Me₂P⁻	—	—	—	−3.5	—	—	—	—	—
F⁻	—	—	—	—	−0.3	—	—	—	—
MeO⁻	—	—	—	—	—	0.5	—	—	—
Me₂N⁻	—	—	—	—	—	—	0.7	—	—
Cl⁻	−5.8	−3.1	−0.5	—	0.8	—	—	1.1	—
MeS⁻	—	—	—	—	—	—	—	—	1.2
Me₃N	−3.5	—	3.0	—	0.5	—	—	1.7	—
Me₂O	—	—	2.6	—	0.4	—	—	2.4	—
Me₂S	−2.1	—	−0.1	—	1.2	—	—	0.7	—
Me₃P	—	—	−1.0	—	0.5	—	—	−0.4	—

[a] Data from Hartman and Miller, 1974; Hartman and Schrobilgen, 1972; Spielvogel and Purser, 1971.

periodic-table positions, one is anxious to see where H falls on the shielding scale because it is difficult to predict by induction from first principles. At 2.2, its electronegativity lies between those for boron and carbon and is identical to the electronegativity of iodine. In combination with trigonal boron, it provides a limiting case in which π bonding is absent. In diborane and in boron hydride clusters, it also exemplifies the three-center bond.

Trigonal boron provides few examples of ligating H. The homogeneous ligand case BH_3 is unavailable because dimerization occurs to give the more stable B_2H_6. An imputed chemical shift of δ 70 for BH_3 is obtained (Noth and Wrackmeyer, 1974) by correlation with the measured ^{13}C shielding for the isoelectronic CH_3^+ carbocation, and this value appears in Fig. 1. We note that this position is close to and slightly upfield of the BC_3 region, consistent with the comparable electronegativities of H and C. The CH_3- and H-substituted bis(organyloxo)boranes have chemical shifts of δ 35 and δ 28, respectively (McAchran and Shore, 1966; Shore et al., 1972), again indicating slightly higher shielding due to H.

Tetrahedrally coordinated boron provides in BH_4^- a limiting example of the homogeneously ligated hydride. At δ −42, it is 18 ppm to high field of BBr_4^- and represents, excepting the anomalous BI_4^-, the most highly shielded of homogeneous ^{11}B environments. This is consistent with the low electronegativity value of H. With less electropositive metal ions, particularly Al^{3+} and transition metal cations, BH_4^- acts as a ligand to form hydrogen-bridged complexes in solution. The most convenient if not the only method for characterizing these complexes is provided by ^1H- and ^{11}B-NMR, and an extensive literature summarized by Noth and Wrackmeyer (1978), has described both the structure and the kinetics of these interesting compounds. Todd and Siedle (1979) have written a comprehensive review that provides NMR parameters for 225 boranes, carboranes, and heteroatom boranes, subdivided according to the number of boron atoms in the cluster.

The MO description of bonding in boron hydride clusters offers distinct advantages over the localized electron pair models, and the topological approach of Lipscomb (1963) based on four structural elements yields parameters with which ^{11}B spectral correlations can be made. There can be three-center B—H—B bonds (labeled s), closed three-center B—B—B bonds (t), conventional B—B bonds (y), and bonds to terminal H (x). These structure elements are illustrated at the top of Table VII. Figure 6 shows that, for boron hydride cluster compounds containing no heteroatoms, ^{11}B absorptions lie in the region between δ 25 and δ −55, coincident both in location and in extent with that for boron tetrahedrally coordinated by other atoms and shown in Fig. 1. One might have expected, conversely, the electron deficiency that characterizes these com-

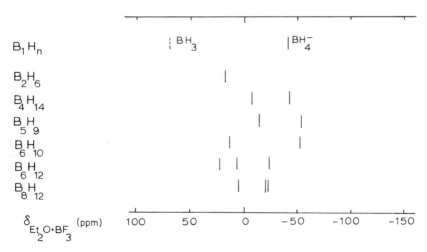

Fig. 6 Boron-11 chemical shifts for boron hydride clusters.

pounds to affect the boron shielding in much the same way that higher states of oxidation affect metal atom resonances by shifting them to lower magnetic fields, but this turns out not to be the case. It is the s and t structure elements that contain the electron deficiency, so if this were to be a factor influencing ^{11}B shielding, the boron atoms in Table III with three of these elements, every boron having at least one terminal hydrogen x element, would absorb in the δ 25 to δ 15 region, and those with only one of the electron-deficient structure elements would be responsible for absorptions in the region δ −40 to δ −55. Table III shows that electron deficiency is not among the factors influencing the shielding of boron in boron hydrides.

TABLE III

Boron-11 Shielding Patterns for Boron Hydride Clusters

Compound	Chemical Shift, Assignment, and $s\ t\ y\ x$ Structure Elements[a]				Reference
	δ23 to δ18	δ8 to δ −13	δ −19 to δ −23	δ −40 to δ −53	
BH_4^-	—	—	—	δ −40 B-1 $4x$	—
B_2H_6	δ 18 B-1,2 $2s + 2x$	—	—	—	Gaines and Schaeffer (1963)
B_4H_{10}	—	δ −7 B-2,4 $2s + 2x$	—	δ −41 B-1,3 $2s + y + x$	Hopkins et al. (1965) Weiss and Grimes (1978)
B_4H_{14}	—	δ −7 B-2,4 $2s + 2x$	—	δ −42 B-1,3 $2s + y + x$	Hopkins et al. (1965)
B_5H_9	—	δ −13 B-2,3,4,5	—	δ −53 B-1 $t + 2y + x$	Gaines, (1969) Odom et al. (1971)
B_6H_{10}	δ 19	δ −7	—	δ −52 B-1 $2t + y + x$	Brice et al. (1973)
B_6H_{12}	δ 23 B-3,6 $2s + t + x$	δ 8 B-1,4 $s + t + 2x$	δ −23 B-2,5 $s + t + y + x$	—	Leach et al. (1970)
B_8H_{12}	—	δ 7 B-3,5,6,8 $2s + t + x$	δ −19 B-4,7 $s + 2t + x$	δ −22 B-1,2 $2t + y + x$	Rietz et al. (1972)

[a] The δ entries refer to chemical shift, B to Assignment and $s\ t\ y\ x$ to Structure Elements.

II. Boron Relaxation

Among quadrupolar nuclei, boron is unique in the extent of coupling data that have been wrested from spectra whose resolution, while poor in comparison to that for spin-$\frac{1}{2}$ nuclei, is higher than expected from a quadrupolar nucleus. The relatively low Q value of 0.036 barn for ^{11}B means that the signals are not excessively broad, and for boron hydrides relaxation rates of ~30 s^{-1} giving line widths of ~10 Hz are typical (Allerhand *et al.*, 1969). The corresponding line width for ^{10}B is ~6 Hz.

A. The Quadrupolar Contribution

Rapid quadrupolar relaxation is favored by a high quadrupole moment Q, a high electric field gradient eq, and a long rotational correlation time τ_c. For molecules of modest size in liquid samples with viscosity ~1 cP, typical correlation times lie in the region 5–50 ps (Boere and Kidd, 1983). Where τ_c has a nominal value of 10 ps, the R value of ~20 s^{-1} observed (Table IV) for certain ^{11}B atoms in a number of the boron hydride molecules reflects a quadrupole coupling constant (QCC) of ~0.7 MHz. By factoring out the ^{11}B quadrupole moment of 0.036 barn, this identifies an electric field gradient of ~2 × 10^{14} esu cm^{-3}.

A quick survey of Table IV reveals that boron atoms bearing two terminal hydrogens have relaxation rates higher than the others by a factor of 4 to 5. This indicates an electric field gradient at these boron sites higher by a factor of 2 to 2.25.

B. The Scalar Coupling Contribution

Where boron is coupled to another nucleus S, the boron resonances appear as a $2n_S I_S + 1$ multiplet provided, inter alia, the S nucleus relaxation rate is significantly less than the coupling constant, $R_1^S < 2\pi J_{BS}$. If nucleus S is quadrupole-relaxed, this inequality is unlikely to hold, and as a result couplings to quadrupolar nuclei are seldom resolved. Boron sees only an average magnetic field from S relaxing rapidly among its discrete spin states. This rapid relaxation modulates the strength of the magnetic coupling to boron with an intensity related to the longitudinal relaxation time of nucleus, S, $T_{1(S)}$.

Equations (1) and (2) identify the contributions made by scalar coupling to the longitudinal and transverse relaxation rates of the boron nucleus.

TABLE IV

Relaxation Parameters for ^{10}B and ^{11}B

Compound	Conditions	T_1 (ms)	R_1 (s^{-1})	T_2 (ms)	R_2 (s^{-1})	$\Delta\nu_{1/2}$ (Hz)[a]	Reference
B$_2$H$_6$	−20°C, ^{11}B	24.2	41	—	—	—	Stampf et al. (1976)
B$_4$H$_{10}$	−20°C, ^{11}B, B-1,3	62.6	16	35	28	4	Stampf et al. (1976)
	−20°C, ^{11}B, B-2,4	13.1	76	11	91	5	Stampf et al. (1976)
	−20°C, ^{10}B, B-1,3	136	7.4	130	7.7	0	Stampf et al. (1976)
	−20°C, ^{10}B, B-2,4	29	34	29	34	0	Stampf et al. (1976)
B$_4$H$_{10}$	−44°C, ^{11}B, B-1,3	36	28	29	35	2	Weiss and Grimes (1978)
	−44°C, ^{11}B, B-2,4	7.4	135	7.4	134	0	Weiss and Grimes (1978)
B$_5$H$_9$	27°C, ^{11}B, B-1	507	2	3.1	326	103	Weiss and Grimes (1978)
	27°C, ^{11}B, B-2–5	63	19	Quartet	—	—	Weiss and Grimes (1978)
	−62°C, ^{11}B, B-1	73	14	3.2	317	96	Weiss and Grimes (1978)
	−62°C, ^{11}B, B-2–5	14	71	4.1	242	54	Weiss and Grimes (1978)
B$_5$H$_{11}$	−20°C, ^{11}B, B-1	55	18	4.9	204	59	Stampf et al. (1976)
	−20°C, ^{11}B, B-2,5	7.5	133	6.8	148	5	Stampf et al. (1976)
	−20°C, ^{11}B, B-3,4	20.7	48	5.2	192	46	Stampf et al. (1976)
B(OMe)$_3$	35°C, ^{11}B	12.9	78	13.0	77	0	Stampf et al. (1976)
	0°C, ^{10}B	12.4	81	12.3	81	0	Stampf et al. (1976)
[(Me$_2$N)$_2$B]$_2$	34°C, ^{11}B	1.9	526	1.8	556	0	Bachmann et al. (1979)
	34°C, ^{10}B	2.7	370	2.4	417	0	Bachmann et al. (1979)

[a] "Additional" broadening $(R_2 - R_1)/\pi$.

Because the $(\omega_B - \omega_S)^2 T_{1(S)}^2$ factor is large, in the range 10^2–10^6, this coupling contributes only to $R_{2(B)}$ and makes no effective contribution to the longitudinal relaxation. In the relaxation region of motional narrowing, only the scalar coupling mechanism makes significantly different contributions to $R_{1(B)}$ and $R_{2(B)}$. This difference provides us with a convenient experimental criterion for evaluating the scalar coupling contribution to the total relaxation process. If the *relaxation*-determined line width $\Delta \nu = R_2/\pi$ in the absence of exchange broadening is greater than the independently measured R_1/π, this indicates the presence of scalar coupling relaxation whose magnitude is given by the "additional" broadening observed.

$$R_1^{sc} = n_S \, \frac{8\pi^2}{3} \, J_{IS}^2 S(S+1) \, \frac{\tau_S}{1 + (\omega_I - \omega_S)^2 \tau_S^2} \tag{1}$$

$$R_2^{sc} = n_S \, \frac{4\pi^2}{3} \, J_{IS}^2 S(S+1) \left[\tau_S + \frac{\tau_S}{1 + (\omega_I - \omega_S)^2 \tau_S^2} \right] \tag{2}$$

Confirmation of the fact that the additional broadening *is* relaxation-determined and not due to unresolved spin–spin coupling can be obtained by measuring the temperature dependence of the $R_2/\pi - R_1/\pi$ difference. Provided R_1 does not contain a significant spin–rotation contribution, its temperature dependence will be normal and R_1 will decrease with increasing temperature. Because $T_{1(S)}^q = \tau$ is directly proportional to the temperature, the scalar component of R_2 is directly proportional to the temperature, and $d(R_2 - R_1)/dT \neq 0$ signals the presence of $R_2^{sc} \neq 0$. The value of J_{BS} is not temperature-dependent, and where unresolved coupling is the source of additional broadening $d(R_2 - R_1)/dT = 0$. A scalar coupling contribution to ^{11}B relaxation in B_4H_{10} has been confirmed, the details of which are given in the following discussion.

C. Sources of Additional Broadening

The relationship among R_1, R_2, and their difference in the region of motional narrowing as reflected in additional line broadening is of pivotal importance in ^{11}B spectroscopy where unresolved couplings and the scalar coupling contribution to relaxation are both evaluated in terms of the difference $R_2 - R_1$. In this context, the term "additional" refers to the component of natural line width not ascribable to quadrupolar relaxation; it does not include broadening introduced by deficiencies in the spectrometer.

The shape of a NMR line in the absence of any spin–spin coupling or chemical exchange effects is Lorentzian, and the full width of the line at half-height $\Delta\nu_{1/2}$ defines the transverse relaxation parameters:

$$\Delta\nu_{1/2} = \frac{R_2}{\pi} = \frac{1}{\pi T_2} \tag{3}$$

When magnetic field inhomogeneity ΔH_0 or any other factor related to spectrometer operation causes instrumental line broadening, then the observed line width defines *apparent* relaxation rates and times, designated by asterisks, which are faster and shorter, respectively, than the natural values of R_2 and T_2.

$$R_2^* = R_2 + \gamma H_0/2 \qquad 1/T_2^* = 1/T_2 + \gamma H_0/2 \tag{4}$$

These instrumental artifacts must be eliminated from R_2^* to obtain the natural R_2 before additional broadening can be evaluated and used to identify unresolved couplings or scalar coupling relaxation.

Weiss and Grimes (1978) have measured T_1 values for the chemically different boron atoms in B_4H_{10}, and these are shown in Table IV. An ^{11}B-enriched sample was used to eliminate ^{10}B contributions to line widths, and values fo 67 and 14 ms were obtained at $-20°$ for B-1,3 and B-2,4 respectively. If T_2 values for these boron atoms were the same, their line line widths would be 4.7 and 8.8 Hz, but widths of 8.6 and 11.1 Hz were observed. In the absence of chemical exchange, this additional line width indicates relaxation by scalar coupling, and the contributions $R_{2(B-1,3)}^{sc} = 12$ s^{-1} and $R_{2(B-2,4)}^{sc} = 13$ s^{-1} comprise 44 and 16%, respectively, of the total R_2 for these borons. At $-44°C$, these contributions reduce to $R_{2(B-1,3)}^{sc} = 7$ s^{-1} (20%) and $R_{2(B-2,4)}^{sc} \sim 0$ s^{-1}, indicating by its direct proportionality to the temperature that the relaxation mechanism is scalar coupling.

Where the additional broadening is due to unresolved spin–spin coupling, the use of signal narrowing techniques that involve an apodization function in Fourier transformation of the raw signal can be implemented. Riley Schaeffer and co-workers (Clouse *et al.*, 1973) have applied this technique to boron hydrides with great success, and the results are discussed in Section III,B.

III. Spin–Spin Coupling to Boron

The ^{11}B isotope has a magnetogyric ratio γ precisely 3.0 times greater than γ_{10B}. This difference is manifest not only in the frequencies at which the two isotopes resonate but also in the relative magnitudes of the scalar couplings to each type of boron. A determination of the coupling constant to one boron isotope is sufficient, since the coupling constant to the other is obtained using the factor 3.0 in the equation

$$J_{^{11}BX} = 3.0 J_{^{10}BX} \tag{5}$$

The designation "indirect" frequently applied to scalar spin–spin coupling indicates that the interaction is transmitted through the valence electrons of the molecule rather than through space. Where the magnitude of the coupling constant observed is used to characterize the structural features of the valence electrons, the reduced coupling constant K, defined by

$$J_{AB} = h(\gamma_A/2\pi)(\gamma_B/2\pi) K_{AB} \tag{6}$$

depends only on electronic structure and is a more useful parameter. The γ-dependent factors C, which yield reduced coupling constants in the equation

$$K_{^{11}BX} \; (nm^{-3}) = (C \times 10^{-2}) J_{^{11}BX} \; (Hz) \tag{7}$$

for isotopes to which boron is most frequently coupled, are 2.59 for ^1H, 24.1 for ^{10}B, 8.08 for ^{11}B, 10.3 for ^{13}C, -25.6 for ^{15}N, 2.76 for ^{19}F, 13.1 for ^{29}Si, 6.41 for ^{31}P, 6.96 for ^{119}Sn, and 12.4 for ^{207}Pb. Some success has been achieved in correlating reduced one-bond coupling constants with the magnitude and type of electron density in the bonding region calculated using various MO models (Kroner and Wrackmeyer, 1976).

A. The Relaxation Curtain

Scalar couplings between ^{13}C and ^1H have proven invaluable in sorting out the structural complexity of a wide range of organic molecules. Were the analogous ^{11}B—^1H couplings to represent as prominent a feature of ^{11}B spectra, structure determinations of boron cluster compounds would become a much less difficult exercise. This, however, is not the case. The relatively efficient relaxation pathway available to both ^{10}B and ^{11}B by virtue of their small nonzero quadrupole moments pulls a relaxation curtain across much of the structural insight that sharply resolved coupling constants would provide.

The criterion determining whether or not under these circumstances the scalar couplings are resolved in the ^{11}B spectrum for a compound rests on the inequalities which can be expressed either in terms of relaxation rate R or in terms of relaxation time T. When observing a spin-$\frac{1}{2}$ nucleus bonded to a quadrupolar nucleus X, the rapid relaxation of X means that coupling between the two is seldom observed. Only when the relaxation rate is sufficiently slow that inequality (8) is satisfied is the coupling constant measurable. When observing a boron isotope, inequality (9) must also be met if evidence of coupling is to be seen, and this second criterion has occasionally been overlooked.

$$R_2(X) < 2\pi J_{^{11}BX} \quad \text{or} \quad T_{2(X)} > (2\pi J_{^{11}BX})^{-1} \tag{8}$$

$$R_2(^{11}B) < 2\pi J_{^{11}BX} \quad \text{or} \quad T_{2(^{11}B)} > (2\pi J_{^{11}BX})^{-1} \tag{9}$$

When X is a spin-$\frac{1}{2}$ nucleus, (8) is always satisfied, and observation of coupling rests on inequality (9). Relaxation rates for ^{11}B are typically slower than 100 s^{-1}, putting coupling constants greater than 15 Hz within the observation window. Tables V and VI indicate that most one-bond couplings exceed 15 Hz.

When X is a quadrupolar nucleus, Eq. (9) will be satisfied only for those with the lowest quadrupole moments occupying situations of reasonably high symmetry. In practice this limits the selection to ^{10}B, ^{11}B, and ^{14}N, along with ^{17}O and ^{2}H in enriched molecules, and of these only boron–boron couplings have been observed. Since $R_{2(X)}$ and $R_{2(^{11}B)}$ both operate independently to interrupt these couplings, the criterion for observation of a resolved coupling pattern is

$$2J_{B_1B_2} > R_{2(B_1)} + R_{2(B_2)} \tag{10}$$

Again using a typical value of 100 s^{-1} as an upper limit for $R_{2(B)}$, J_{BB} constants greater than 30 Hz should be at least partially resolved, and this is borne out by the 1:1:1:1 quartet pattern observed by Onak et al. (1976) from a number of boron hydrides.

B. Coupling to ^1H

Hydrogen can form bonds to boron in either the terminal or the bridging mode. The large difference in $J_{^{11}B^1H}$ in the terminal and the bridging modes assists in both spectral assignments and structure determinations. It can be seen from Table V that, in boron cluster compounds, couplings to terminal hydrogens occur in the region 120–190 Hz, with couplings to different terminal hydrogens in the same molecule spanning a range up to 30 Hz, whereas couplings to bridge hydrogens occur in the region 30–60 Hz and couplings to different bridge hydrogens in the same molecule can span most of this 30-Hz range.

All the available evidence suggests that it is the Fermi contact term that makes the dominant or only contribution to boron–hydrogen couplings, with the result that coupling constants provide a measure of s-electron density at specific locations within the molecule. Onak and co-workers in California, using a semiempirical partial retention of diatomic differential

TABLE V

Typical $^1J_{^{11}B^1H}$ Coupling Constants

Compound	$^1J_{^{11}B^1H_t}$ (Hz)	$^1J_{^{11}B^1H_b}$ (Hz)	Reference
BH_4^-	82	—	Goldstein and Hobgood (1964)
B_2H_6	133	46	Onak et $al.$ (1976)
B_4H_{10}	125–155	30–55	Onak et $al.$ (1976)
B_5H_9	166–175	33	Onak et $al.$ (1976)
B_5H_{11}	127–160	36–60	Onak et $al.$ (1976)
B_6H_{10}	155	—	Onak et $al.$ (1976)
$2,3$-$C_2B_4H_8$	154–181	44	Onak et $al.$ (1976)
$2,4$-$C_2B_5H_7$	167–180	—	Onak et $al.$ (1976)

overlap (PRDDO) MO description for the molecule, found $J_{^{11}B^2H}$ to be linearly correlated with the product of s-orbital populations on B and H atoms. Empirical evaluation of the proportionality constant in Eq. 11 yields $K_{BH_t} = 320$ Hz electron^{-2} and $K_{BH_b} = 533$ Hz electron^{-2}, permitting the determination of s-electron density from measured coupling constants. Noth and Wrackmeyer (1974) in Munich have independently observed the same correlation but use a CNDO/S MO model to represent s-electron density proportionality constants. The linearity of the correlation over a broad range of compounds, however, is confirmed.

$$J_{^{11}B^1H} = K_{BH}S_BS_H \tag{11}$$

C. Coupling to Other $I = \frac{1}{2}$ Nuclei

Because ^{11}B undergoes quadrupolar relaxation, the observation of its scalar couplings is never the routine business it has become in the NMR spectroscopies of ^1H and ^{13}C. Table VI lists representative values for $J_{^{11}BX}$, where X has a spin of $\frac{1}{2}$. There is the usual pronounced trend toward higher coupling constants with increasing size of the coupled atom. In the column for reduced coupling constants in Table VI, where the effects due to variations in nuclear magnetic moment have been eliminated and differences can be ascribed to variations in bond character, maximum constants in the region of 6 nm^{-3} are observed for coupling to first-row atoms, 12 nm^{-3} for second-row atoms, 70 nm^{-3} for fourth-row atoms, and 170 nm^{-3} for fifth-row atoms. These range values with which one can approximate unobserved coupling constants are extremely useful when estimating the potential contribution of scalar coupling to the total relaxation rate of a nucleus. Also useful is the recognition that they vary

TABLE VI

$^1J_{11BX}$ Spin–Spin Coupling Constants

Compound	X	$^1J_{11BX}$ (Hz)	$^1K_{11BX}$ (nm^{-3})	Reference
Me$_2$BCH$_3$	^{13}C	47	4.8	Fußstetter *et al.* (1977)
(Me$_2$N)$_2$BCH$_3$	^{13}C	59	6.1	Fußstetter *et al.* (1977)
Me$_3$BCH$_3^-$	^{13}C	22	2.3	Fußstetter *et al.* (1977)
Cl$_3$BNMe$_3$	^{15}N	14	3.6	Fußstetter *et al.* (1977)
BF$_3$X$^-$	^{19}F	4.5 (X = CF$_3$COO$^-$) to 26.8 (X = CN$^-$)	0.1–0.7	Brownstone and Latremouille (1978)
BF$_2$X$_2^-$	^{19}F	9.0 (X = CF$_3$COO$^-$) to 76.1 (X = Br$^-$)	0.2–2.1	Brownstone and Latremouille (1978)
BFBr$_3^-$	^{19}F	113.3	3.1	Brownstone and Latremouille (1978)
(Me-N–CH$_2$CH$_2$–N-Me)B—SiMe$_3$	^{29}Si	-97	-12.7	Fußstetter *et al.* (1977)
H$_3$B-P(OMe)$_3$	^{31}P	100	6.4	McFarlane *et al.* (1972)
Me$_2$B—P(N-Me ... N-Me)	^{31}P	44.5	2.9	Fußstetter *et al.* (1977)
F$_3$BPR$_3$	^{31}P	165	10.6	Muylle *et al.* (1976)
Cl$_3$BPR$_3$	^{31}P	154–157	9.9–10.1	Muylle *et al.* (1976)
Br$_3$BPR$_3$	^{31}P	145–147	9.3–9.4	Muylle *et al.* (1976)
Me$_2$N(Cl)BSnMe$_3$	^{119}Sn	-1007	70.1	Fußstetter *et al.* (1977)
(Me$_2$N)$_2$BSnMe$_3$	^{119}Sn	-930	64.7	Fußstetter *et al.* (1977)
Me$_2$NB-(SnMe$_3$)$_2$	^{119}Sn	-657	45.7	Fußstetter *et al.* (1977)
(Me-N–CH$_2$CH$_2$–N-Me)B—PbMe$_3$	^{207}Pb	1330	165.0	Fußstetter *et al.* (1977)

by a factor of about 30 between the upper and the lower members of a periodic group.

The coupling constants for ^{19}F are noteworthy in that they confirm the conclusion already drawn from ^{19}F chemical shifts concerning the inadequacy of a first-order theoretical model for rationalizing the structural changes that occur on substitution. The successive substitution of F^- for Br^- in $BFBr_3^-$ reduces the ^{11}B coupling to the adjacent F^- by about 50 Hz with each substitution, indicating that substitutions one bond removed from the fluorine atom under study exert a profound effect on the electronic character of neighboring B—F bonds.

D. Boron–Boron Coupling

Because of the presence of a quadrupolar relaxed nucleus at *both* ends of the coupled system, boron–boron couplings are never fully resolved in boron NMR spectra. The manifestation of coupling is seen at best as a poorly resolved multiplet and at worst as a broad single line, neither of which is improved by the standard line-narrowing techniques (Ernst, 1966). Under these circumstances coupling constants can, in favorable cases, be estimated from computer-simulated spectra.

The success of this spectral matching technique requires that unresolved couplings be the only factors beyond quadrupole relaxation contributing to the boron line width. The demonstration (Weiss and Grimes, 1978) that relaxation by scalar coupling (q.v.) contributes to the line width of nucleus I where R^q of coupled nucleus S is sufficiently rapid to collapse a resolved multiplet, indicates the need for caution in carrying out a line fit analysis. In assessing the possibility of scalar coupling, the proportionality between R^{sc} and $(J_{BB})^2$ must be borne in mind. Since $J_{^{11}B^{11}B} = 3.0$ $J_{^{11}B^{10}B}$, a demonstration that $J_{^{11}B^{10}B}$ does not contribute to R^{sc} is insufficient.

The number of J_{BB} constants evaluated is relatively few, and typical examples are shown in Table VII. Particularly significant are the couplings between equivalent borons in $(Me_2N)_2BB(NMe_2)_2$ and in B_4H_{10} (Fig. 7), couplings that are unobservable under normal circumstances. Coupling between the two carbons in H_3CCH_3, for example, has not been observed because identical atoms occupy magnetically equivalent sites in the molecule. With boron, the existence of two magnetic isotopes unlocks the door to these elusive couplings. For boron isotopes present in their natural ratio, a pair of magnetically equivalent sites contain a ^{10}B atom and a ^{11}B atom in 31% of the molecules, and $J_{^{10}B^{11}B}$ is observable in these molecules. Multiplication by the $\gamma_{^{11}B}/\gamma_{^{10}B} = 2.98$ ratio yields the unobservable $J_{^{11}B^{11}B}$.

According to localized molecular orbital views of bonding in boron hydrides (Lipscomb, 1973) two-center two-electron bonds (*y* type) between borons involve an sp^3 hybrid atomic orbital (AO) from each, and maximum $J_{^{11}B^{11}B}$ values of $\sim$20 Hz are observed in these cases. In

TABLE VII

Boron–Boron Coupling Constants

Compound	Bond Type[a]	$J_{^{11}B^{11}B}$ (Hz)	Method	Reference
Me_2BBMe_2	y	79	Calculated from orbital populations	Kroner *et al.* (1976)
$(Me_2N)_2BB(NMe_2)_2$	y	$\leq 75 \pm 6$	T_1 and T_2 spectral fitting	Bachmann *et al.* (1979)
B_2H_6	s	5 ± 2	Spectral fitting	Farrar *et al.* (1968)
B_2H_6	s	<1.1	Spectral fitting	Odom *et al.* (1973)
B_4H_{10}	y	20.4 ± 1.5	$J_{^{10}B^{11}B}$ spectral fitting	Onak *et al.* (1976)
	s	<0.3	Triple resonance spectral fitting	Onak *et al.* (1976)
B_5H_9	ty^2 hybrid	19.4	—	Lowman *et al.* (1973)
B_5H_{11}	t	17	—	Clouse *et al.* (1973)
B_6H_{10}	y	12–15	—	Onak *et al.* (1976)
B_6H_{12}	y	20	—	Onak *et al.* (1976)

[a] Bond types:

B $\overset{H}{\frown}$ B	B$\big<^{B}_{B}$	B—B	B—H
s	t	y	x

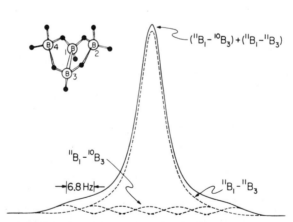

Fig. 7 $J_{^{11}B^{10}B} + J_{^{11}B^{11}B}$ between magnetically equivalent sites in B_4H_{10}. (Reproduced from T. Onak, J. B. Leach, S. Anderson, M. J. Frisch, and D. Marynick (1976) *J. Magn. Reson.* **23**, 237 by permission of Academic Press.)

$(Me_2N)_2BB(NMe_2)_2$, where the B—B bond involves sp^2 hybrid AOs with their higher s-character, the observed coupling constant is three to four times larger in the region of the 79 Hz calculated for Me_2BBMe_2 from s-orbital populations in a localized MO model (Kroner and Wrackmeyer, 1976). Coupling between borons joined by a t-type three-center two-electron bond is slightly lower but of magnitude comparable to that for y-bonded borons. Coupling between borons joined by the traditional electron-deficient hydrogen-bridged structure, once thought to be ~5 Hz, is now shown to be less than 1 Hz.

IV. Conspectus

Diborane occupies two pivotal positions in the chemical firmament; it is the compound that forced chemists to break out of the theoretical confines of the Lewis electron-pair bond and accommodate the idea of a three-center σ-bond. It is also the progenitor of the infinitely wonderful and extremely complex class of cluster compounds that includes boranes, carboranes, and heteroatom boranes. The isotopic mix of boron nuclei, the high ratio of near-neighbor protons, the variety of bonding interconnections, and the fluxional behavior of many clusters all contribute to making their ^{11}B spectra incredibly complex. By the same token, NMR spectroscopy in many cases provides the only tool capable of unraveling the structural puzzle, and simplifying devices such as isotopic enrichment or substitution, very high magnetic fields, intricate decoupling sequences, and hard work have led to successful assignments of chemical shifts and coupling constants for the smaller members of the class. Current developments in spectrometer design are bringing even larger boron compounds within the purview of the NMR spectroscopist, and the future of boron chemistry is inextricably linked with that of NMR.

References

Akitt, J. W. (1972). "Annual Reports on NMR Spectroscopy" (G. A. Webb, ed.), Vol. 5A, p. 465. Academic Press, London.

Akitt, J. W., Greenwood, N. N., and Storr, A. (1965). J. Chem. Soc. 4410.

Allerhand, A., Odom, J. D., and Moll, R. E. (1969). J. Chem. Phys. 50, 5037.

Bachmann, F., Noth, H., Pommerening, H., Wrackmeyer, B., and Wirthlin, T. (1979). J. Magn. Reson. 34, 237.

Boere, R. T., and Kidd, R. G. (1983). "Annual Reports on NMR Spectroscopy," Vol. 13, pp. 319–385. Academic Press, London.

Brice, V. T., Johnson, H. D., and Shore, S. G. (1973). *J. Am. Chem. Soc.* **95,** 6629.
Brownstone, S., and Latremouille, G. (1978). *Can. J. Chem.* **56,** 2764.
Chan, S. O., and Reeves, L. W. (1974). *J. Am. Chem. Soc.* **96,** 404.
Clouse, A. O., Moody, D. C., Reitz, R. R., Roseberry, T., and Schaeffer, R. (1973). *J. Am. Chem. Soc.* **95,** 2496.
Ernst, R. R. (1966). *Adv. Magn. Reson.* **2,** 1.
Farrar, T. C., Johanneson, R. B., and Coyle, T. D. (1968). *J. Chem. Phys.* **49,** 281.
Fußstetter, H., Noth, H., Wrackmeyer, B., and McFarlane, W. (1977). *Chem. Ber.* **110,** 3172.
Gaines, D. F. (1969). *J. Am. Chem. Soc.* **91,** 1230.
Gaines, D. F., and Schaeffer, R. (1963). *J. Am. Chem. Soc.* **85,** 3592.
Goldstein, J. H., and Hobgood, R. T. (1964). *J. Chem. Phys.* **40,** 3592.
Grinter, R., and Mason, J. (1970). *J. Chem. Soc.* (A) 2196.
Hall, J. H., Marynick, D. S., and Lipscomb, W. N. (1974). *J. Am. Chem. Soc.* **96,** 770.
Hartman, J. S., and Miller, J. M. (1974). *Inorg. Chem.* **13,** 1467.
Hartman, J. S., and Schrobilgen, G. J. (1972). *Inorg. Chem.* **11,** 940.
Hopkins, R. C., Baldeschwieler, J. D., Schaeffer, R., Tebbe, F. N., and Norman, A. D. (1965). *J. Chem. Phys.* **43,** 975.
Kidd, R. G. (1967). *Can. J. Chem.* **45,** 605.
Kidd, R. G. (1980). "Annual Reports on NMR Spectroscopy." (G. A. Webb, ed.), Vol. 10a, p. 6. Academic Press, London.
Kidd, R. G., and Spinney, H. G. (1973a). *J. Am. Chem. Soc.* **95,** 88.
Kidd, R. G., and Spinney, H. G. (1973b). *Inorg. Chem.* **12,** 1967.
Kidd, R. G., and Truax, D. R. (1968). *J. Am. Chem. Soc.* **90,** 6867.
Koppel, H., Dallorso, J., Hoffman, G., and Walther, B. (1976). *Z. Anorg. Allg. Chem.* **427,** 24.
Kroner, J., and Wrackmeyer, B. (1976). *J. Chem. Soc. Faraday Trans. 2* **72,** 2283.
Leach, J. B., Onak, T., Spielman, J., Rietz, R. R., Schaeffer, R., and Sneddon, L. G. (1970). *Inorg. Chem.* **9,** 2170.
Lipscomb, W. N. (1963). "Boron Hydrides." Benjamin, New York.
Lipscomb, W. N. (1973). *Accounts Chem. Res.* **6,** 257.
Lowman, D. W., Ellis, P. D., and Odom, J. D. (1973). *Inorg. Chem.* **12,** 681.
McAchran, G. E., and Shore, S. G. (1966). *Inorg. Chem.* **5,** 2044.
McFarlane, H. C. E., McFarlane, W., and Rycroft, D. S. (1972). *J. Chem. Soc. Faraday Trans. 2* **72,** 2283.
McFarlane, W. (1973). *J. Magn. Reson.* **10,** 98.
Mann, B. E. (1978). "NMR and the Periodic Table." (R. K. Harris and B. E. Mann, eds.), Ch. 4. Academic Press, London.
Muylle, E., van der Kelen, G. P., and Claeys, E. G. (1976). *Spectrochim. Acta* **32a,** 1149.
Noth, H., and Wrackmeyer, B. (1974). *Chem. Ber.* **107,** 3070, 3089.
Noth, H., and Wrackmeyer, B. (1978). "Nuclear Magnetic Resonance Spectroscopy of Boron Compounds," Vol. 14. Springer-Verlag, Berlin.
Odom, J. D., Ellis, P. D., and Walsh, H. C. (1971). *J. Am. Chem. Soc.* **93,** 3529.
Odom, J. D., Ellis, P. D., Lowman, D. W., and Gross, M. H. (1973). *Inorg. Chem.* **12,** 95.
Onak, T., Leach, J. B., Anderson, S., Frisch, M. J., and Marynick, D. (1976). *J. Magn. Reson.* **23,** 237.
Ramsey, N. F. (1950). *Phys. Rev.* **78,** 699.
Rietz, R. R., Schaeffer, R., and Sneddon, L. G. (1972). *Inorg. Chem.* **11,** 1242.
Shore, S. G., Christ, J. L., and Long, D. R. (1972). *J. Chem. Soc. Dalton Trans.* 1123.
Spielvogel, B. F., and Purser, J. M. (1971). *J. Am. Chem. Soc.* **93,** 4418.

Spielvogel, B. F., Nutt, W. R., and Izydore, R. A. (1975). *J. Am. Chem. Soc.* **97,** 1609.
Stampf, E. J., Garber, A. R., Odom, J. D., and Ellis, P. D. (1976). *J. Am. Chem. Soc.* **98,** 6550.
Todd, L. J., and Siedle, A. R. (1979). *Prog. Nucl. Magn. Reson. Spectrosc.* **13,** 87.
Vladimiroff, T., and Malinowski, E. R. (1967). *J. Chem. Phys.* **46,** 1830.
Weiss, R., and Grimes, R. N. (1978). *J. Am. Chem. Soc.* **100,** 1401.
Witanowski, M., and Webb, G. A. (1973). "Nitrogen NMR." Plenum, London.

4 Oxygen-17 NMR

J. P. Kintzinger

Institut de Chimie
Université Louis Pasteur
Strasbourg, France

I. Introduction

It is primarily the extremely low natural abundance of its one magnetically active isotope (^{17}O, 0.037%) that has made oxygen a little-studied atom from the NMR point of view. Despite good sensitivity, its receptivity is still only 6.11×10^{-2} that of ^{13}C. This difficulty can often be overcome, because for many functional groups containing oxygen direct exchange with ^{17}O-enriched water or synthesis via an enriched precursor is feasible (^{17}O is a relatively inexpensive isotope). However, oxygen is a quadrupolar nucleus (spin number $I = \frac{5}{2}$, quadrupole moment $eQ = -0.0263 \times 10^{-24}$ ecm^2) whose Larmor frequency ν_0 is relatively low ($\nu_0 = 13.5$ MHz for a magnetic field $B_0 = 2.34$ T). An unfortunate consequence is that high-resolution NMR probes are subject to acoustic ringing, which spoils the beginning of the free induction decay (FID). Acoustic ringing can be eliminated by introducing a delay time (up to hundreds of microseconds) between the end of the pulse and the beginning of the acquisition. This reduces the signal/noise ratio and introduces phase errors in multiline spectra (Canet *et al.,* 1976). For broad lines, this reduction may be unacceptable, and several pulse sequences have been proposed to overcome this problem (Chapter 1, Vol. I). From the author's experience and from spectra in the literature, it is apparent that this problem disappears for frequencies higher than 50 MHz. Despite all these difficulties, the first oxygen NMR signals were reported by Alder and Yu (1951) early in the history of NMR, and this field is now a quickly growing one.

II. Characteristic Parameters

The basic parameters, namely, chemical shifts, line widths, and indirect coupling constants, have been reviewed (Amour and Fiat, 1980; Kintzinger, 1981; Klemperer, 1978). Hence only a brief overview will be given here, and the discussion will concentrate on the most recent publications.

A. Chemical Shifts

As an origin of the chemical shift scale, it is common to select the water signal, although it is a poor reference because its line width and absolute frequency are dependent on temperature (Florin and Alei, 1967), pH (Meiboom, 1961), ionic content (Hertz and Klute, 1970), the protium/deuterium ratio (Lutz and Oehler, 1977), and hydrogen bonding (Alei and Florin, 1969; Reuben, 1969).

The few parts per million of inaccuracy due to these factors are generally negligible when compared to the overall scale of chemical shifts (more than 1000 ppm) and to the inaccuracy of the shift measurements for broad lines. Figure 1 shows the ^{17}O chemical shifts for oxygen in simple

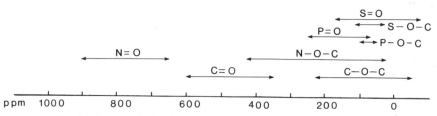

Fig. 1   Oxygen-17 chemical shifts ranges for different X—O bonds.

carbon-, nitrogen-, phosphorus-, and sulfur-containing compounds. Table I contains selected values for the first members of the most common oxygen functions. From measurements on series of homologous compounds, the following semiempirical rules have been established.

(a)   Correlation between δ and the mean excitation energy (Figgis *et al.*, 1962).

(b)   Correlation between δ and the π-bond order (Kidd, 1967). In the case of Mo—O bonds, this relationship causes also a correlation between δ_0 and the force constant for Mo—O stretching and Mo—O bond length (Miller and Wentworth, 1979).

(c)   Correlation between δ_X and δ_O in X=O bonds, X being nitrogen (Andersson and Mason, 1974) or carbon (Delseth and Kintzinger, 1976).

(d)   Correlation between δ_O and δ_C in X—O and X—C fragments (Delseth and Kintzinger, 1978; Eliel *et al.*, 1979; Nguyên *et al.*, 1980; Sardella and Stothers, 1969).

Additivity rules for oxygen shifts are based on the preceding correlations. As for ^{13}C shifts, one can define α, β, γ, ... effects for disubstituted oxygen and β^π, γ^π, ... effects for monosubstituted oxygen. Quite surprisingly, the α effect (OH $\rightarrow$ OCH$_3$) is strongly shielding (-40 ppm), whereas the β (OCH$_3$ $\rightarrow$ OCH$_2$CH$_3$) and γ (OCH$_2$CH$_3$ $\rightarrow$ OCH$_2$CH$_2$CH$_3$) effects are deshielding (up to 30 ppm) and shielding (up to -12 ppm), as expected. For carbonyl compounds, downfield shifts are caused by π acceptors, and upfield shifts by π donors. Cyclization effects on ^{17}O shifts are puzzling. Monocyclic ethers are only slightly shifted with respect to the alicyclic, but there is a strong deshielding for bicyclic ethers (Iwamura *et al.*, 1979; Nguyên *et al.*, 1980). Small cyclic ketones are shielded compared to the alicyclic (Delseth and Kintzinger, 1976), but surprising results are obtained for lactones (Delseth and Kintzinger, 1982), cyclic sulfones, and cyclic sulfoxides (Block, 1980). In γ-propiolactone, the C=O group resonates at the expected position (δ 340), but the ether oxygen is strongly deshielded (δ 235 compared to δ ~170 ppm for homologous compounds). In the sulfoxide and sulfone series, the six-membered rings have a normal shift (-3 and 142 ppm, respectively), the four-membered rings are deshielded (61 and 182 ppm), and the three-membered rings are shielded (-71 and 111 ppm) (Block *et al.*, 1980). Stereoelectronic effects on the chemical shifts have also been characterized in geminal dioxo groups for the series CH$_{4-n}$(OCH$_3$)$_n$ (Kintzinger *et al.*, 1980) and in 2-alkoxytetrahydropyrans (McKelvey *et al.*, 1981). Magnetic nonequivalence of diastereotopic oxygen atoms has been observed with

TABLE I

Selected Chemical Shifts and Line Widths (or Spin Coupling Constants) of Oxygen-Containing Compounds

Compound	δ (ppm)	$W_{1/2}$	or J (Hz)	Reference[a]
H_3O^+	9	—	106	Mateescu and Benedikt (1979); Olah et al. (1982)
CH_3OH	−37	100	85.5	Crandall and Centeno (1979); Versmold and Yoon (1972)
CH_3OCH_3	−52.5	30	—	Delseth and Kintzinger (1978); Sugawara et al. (1978)
$CH_2(OCH_3)_2$	3	—	—	Kintzinger et al. (1980)
$(CH_3)_2CO$	569	41	—	Crandall et al. (1979); Delseth and Kintzinger (1976)
$(CH_3)_2COH^+$	310	—	74	Olah et al. (1982)
$CH_2{=}CH{-}OCH_3$	59	45	—	Kalabin et al. (1982)
$CH_2{=}CH{-}CHO$	579	26	—	Delseth et al. (1980)
CH_3COOH	251	175	—	Christ et al. (1961)
CH_3COCl	502	30	—	Delseth et al. (1980)
CH_3CONH_2	303	137	—	Burgar et al. (1981)
$CH_3CO{-}OCH_3$				
—O—	137	120	—	Christ et al. (1961)
C=O	355	120	—	
$(CH_3CO)_2O$				
—O—	259	170	—	Christ et al. (1961)
C=O	393	170	—	
CO_3^{2-}	192	—	—	Figgis et al. (1962)
$Ni(CO)_4$	362	—	—	Bramley et al. (1962)
$Co_4(CO)_{12}$				
Bridging	501.2	—	—	Aime et al. (1981)
Terminal	381.5, 375.9, 365.8	—	—	
$(C_2H_5O)_4Si$	9	60	—	Harris and Kimber (1975)
CH_3NO_2	605	—	—	Christ and Diehl (1963); Lippmaa et al. (1972); Mägi et al. (1971)
CH_3ONO				
—O—	420	—	—	Andersson and Mason (1974)
NO	790	—	—	
$C_2H_5ONO_2$				
—O—	340	—	—	Weaver et al. (1955)
NO_2	470	—	—	
NO_3^-	410	—	—	Andersson and Mason (1974)
H_3PO_4	79	—	—	Christ and Diehl (1963)
$(CH_3O)_3PO$				
—O—	68	—	90	Christ and Diehl (1963)

(*continued*)

TABLE I (*Continued*)

Compound	δ (ppm)	$W_{1/2}$	or J (Hz)	Reference[a]
P=O	19	—	165	Christ and Diehl (1963)
$[(CH_3)_2N]_3PO$	73	—	145	Gray and Albright (1977); Grossman *et al.* (1976)
$(CH_3)_2SO$	20	—	—	Block *et al.* (1980)
$(CH_3)_2SO_2$	163	—	—	Block *et al.* (1980)
$(CH_3O)_2SO_2$				
—O—	102	—	—	Christ *et al.* (1961)
SO_2	150	—	—	
SO_4^{2-}	167	—	—	Lutz *et al.* (1976a)
SeO_4^{2-}	204	—	—	Figgis *et al.* (1962)
$Te(OH)_6$	120	—	—	Luz and Pecht (1966)
ClO_3F	302	—	106[b]	Virlet and Tantot (1976)
ClO_3^-	290	—	—	Alei (1965)
ClO_4^-	290	—	85.5	Alei (1965)
BrO_3^-	297	—	—	Figgis *et al.* (1962)
IO_3^-	206	—	—	Dwek *et al.* (1971)
VO_4^{3-}	568	—	61.6	Lutz *et al.* (1976b)
CrO_4^{2-}	835	—	10	Figgis *et al.* (1962)
MoO_4^{2-}	831	—	40.5	Vold and Vold (1974)
WO_4^{2-}	420	—	—	Figgis *et al.* (1962)
MnO_4^-	1230	—	28.9	Jackson and Taube (1965)
TcO_4^-	749	—	—	Figgis *et al.* (1962)
ReO_4^-	569	—	—	Figgis *et al.* (1962)
RuO_4	1106	—	23.4[c]	Brevard and Granger (1981)
OsO_4	796	—	—	Brevard and Granger (1981)
$Cr_2O_7^{2-}$				
—O—	345	—	—	Kidd (1967)
GrO	1129	—	—	
$Mo_2O_7^{2-}$				
—O—	248	—	—	Day *et al.* (1977a)
MoO	715	—	—	

[a] References given refer not only to the selected compound but also to parent compounds.
[b] Coupling to ^{35}Cl.
[c] Coupling to ^{99}Ru.

^{17}O NMR for sulfones (Kobayaski *et al.*, 1981) in the range 5–50 ppm and for cyclic 2'-deoxyadenosine 3'-5'-monophosphate (Coderre *et al.*, 1981).

Numerous chemical shifts in oxyanions have also been reported (Filowitz *et al.*, 1979): They are within the range 0–1200 ppm, and one can note a rough correlation between the ^{17}O high-field shift and the increased number of metal atoms coordinated to the oxygen atom. Because of small line widths (about 25 Hz compared to line widths of 200 to 300 Hz for organic compounds having comparable molecular weights), metal carbonyl compounds are easily observed (Cozak et al., 1979; Kawada et al., 1979; Kump and Todd, 1980). The resonances of ^{17}O can be observed even more easily than those of ^{13}C because there is no interference from long relaxation times or from scalar coupling to quadrupolar nuclei (Aime et al., 1981).

Among the biologically relevant molecules, only the building blocks— amino acids (Gerothanassis et al., 1982b; Hunston et al., 1982; Valentine et al., 1980), small peptides (Irving and Lapidot, 1976), monosaccharides (Gerothanassis et al., 1982a; Gorin and Mazurek, 1978), polyphosphates (Gerothanassis and Sheppard, 1982), and some nucleosides (Schwartz et al., 1980)—have been submitted to ^{17}O-NMR analysis. For amino acids, the chemical shift and line width variations versus pH are classic titration curves. Some anomalies detected in the line widths (Valentine et al., 1980) are due to the presence of paramagnetic impurities (Gerothanassis et al., 1982c). In the nucleoside series (Schwartz et al., 1980) only the carbonyl group of the pyrimidine ring has been ^{17}O-enriched and detected by NMR, whereas the oxygen atoms from the furanose ring are not observable. For observation of monosaccharides at natural abundance (Gerothanassis et al., 1982a) one must use high-quality ^{17}O-depleted water and work at an elevated temperature (80°C) on a high-field superconducting spectrometer with fast pulsing. Furthermore, the addition of aqueous shift reagents (Frahm and Füldner, 1980; Jackson et al., 1960) and strong Lorentzian-to-Gaussian deconvolution are useful either in shifting the water line or increasing the resolution. For macromolecules, the lines are so broad that even at 100% enrichment they are quite unobservable. For instance, no ^{17}O resonance could be found for either a hemoglobin–oxygen complex (Irving and Lapidot, 1971) or, more surprisingly, Vaska complexes (Lapidot and Irving, 1972).

B. Relaxation, Line Widths, Quadrupolar Coupling Constants

1. Relaxation Mechanism

The quadrupolar relaxation mechanism is overwhelmingly dominant in ^{17}O NMR, at least for diamagnetic compounds. When the conditions of extreme narrowing are fulfilled and in the absence of chemical exchange, the longitudinal (T_1) and transverse (T_2) relaxation times are equal and are given by

$$\frac{1}{T_1} = \frac{1}{T_2} = \frac{3}{125}\left(1 + \frac{\eta^2}{3}\right)\left(\frac{e^2qQ}{\hbar}\right)^2 f(\Omega,D) \tag{1}$$

where $e^2qQ/\hbar$ is the quadrupolar coupling constant (QCC, in radians per second), η the asymmetry parameter (dimensionless), and $f(\Omega,D)$ the correlation function depending on the rotational diffusion constants D and the relative orientation of the field gradient components with respect to the diffusion constants. When isotropic reorientation is assumed, $f(\Omega,D)$ reduces to the correlation time τ_c. As oxygen is a spin-$\frac{5}{2}$ nucleus, nonexponential relaxation behavior is expected for nonextreme narrowing situations (Hubbard, 1970). The longitudinal and transverse decays are then the sum of three exponential decays. For practical cases describing fast exchange between two states with one state under extreme narrowing conditions (Bull *et al.*, 1979), the longitudinal magnetization decays as a single exponential, whereas the transverse magnetization decays as the sum of three exponentials.

2. Linewidths

If the resonance line is a Lorentzian curve, the measurement of the line width at half-height $W_{1/2}$ is a direct measure of the relaxation time because

$$W_{1/2} = 1/\pi T_2 \tag{2}$$

To apply Eq. (2) safely, one has to be sure that

(a) No unresolved indirect spin coupling is present. (Which can be verified by line shape analysis (Delseth *et al.*, 1978) or more simply by broad-band decoupling)

(b) For labile protons directly bonded to oxygen, the rate of proton exchange is fast compared to the reciprocal of the coupling (by modifying the pH value or broad-band decouple)

(c) No paramagnetic impurities are present, which is particularly important in studies with amino acids (Gerothanassis *et al.*, 1982c).

Considering that the actually known QCC values (Section II,B,3) range from 1.3 to 17 MHz and that the correlation times range from 1 to 100 ps for molecules having a molecular weight below 1000, one calculates line widths in the range 0.5–9000 Hz.

Line widths can be used to extract either values for the quadrupolar interaction or correlation times for molecular reorientation, depending on which of the two parameters appearing in Eq. (1) is known. Thus far line widths and relaxation times have been of little use in molecular dynamics studies. Only the reorientations of small molecules, such as water (Hindman *et al.*, 1973) and formamide (Burgar *et al.*, 1981; Wallach and Huntress, 1969) as pure liquids or small molecules in binary mixtures, have been studied (Goldammer and Hertz, 1970). Generally the reverse procedure is used to determine QCC values in the liquid state.

3. Quadrupolar Coupling Constants

The electrostatic interaction between the nuclear quadrupolar electric moment eQ and the electric field gradient tensor eq_{ij} is the quadrupolar interaction. An appropriate reference axis system can be chosen to reduce this interaction to two values:

$$\chi = e^2q_{zz}Q/h \qquad \text{(the main component)} \tag{3}$$
$$\eta = (e^2q_{xx}Q - e^2q_{yy}Q)/e^2q_{zz}Q \qquad \text{(the asymmetry parameter)} \tag{4}$$

where $|e^2q_{zz}Q| \geqslant |e^2q_{yy}Q| \geqslant |e^2q_{xx}Q|$.

Within the framework of the Townes–Dailey theory (Townes and Dailey, 1949), the three components of the field gradient are easily related to the electronic populations in the orbitals describing the oxygen bonding and to eq_0, the field gradient created by an electron in a $2p$ atomic orbital. From spectroscopic measurements on atomic oxygen, a χ_0 value corresponding to the field gradient eq_0 is obtained (Kamper et al., 1957):

$$\chi_0 = 20.88 \quad \text{MHz}$$

a. Monocoordinated Oxygen. The simplest hybrid orbital set requires only one sp hybridization parameter for the X—O bond.

$$\overset{\longleftarrow}{\underset{z}{}} \quad \text{X—O}$$

$$\psi_1 = (1 - \lambda^2)^{1/2}s + \lambda p_z$$
$$\psi_2 = \lambda s - (1 - \lambda^2)^{1/2}p_z \tag{5}$$
$$\psi_3 = p_x \qquad \psi_4 = p_y$$

Considering that in the oxygen molecule the ψ_i form, respectively, σ-bonding (ψ_1, population 1), nonbonding (ψ_2, population 2), π-bonding, and π-antibonding (ψ_3, ψ_4, population 1.5) orbitals, the Townes–Dailey theory gives the following field gradients (Kintzinger, 1981):

$$eq_{xx} = (-0.25 + \lambda^2/2)eq_0$$
$$eq_{yy} = (-0.25 + \lambda^2/2)eq_0 \tag{6}$$
$$eq_{zz} = (0.5 - \lambda^2)eq_0$$

The main component is directed along the z axis and depends only on λ^2. Identifying eq_{zz} with the experimental value $\chi_{zz} = -8.42$ MHz, Eq. (6)

becomes

$$\chi_{zz} = -8.42 = (0.5 - \lambda^2)\chi_0$$

The hybridization parameter λ^2 is found to be equal to 0.9.

Formaldehyde could be discussed in relation to the same set of hybrid orbitals and with the following distributions: ψ_1, bonding, population a; ψ_2, nonbonding, population 2; ψ_3, bonding, population b; ψ_4, nonbonding, population 2. With $\lambda^2 = 0.9$, the different components are

$$eq_{xx} = (-1.1 - 0.45a + b)eq_0$$
$$eq_{yy} = (1.9 - 0.45a - b/2)eq_0 \qquad (7)$$
$$eq_{zz} = (-0.8 + 0.9a - b/2)eq_0$$

Equivalent expressions are obtained by following the treatment of Cheng and Brown (1979). As shown by Cheng and Brown (1979), and for reasonable values of a, b, and c, the main component is either xx or yy and the η values can be large.

   $b.$ *Dicoordinated Oxygen.* For the water molecule, it is convenient to take a variable hybrid set of orbitals similar to those used in the descrip-

tion of strained cycloalkanes (Coulson and Moffitt, 1949). The yz plane is the molecular plane, and the nonbonding orbitals are in the xz plane:

$$\psi_a = l[s + \lambda(p_y \cos \xi + p_z \sin \xi)]$$
$$\psi_b = l[s + \lambda(-p_y \cos \xi + p_z \sin \xi)]$$
$$\psi_c = m[s + \mu(p_x \cos \eta - p_z \sin \eta)] \qquad (8)$$
$$\psi_d = m[s + \mu(-p_x \cos \eta - p_z \sin \eta)]$$

Coefficients l, m, λ, μ, ξ, and η are interrelated through the orthonormalization relations. They allow the description of all types of sp^x hybridization, and the knowledge of one of them defines the others. If r is the population of the bonding ψ_a and ψ_b orbitals, and the populations of the nonbonding orbitals are equal to 2, the components of the field gradient are equal to

$$eq_{xx} = l^2\lambda^2(2 - r)eq_0$$
$$eq_{yy} = m^2\lambda^2(r - 2)eq_0 \qquad (9)$$
$$eq_{zz} = l^2\lambda^2 - m^2\mu^2(r - 2)eq_0$$

A perfect sp^3 hybridization (angle HOH $= 109°28'$) corresponds to an angle $\xi = 35°16'$ and to the condition $l^2\lambda^2 = m^2\mu^2$, because from orthonormality relationships one can show that $m^2\mu^2/l^2\lambda^2 = 3\cos\xi - 1$. In this situation one obtains $eq_{xx} = -eq_{yy}$; $\eta = 1$.

c. *Experimental values.* Experimental values for χ and η are given in Table II. The QCC values cover a range from -8 to 17 MHz, and the asymmetry parameter goes from zero almost to the maximum possible value of unity. The sign of the QCC is irrelevant to the NMR experiment, but from the theoretical point of view it gives indications about the orientation of the field gradient. It is interesting to note that, for water hydrates (Achlama, 1977; Sphorer and Achlama, 1976), the sign of X is reversed with respect to that of water in different ices (Edmonds *et al.,* 1977). This change is easily explained by taking into account Eq. (9) and the change in the HOH angle in these hydrates. Measurements on ice Ih (Lumpkin and Dixon, 1979a) show that, because of proton tunneling along one O...O direction, the field gradient tensor has axial symmetry and the quadrupolar interaction in this solid phase is very close to the value measured in the gas phase (Verkoeven *et al.,* 1969). For the other forms of ice (Edmonds *et al.,* 1977) one observes a much smaller quadrupolar interaction, as expected for hydrogen-bonded systems. The measured QCC values for a series of solid state compounds with C=O, —O—, PO, NO, and SO bonds show some interesting trends:

(a) For a substituted aromatic function [nitrobenzenes (Cheng and Brown, 1980), nitrophenols, hydroxybenzaldehydes (Butler and Brown, 1981), nitrobenzaldehydes (Cheng and Brown, 1979)] the QCCs are nearly invariant with the electromeric properties of the substituent but are reduced by hydrogen bonding.

(b) Strong hydrogen bonding reduces the QCC up to 2.5 MHz (Butler and Brown, 1981). In summarizing very roughly the results available at present, the accompanying scale for decreasing QCC can be constructed.

$$O—O > N—O > NO_2 > C=O > P—OR \simeq S=O > ROH > SO_2 > P=O \sim C=O > XO_n{}^{m-}$$

| 16–17 | 16 | 13 | 11–9 | 9.5–9 | | 9–8 | 6.7 | 4 | 1 |

C. Indirect Spin Coupling Constants

Despite the fact that oxygen is a quadrupolar nucleus, many spin–spin coupling constants, even with other quadrupolar nuclei, are known. A selection of ^1J values is given in Table I, where one can observe that, although the general trend is an increase in the reduced coupling constants on moving from the upper left to the lower right corner of the

TABLE II

Oxygen-17 Nuclear Quadrupole Coupling Constants

Compound	State	e^2qQ/h (MHz)	η (%)	Reference
O_2	Gas	-8.42	0	Gerber (1972)
CO	Gas	4.43	0	Rosenblum and Nethercot (1957)
CO	Solid α	3.85	0	Li et al. (1980)
OCS	Gas	-1.32	0	Geschwind et al. (1952)
HDO	Gas	10.175	75	Verhoeven et al. (1969)
H_2O	Ice, Ih, 4.2 K	11.33	6	Lumpkin and Dixon (1979a)
H_2O	Ice, II	6.893	86.5	Edmonds et al. (1976)
D_2O	Ice	6.66	93.5	Spiess et al. (1969)
D_2O, H_2O	Liquid	7.6^a	—	Garrett et al. (1967)
$Ba(ClO_3)_2H_2O$	Solid	-7.61	94	Shporer and Achlama (1976)
$NaAuCl_4$, $2H_2O$	Solid	-7.79	87	Achlama (1977)
HOOH	Solid	16.31	68	Lumpkin and Dixon (1979b)
Vaska's iridium compound	Solid	16.9	65	Lumpkin et al. (1979)
		15.6	91	
KH_2PO_4	Solid, 190 K	5.16	55	Blinc et al. (1974)
	Solid, 77 K	5.96, 4.85	72, 18	
ClO_3F	Liquid	14^a	—	Virlet and Tantot (1976)
MoO_4^{2-}	Liquid	0.7	0	Vold and Vold (1974)
RuO_4	Liquid	1.12	0	Brevard and Granger (1981)
OsO_4	Liquid	0.97	0	Brevard and Granger (1981)
Triphenyl phosphate	Solid, P=O	3.82	10	Cheng and Brown (1980)
	Solid, P—O—C	9.32	51	
		9.06	68	
		9.15	74	
Diphenyl sulfoxide	Solid	9.44	23	Cheng and Brown (1980)
Diphenyl sulfone	Solid	9.73	21	Cheng and Brown (1980)

(continued)

TABLE II (*Continued*)

Compound	State	e^2qQ/h (MHz)	η (%)	Reference
Nitrobenzene	Solid	13.09	57.6	Cheng and Brown (1980)
p-Nitroaniline	Solid	12.57	74	Cheng and Brown (1980)
NO_3^-	Solid	12.57	75	Cheng and Brown (1980)
γ-Picoline *N*-oxide	Solid	15.63	33	Cheng and Brown (1980)
Formaldehyde	Gas	12.37	69.5	Flygare and Lowe (1965)
Benzophenone	Solid	10.883	36.9	Cheng and Brown (1979)
9-Fluorenone	Solid	10.41	39	Butler *et al.* (1981)
Dichloroacetone	Solid	10.725	45.4	Kado and Takarada (1978)
Cyclohexanone	Liquid	10.3[a]	—	Delseth and Kintzinger (1982)
Benzoyl chloride	Solid	8.613	37.6	Cheng and Brown (1979)
Benzoyl fluoride	Solid	8.507	35	Cheng and Brown (1979)
m-Chlorobenzaldehyde	Solid	10.708	44.3	Kado and Takarada (1978)
o-Hydroxybenzaldehyde	Solid, C=O	9.891	28	Butler *et al.*
	Solid, C—OH	8.304	53	(1981)
m-Hydroxybenzaldehyde	Solid, C=O	9.923	32	Butler *et al.*
	Solid, C—OH	8.736	66	(1981)
p-Hydroxybenzaldehyde	Solid, C=O	9.681	21	Butler *et al.*
	Solid, C—OH	8.468	50	(1981)
Maleic anhydride	Solid, C=O	9.322	32	Cheng and Brown
	Solid, —O—	7.491	54	(1979)
Phtalide	Solid, C=O	8.354	5	Cheng and Brown
	Solid, —O—	8.931	53	(1979)
Phenylbenzoate	Solid, C=O	9.504	44	Cheng and Brown
	Solid, —O—	9.791	14	(1979)
γ-Butyrolactone	Liquid, C=O	9.2[a]	—	Delseth and
	Liquid, —O—	11.2[a]	—	Kintzinger (1982)
Formic acid	Solid, C=O	7.818	7	Brosnan *et al.*
	Solid, COH	−6.90	8	(1981)

(*continued*)

TABLE II (Continued)

Compound	State	e^2qQ/h (MHz)	η (%)	Reference
1,8-Dihydroxyanthra-	Solid, C=O	8.653	3.8	Butler and Brown
quinone	Solid, COH	8.215	45.1	(1981)
		8.121	46.9	
2,6-Dichloro-p-benzo	Solid	11.240	42.5	Hsieh et al.
quinone				(1972)
2,5-Dichlorohydroquinone	Solid	8.903	88	Hsieh et al.
				(1972)
Xanthene	Solid	9.689	61	Hsieh et al.
				(1972)
Dibenzyl ether	Solid	10.855	92	Butler et al.
				(1981)
Tetrahydropyran	Liquid	12.7[a]	—	Delseth and
				Kintzinger
				(1982)
Ethane diol	Solid	8.895	97	Brosnan and
				Edmonds (1980)
Sodium formate	Solid	7.311	45.6	Cheng and Brown
				(1979)

[a] Value for the product $(e^2qQ/h)(1 + \eta^2/3)^{1/2}$.

periodic table, there are notable exception such as the isoelectronic series VO_4^{3-}, CrO_4^{2-}, MnO_4^-. In some cases, the existence of a coupling allows experiments that otherwise would be impossible. It should be remembered that measurement of the proton exchange rate in water is based on broadening of the ^1H and ^{17}O signals due to partial averaging of $^1J_{OH}$ (Meiboom, 1961). Clear proof of the structure of the hydronium cation has been obtained with ^{17}O NMR (Mateescu and Benedikt, 1979). In a SO_2 solution at $-15°C$, the ^{17}O signal of H_3O^+ is a quartet with $J = 106 \pm 1.5$ Hz, indicating that it should be a planar cation because the increase in J (compared to liquid water) has the expected value for a hybridization change from sp^3 to sp^2. The J value for protonated acetone is reported as 74 Hz (Olah et al., 1982) or 85 Hz (Mateescu and Benedikt, 1979), which should support the hydroxycarbenium structure. This finding, however, contradicts the chemical shift measurement which shows an important π-bond order for the C—O bond (Olah et al., 1982). Among the more exotic couplings it is worthwhile to mention the ^{17}O sextuplet in RuO_4 due to coupling between ^{17}O and ^{99}Ru, demonstrating that ^{99}Ru is a spin-$\frac{5}{2}$ nucleus (Brevard and Granger, 1981). In favorable cases $^2J_{OH}$ (Delseth et al., 1978) and $^2J_{OF}$ (Reuben and Brownstein, 1967) couplings have been resolved, but generally ^2J and ^3J values are too small to be directly measured on the ^{17}O spectrum and one has to computer-fit the line shape to obtain their values.

D. Isotopic Shifts and Quadrupolar Effects

1. Isotopic Shifts

Isotopic shifts arising from the replacement of ^{16}O by ^{17}O are expected in the spectrum of a nucleus X directly bonded to oxygen. They are more easily measured by the substitution of ^{18}O for ^{16}O and are conveniently expressed in parts per billion. Following the general rule, diamagnetic shifts are observed. The best documented studies concern carbon NMR with values of about -20 ppb per oxygen for dicoordinated oxygen as measured for alcohols (Diakur et al., 1980; Risley and Van Etten, 1979), esters (Risley and Van Etten, 1980a), orthocarbonates (Risley and Van Etten, 1980b), and acetals (Moore et al., 1981), -25 ppb per oxygen for carboxylic acids and salts (Risley and Van Etten, 1981), and -50 ppb for monocoordinated oxygen in ketones and esters (Diakur et al., 1980; Vederas, 1980). In ^{31}P NMR, isotopic shifts are known for the simplest phosphates and also for AMP, ADP, and ATP (Lowe et al., 1979). For ATP, the shifts are slightly dependent on the position of the phosphorus atom and on the site of substitution: Bridging oxygen produces smaller shifts (-16 to -21 ppb) than terminal oxygen (-22 to -28 ppb) (Cohn and Hu, 1980). The existence of these isotopic shifts is of prime importance in biosynthetic studies, because in labeling experiments an O-labeled site can be determined by ^{13}C or ^{31}P NMR. An isotopic shift of -138 ppb per oxygen is reported in ^{15}H NMR for the nitrite ion (Van Etten and Risley, 1981), and shift values are also known in ^1H, ^{55}Mn, and ^{95}Mo NMR (Buckler et al., 1977).

2. Quadrupolar Effects in the NMR Spectra of Other Nuclei

It is well known that the spectrum of a spin-$\frac{1}{2}$ nucleus is dependent on the relaxation time of a quadrupolar nucleus spin coupled to it (Susuki and Kubo, 1964). This property has been used to extract the ^{17}O relaxation time in alcohols from the line shape of the proton spin coupled to it (Versmold and Yoon, 1972). Consider a P=O group where the oxygen atom is 50% labeled with ^{17}O. If the isotopic shift is ignored, the spectrum of ^{31}P will be the superposition of a singlet arising from P$=^{16}O$ and of a sextuplet with $J \sim 150$ Hz arising from P$=^{17}O$. The line widths of the sextuplet's components are related to the quadrupolar relaxation time Tq by $0.597/Tq$, $0.915/Tq$, $0.716/Tq$, $0.716/Tq$, $0.915/Tq$, and $0.597/Tq$. Even for a *long* ^{17}O relaxation time of 10 ms, the line width of the different components is in the range 60–90 Hz. The intensity of the sextuplet components is then less than 0.2 compared to 50 for P$=^{16}O$. These components are hard to observe, and integration over an expanded scale even cannot account for them (Tsai, 1979). Shortening the ^{17}O relaxation time transforms the sextuplet into a broad singlet when the product TqJ is less than unity. For the condition $TqJ \ll 1$ the line width of the ^{31}P signal is given by

$$W_{1/2} = \tfrac{35}{3} \pi^2 TqJ^2 \tag{10}$$

For a phosphorus atom ^1J-coupled to ^{17}O, the residual width will be greater than 25 Hz as long as the oxygen relaxation time is longer than 10 μs (Tsai *et al.*, 1980). Then, when the ^{17}O relaxation time is between 10 ms and 10 μs, (line width between 30 and 30,000 Hz) the ^{31}P-NMR spectrum will have a lower integrated intensity. From the measured reduction, it is easy to obtain the amount and the site of labeling for molecules such as ATP which contain several types of phosphorus and oxygen atoms (Tsai, 1979; Tsai *et al.*, 1980).

III. Chemical and Biochemical Applications

Because of its great chemical shift range, the ^{17}O nucleus is a good nucleus for structural studies. However, as long as the molecule also contains a nucleus such as ^1H, ^{13}C, or ^{31}P, identification is probably easier using one of these spectroscopies. This remark does not discredit ^{17}O as a useful nucleus; the numerous recent publications in the field of oxyanion chemistry demonstrate the usefulness of ^{17}O as an analytical tool, and for metal carbonyls ^{17}O NMR sometimes appears to be superior to ^{13}C NMR. From previous remarks concerning chemical shifts and the relaxation of ^{17}O, it is clear that ^{17}O NMR will be a powerful tool in studies concerned with π-bond order, hydrogen bonding, labeling experiments, and the dynamic NMR of small molecules.

A. Bond Polarization and π-Bond Order

Theoretical arguments and experimental results support at least qualitatively a relation between π-bond order and δ_{17O}. A value of -530 ppm has been proposed for the shift experienced by an oxygen atom when its charge is increased by the addition of one $2p$ electron (Delseth et al., 1980). A diamagnetic shift is therefore expected for structures like $X^+\!-\!O^-$ when compared to $X{=}O$. For example, orthosubstitution in nitrobenzenes (Christ and Diehl, 1963) and acetophenones (Amour et al., 1981; Sardella and Stothers, 1969) generally produces paramagnetic shifts with respect to the unsubstituted parent compounds because the non-planarity of the substituted molecules favors the less polar structures.

For the series of cyclic ketones the transannular N–C interaction in the molecule with $m = 4$ and $n = 3$ shifts the oxygen by -90 ppm compared to

the parent compound with $m = n = 2$ (Dahn et al., 1972). Observation of a chemical shift δ 310 of oxygen in protonated acetone (Olah et al., 1982) proves that a partial π bond exists between C and O, in contrast to the previously held idea of either a total double-bond character based on ^{1}H and ^{13}C NMR or a zero π-bond order based on the value of $^1J_{OH}$ (Mateescu and Benedikt, 1979). The advantage of the large chemical shift difference between a formal C$=$O and a formal C—O—H structure has been fully used in determination of the equilibrium constant of enols of types A and

B. Because of rapid proton transfer the chemical shifts of the oxygen atoms labeled α and β are given by (Gorodetsky et al., 1967)

$$\delta^\alpha = A\delta_A^{CO} + B\delta_B^{COH} \qquad \delta^\beta = A\delta_A^{COH} + B\delta_B^{CO}$$

Small differences between δ_A^{CO} and δ_B^{CO} or δ_A^{COH} and δ_B^{COH} are negligible with respect to the large difference $\Delta = \delta^{CO} - \delta^{COH}$, so that the equilibrium constant is simply related to the chemical shift differences $\delta = \delta^\alpha - \delta^\beta$ and Δ through

$$K = B/A = (\Delta - \delta)/(\Delta + \delta)$$

B. Hydrogen Bonding Studies

The ^{17}O chemical shifts in water are temperature-dependent (Florin and Alei, 1967), with a diamagnetic shift of -9 ± 1 ppm between 25 and 215°C. Furthermore, the signal of the vapor is shielded by -36 ppm. The dilution of water with various solvents also produces diamagnetic shifts (Reuben, 1969). These shifts arise from hydrogen bond breaking, and it has been proposed that, starting from an isolated water molecule, proton donation has an effect of 12 ppm and proton acceptance an effect of 6 ppm on the ^{17}O shift (Reuben, 1969). Ammonia produces a paramagnetic shift of 9 ppm, whereas trimethylamine creates a shielding of -15 ppm (Alei and Florin, 1969). Acetone at infinite dilution in water is also diamagnetically shifted by -52 ppm with respect to pure acetone (Christ and Diehl, 1963; Reuben, 1969). Such a diamagnetic shift is indicative of a higher contribution from the polar structure C^+—O^- in water. The dilution of acetone in alkanes creates a low-field shift of about 16 ppm (Tiffon, 1980). Mixtures of formamides [formamide, N-methylformamide, N,N-dimethylformamide (DMF)] in water or acetone have been examined (Burgar et al., 1981). For acetone systems, nearly no shift is detected for either acetone or for DMF over the entire composition range, and the greatest variations are obtained with formamide (acetone is -25 ppm shielded, and formamide is 30 ppm deshielded). For water systems, the greatest high-field shifts are observed for DMF in water (-53 ppm for DMF and -10 ppm for water), whereas the smallest shifts are obtained for formamide (-20 ppm for the amide and no shift for water). These variations reflect the hydrogen bonding properties of amides, acetone, and water. For the three amides studied, the overall upfield shift on going from infinite dilution in acetone to infinite dilution in water is nearly constant and equal to -53 ppm. In a series of substituted acetophenone and benzaldehydes, intramolecular hydrogen bonding in the ortho amino and ortho hydroxy molecules produces high-field shifts (Amour et al., 1981).

C. Dynamic NMR

Numerous types of dynamic experiments have been performed using ^{17}O NMR. Line shape analysis is difficult because the static line width itself is temperature-dependent but, as shown in the pioneering work of Diehl *et al.* (1962) on benzofuroxan, it is always possible to measure at least a temperature of coalescence and from that to calculate the free energy of activation. Aime *et al.* (1981) have shown that dynamic ^{17}O NMR is a convenient alternative to ^{13}C NMR for metal carbonyl compounds, especially when the metal has a quadrupolar nucleus. A solution of $HFeCO_3(CO)_{12}$ at $-89°C$ gives only two ^{13}CO signals in a ratio of $1:2$ instead of the expected four signals in a $1:1:1:1$ ratio. In ^{17}O NMR, the four signals are easily observed at $-11°C$ and a two-step exchange process averages all the sites producing a single line at $108°C$.

Dynamic ^{17}O NMR has also been used in polymolybdate chemistry (Day *et al.*, 1977b). The microdynamic behavior of liquid water has been analyzed over a wide range of temperatures (-16 to $145°C$) by relaxation time measurements, and the non-Arrhenius temperature dependence is interpreted in terms of an equilibrium between "lattice" water and "free" water (Hindman *et al.*, 1973). Little work has been done with other liquids because of the lack of known QCCs until recently, and the previous results obtained by Wallach and Huntress (1969) on formamide are questionable in view of more recent measurements (Burgar *et al.*, 1981). In the case of paramagnetic systems, the quadrupolar line width is negligible with respect to both the paramagnetic broadening and the exchange broadening. It is then possible to study the rate of exchange of water between the coordination sphere of a cation and the bulk water (Achlama and Fiat, 1974). Exchange studies on diamagnetic systems like $Be^{2+}(H_2O)_4$ are also feasible, but first the bulk water line must be shifted using shift reagents and then the variation in line width with temperature must be accounted for (Frahm and Füldner, 1980).

D. Kinetics of Oxygen Exchange

The application of ^{17}O NMR to measurement of the rate of oxygen exchange at a given site—generally with oxygen in water—is obvious. The only experimental requirement is to ensure that there is only a small change in the ^{17}O content of the site during the observation period. The exchange rate can be placed in this range adjusting the temperature and the pH (Luz and Silver, 1966). Aldehydes and ketones have been well studied (Amour *et al.*, 1981; Dahn *et al.*, 1972; Greenzaid *et al.*, 1968), and other slow-exchange studies with $Te(OH)_6$ (Luz and Pecht, 1966), cobaloximes (Curzon *et al.*, 1982), and nitritopentamminecobalt(III) (Jackson *et al.*, 1982) are known. Rapid-exchange reactions can be studied via line broadening, as shown by the periodate–water system (Pecht and Luz, 1965).

E. Water Orientation in Lyotropic Phases and Bilayers

Water molecules are important constituents of lyotropic phases and bilayers. Ordering of the water molecule can be studied with deuterium or with oxygen NMR, the quadrupolar splitting Δ being related to the order parameter by

$$\Delta = \frac{1}{2I(2I - 1)} \chi[3S_{zz} + \eta(S_{xx} - S_{yy})] \tag{11}$$

The discrepancies found by Fujiwara *et al.* (1974) between the results obtained with ^{17}O and ^{2}H NMR were due to the fact that an unique order parameter was defined (Niederberger and Tricot, 1977). It has been shown that dynamic information is easily obtained from ^{17}O line width data and complements the ordering results from deuterium NMR (Tricot and Niederberger, 1979)

F. Biological Studies

As discussed previously, ^{17}O NMR of biologically relevant molecules is possible only for medium-sized molecules. In a study on the conversion of aminoadipyl-cysteinyl-valine to isopenicillin, it was shown with ^{17}O NMR that the oxygen atoms of the isopenicillin came from the tripeptide and not from the medium (Adlington *et al.*, 1982). The success of this study relies on the fact that the labeled oxygen is in COO$^-$, C=O, or OMe groups which are well separated on the chemical shift scale. In the case of the ^{17}O-labeled ATP molecule (Tsai *et al.*, 1980) only a unique broad resonance is observed for the different oxygen atoms, and the decreased intensity in ^{31}P resonances is used to evaluate the extent of labeling.

Because water plays a fundamental role in the conformation and activity of biological macromolecules, it is clear that the water molecule itself is a perfect target for NMR studies. Preliminary studies (Glasel, 1966; 1968; Koenig *et al.*, 1975) have demonstrated the superiority of ^{17}O NMR compared to ^{1}H and ^{2}H NMR for protein hydration studies. Based on the statements of the Lund NMR group (Halle *et al.*, 1981) one can summarize the advantages as follows.

(a) The relaxation effect is large, permitting studies at low protein concentrations.

(b) The ^{17}O relaxation in water is generally dominated by the quadrupolar mechanism (Glasel, 1966, 1968). The intramolecular origin of the electric field gradient makes the quadrupolar interaction virtually independent of the molecular environment.

(c) The ^{17}O relaxation is not affected by proton (or deuteron) exchange with prototropic residues on the protein. Only T_2 can be affected by exchange within a narrow pH range near neutrality.

(d) Cross-relaxation that contributes to ^{1}H relaxation is unimportant for ^{17}O (Koenig et al., 1978).

In order to extract quantitative results from relaxation measurements, one must use an elaborate theory of the relaxation of quadrupolar nuclei under nonextreme narrowing conditions with exchange (Bull et al., 1979). Based on T_1 and T_2 measurements at four frequencies and for seven proteins at different pH values and temperature, the Lund group presented a very convincing picture of protein hydration (Halle et al., 1981). A two-state model with a fast exchange between bulk water and a single class of hydration water explains the experimental results. Because of the influence of the protein surface, reorientation of the hydration water is anisotropic. The variation in the extent of hydration of different proteins results from the relative number of charged residues, particularly carboxylate groups which are more extensively hydrated. There are approximately two layers of hydration having a reorientational correlation time of about 20 ps (about eight times slower than that of bulk water and independent of the nature of the protein). This rapid local motion has a small anisotropic component (corresponding to an order parameter of 0.06) which is averaged out by protein reorientation with a correlation time on the order of 10 μs.

Binding of water to the paramagnetic metallic center of a protein is quite easily detected with ^{17}O NMR. In the case of concanavalin A (Con A) (Meirovitch and Kalb, 1973) T_2 for water is reduced from 4.3 ms in water itself to 1.4 ms in the presence of Mn^{2+}–Con A. Demetallating concanavalin A reduces T_2 to 2.3 ms. A mean residence time of 3×10^{-8} s for the water molecule at the Mn^{2+} site is computed from these results. The binding of water to the copper(II) center of the superoxide dismutase did not modify the T_1 value in the pH range 7.5–11.7; however, there was an appreciable increase in the line width with increasing pH with an apparent pK_a of 11.3 (Bertini et al., 1981). These data are interpreted as being due to the binding of a OH^- anion to the copper ion at high pH values. Binding to diamagnetic centers is more difficult to detect. Oxygen-17 NMR studies on carbonic anhydrase (Rose and Bryant, 1980) reveal that there is no difference in the relaxation rates of the apoenzyme and the zinc-containing enzyme.

The permeability of phospholipid vesicle membranes to water has been measured with ^{17}O NMR (Haran and Shporer, 1976). First, the relaxation time is not affected by the presence of vesicles in suspension in water.

Second, the addition of Mn^{2+} permits resolution of the magnetization into two exponential components, a fast one arising from external water and a slow one (identical to that of pure water) arising from the intravesicular fluid. From the rates of relaxation, the mean lifetime of water within the vesicles is calculated to be 1 ms at 22°C. The ^{17}O longitudinal relaxation time of intracellular water in human erythrocytes is four to five times shorter than in the supernatant solution. This reduction is attributed to interaction with hemoglobin (Shporer and Civan, 1975). Even more complicated systems—frog skeletal muscle—have been submitted to ^{17}O NMR in order to understand the nature of water (Civan and Sphorer, 1972, 1974; Swift and Barr, 1973). The results are not only more qualitative than the preceding ones, but unfortunately different conclusions were derived from the same types of experiments. For example, Civan and Shporer (1972) found a decrease in the intensity of the water signal in the muscle when it was compared to the signal obtained from water collected by vacuum distillation of the muscle, but no such reduction was found by Swift and Barr (1973). The T_1 and T_2 relaxation times of ^1H and ^{17}O have also been measured for an aqueous suspension of dark-adapted chloroplasts (Wydrzynski et al., 1978). Relaxation times for ^1H are determined largely by the loosely bound Mn present in the chloroplast membranes. The ^{17}O T_1 and T_2 data for suspensions before and after treatment with a detergent are consistent with the location of the manganese in the interior of the thylakoids. An analysis of the relaxation rates shows that the average lifetime of a water molecule inside a thylakoid is >1 ms.

IV. Conclusions

Quite often oxygen atoms are located at the reactive sites of a molecule. Oxygen NMR is then a convenient tool either for following the course of a reaction or for correlating reactivity with spectroscopic properties. Sometimes technical limitations are frustrating, for example, when the line width of the resonance exceeds the observation capacity of the spectrometer. During a study on complexation shifts induced by alkali cations in a series of crown ethers, it was quite easy to observe at 12.2 MHz the ^{17}O signal from the crown ether 12C6 (Popov et al., 1980). On complexation with K^+, the signal disappeared. Relaxation time measurements for ^{13}C showed that the correlation time increased on complexation. The concomitant line broadening made the ^{17}O line unobservable at 12.2 MHz, but on going to 27.15 MHz it was again possible to detect the signal. Remember also that no oxygen signal was found in highly enriched Vaska complexes (Lapidot and Irving, 1972). Because the line width should not exceed a few kilohertz and because Dundon (1981) actually has observed a resonance of about 50 kHz in liquid oxygen, it is likely that the missing line will be detected in the near future.

The future of ^{17}O NMR appears bright. The combination of both scientific interest in the reactivity of small oxygen-containing molecules and the actual possibilities of high-field spectrometers will create increased interest in this hither to capricious spectroscopy. In the field of molecular biology, the state of the art appears to have been achieved with studies on saccharides at natural abundance in aqueous solutions (Gerothanassis *et al.*, 1982a), but a vast field of studies remain possible in investigation of the interaction of small molecules with macromolecules (Halle *et al.*, 1981).

References

Achlama, A. M. (1977). *Chem. Phys. Lett.* **48**, 501–504.

Achlama, A. M., and Fiat, D. (1974). "Nuclear Magnetic Resonance Spectroscopy of Nuclei Other than Protons", pp. 143–152. Wiley, London.

Adlington, R. M. *et al.* (1982). *J. Chem. Soc. Chem. Commun.*, 137–139.

Aime, S., Osella, D., Milone, L., Hawles, G. E., and Randall, E. W. (1981). *J. Amer. Chem. Soc.* **103**, 5920–5922.

Alder, F., and Yu, F. C. (1951). *Phys. Rev.* **81**, 1067–1068.

Alei, M. (1965). *J. Chem. Phys.* **43**, 2904–2906.

Alei, M., and Florin, A. E. (1969). *J. Phys. Chem.* **73**, 863–867.

Amour, T. E. St., and Fiat, D. (1980). *Bull. Magn. Reson.* **1**, 118–129.

Amour, T. E. St., Burgar, M. I., Valentine, B., and Fiat, D. (1981). *J. Am. Chem. Soc.* **103**, 1128–1136.

Andersson, L. O., and Mason, J. (1974). *J. Chem. Soc. Dalton Trans.*, 202–205.

Bertini, F., Luchinat, C., and Messori, L. (1981). *Biochem. Biophys. Res. Commun.* **101**, 577–583.

Blinc, R., Seliger, J., Osredkar, R., and Mali, M. (1974). *Phys. Lett.* **47a**, 131–132.

Block, E. *et al.* (1980). *J. Org. Chem.* **45**, 4810–4812.

Bramley, R., Figgis, B. N., and Nyholm, R. S. (1962). *Trans. Faraday Soc.* **58**, 1893–1896.

Brevard, C., and Granger, P. (1981). *J. Chem. Phys.* **75**, 4175–4177.

Brosnan, S. G. P., and Edmonds, D. T. (1980). *J. Magn. Reson.* **38**, 47–63.

Brosnan, S. G. P., Edmonds, D. T., and Poplett, I. J. F. (1981). *J. Magn. Reson.* **45**, 451–460.

Buckler, K. U., Haase, A. R., Lutz, O., Muller, M., and Nolle, A. (1977). *Z. Naturforsch.* **32A**, 126–130.

Bull, T. E., Forsen, S., and Turner, D. L. (1979). *J. Chem. Phys.* **70**, 3106–3111.

Burgar, M. I., Armour, T. E. St., and Fiat, D. (1981). *J. Phys. Chem.* **85**, 502–510.

Butler, L. G., and Brown, T. L. (1981). *J. Am. Chem. Soc.* **103**, 6541–6549.

Butler, L. G., Cheng, C. P., and Brown, T. L. (1981). *J. Phys. Chem.* **85**, 2738–2740.

Canet, D., Goulon-Ginet, C., and Marchal, J. P. (1976). *J. Magn. Reson.* **22**, 337–342; **25**, 397.

Cheng, C. P., and Brown, T. L. (1979). *J. Am. Chem. Soc.* **101**, 2327–2334.

Cheng, C. P., and Brown, T. L. (1980). *J. Am. Chem. Soc.* **102**, 6418–6421.

Christ, H. A., and Diehl, P. (1963). *Helv. Phys. Acta* **36**, 170–182.

Christ, H. A., Diehl, P., Schneider, H. R., and Dahn, H. (1961) *Helv. Chim. Acta* **44**, 865–880.

Civan, M. M., and Shporer, M. (1972). *Biophys. J.* **12**, 404–413.

Civan, M. M., and Shporer, M. (1974). *Biochim. Biophys. Acta* **343**, 399–408.

Coderre, S. A. *et al.* (1981). *J. Am. Chem. Soc.* **103**, 1870–1872.

Cohn, M., and Hu, A. (1980). *J. Am. Chem. Soc.* **102**, 913–916.

Coulson, C. A., and Moffit, W. E. (1949). *Philos. Mag.* **40**, 1–35.

Cozak, D., Butler, I. S., Hickey, J. P., and Todd, L. J. (1979). *J. Magn. Reson.* **33**, 149–157.

Crandall, J. K., and Centeno, M. A. (1979). *J. Org. Chem.* **44**, 1183–1184.

Crandall, J. K., Centeno, M. A., and Borresen, S. (1979). *J. Org. Chem.* **44**, 1184–1186.

Curzon, E. H., Golding, B. T., and Wong, A. K. (1982). *J. Chem. Soc. Chem. Commun.* 63–65.

Dahn, H., Schlumke, H. P., and Temler, J. (1972). *Helv. Chim. Acta* **55**, 907–916.

Day, V. W., Fredrich, M. F., Klemperer, W. G., and Shum, W. (1977a). *J. Am. Chem. Soc.* **99**, 6146–6148.

Day, V. W., Fredrich, M. F., Klemperer, W. G., and Shum, W. (1977b). *J. Am. Chem. Soc.* **99**, 952–953.

Delseth, C., and Kintzinger, J. P. (1976). *Helv. Chim. Acta* **59**, 466–475; **59**, 1411.

Delseth, C., and Kintzinger, J. P. (1978). *Helv. Chim. Acta* **61**, 1327–1334.

Delseth, C., and Kintzinger, J. P. (1982). *Helv. Chim. Acta* **65**, 2273–2279.

Delseth, C., Kintzinger, J. P., Nguyên, T. T. T., and Niederberger, W. (1978). *Org. Magn. Reson.* **11**, 38–39.

Delseth, C., Nguyên, T. T. T., and Kintzinger, J. P. (1980). *Helv. Chim. Acta* **63**, 498–503.

Diakur, J., Nakashima, T. T., and Vederas, J. C. (1980). *Can. J. Chem.* **58**, 1311–1315.

Diehl, P., Christ, H. A., and Mallory, F. B. (1962). *Helv. Chim. Acta* **45**, 504.

Dundon, J. M. (1982). *J. Chem. Phys.* **76**, 2171–2172.

Dwek, R. A., Luz, Z., Peller, S., and Shporer, M. (1971). *J. Am. Chem. Soc.* **93**, 77–79.

Edmonds, D. T., Goren, S. D., Mackay, A. L., and White, A. A. L. (1976). *J. Magn. Reson.* **23**, 505–514.

Edmonds, D. T., Goren, S. D., White, A. A. L., and Sherman, W. F. (1977). *J. Magn. Reson.* **27**, 35–44.

Eliel, E. L., Pietrusiewicz, K. M., Jewell, L. M., and Kenan, W. R. (1979). *Tetrahedron Lett.*, pp. 3649–3652.

Figgis, B. N., Kidd, R. S., and Nyholm, R. S. (1962). *Proc. R. Soc. A* **269**, 469–480.

Filowitz, M., Ho, R. K. C., Klemperer, W. G., and Shum, W. (1979). *Inorg. Chem.* **18**, 93–103.

Florin, A. E., and Alei, M. (1967). *J. Chem. Phys.* **47**, 4268–4269.

Flygare, W. H., and Lowe, J. T. (1965). *J. Chem. Phys.* **45**, 3645–3653.

Frahm, J., and Füldner, H. H. (1980). *Ber. Bunsenges. Phys. Chem.* **84**, 173–176.

Fujiwara, F., Reeves, L. W., Tracey, A. S., and Wilson, L. A. (1974). *J. Am. Chem. Soc.* **96**, 5249–5250.

Garrett, B. B., Denisson, A. B., and Rabideau, S. W. (1967). *J. Phys. Chem.* **71**, 2606–2611.

Gerber, P. (1972). *Helv. Phys. Acta* **45**, 655–682.

Gerothanassis, I. P., and Sheppard, N. (1982). *J. Magn. Reson.* **46**, 423–439.

Gerothanassis, I. P., Lauterwein, J., and Sheppard, N. (1982a). *J. Magn. Reson.*, **48**, 431–446.

Gerothanassis, I. P., Hunston, R., and Lauterwein, J. (1982b). *Helv. Chim. Acta* **65**, 1764–1773.

Gerothanassis, I. P., Hunston, R., and Lauterwein, J. (1982c). *Helv. Chim. Acta* **65**, 1774–1784.

Geschwind, S., Gunther-Mohr, G. R., and Silvey, G. (1952). *Phys. Rev.* **85**, 474–477.

Glasel, J. A. (1966). *Proc. Natl. Acad. Sci. USA* **55**, 479–485.

Glasel, J. A. (1968). *Nature* **218**, 953–955.

Goldammer, E. V., and Hertz, H. G. (1970). *J. Phys. Chem.* **74**, 3734–3755.

Gorin, P. A. J., and Mazurek, M. (1978). *Carbohydr. Res.* **67**, 479–483.

Gorodetsky, M., Luz, Z., and Mazur, Y. (1967). *J. Am. Chem. Soc.* **89**, 1183–1189.

Gray, G. A., and Albright, T. A. (1977). *J. Am. Chem. Soc.* **99**, 3243–3250.

Greenzaid, P., Luz, Z., and Samuel, D. (1968). *Trans. Faraday Soc.* 2780–2787; 2787–2793.

Grossman, G., Gruner, M., and Seifert, G. (1976). *Z. Chem.* **16**, 362–363.

Halle, B., Andersson, T., Forsén, S., and Lindman, B. (1981). *J. Am. Chem. Soc.* **103**, 500–508.

Haran, N., and Shporer, M. (1976). *Biochim. Biophys. Acta* **426**, 638–646.

Harris, R. K., and Kimber, B. J. (1975). *Org. Magn. Reson.* **7**, 460–464.

Hertz, H. G., and Klute, R. (1970). *Z. Phys. Chem.* **69**, 101–107.

Hindman, J. C., Svirmickas, A., and Wood, M. (1973). *J. Chem. Phys.* **59**, 1517–1522.

Hsieh, Y., Koo, J. C., and Hahn, E. L. (1972). *Chem. Phys. Lett.* **13**, 563–566.

Hubbard, P. S. (1970). *J. Chem. Phys.* **53**, 985–987.

Hunston, R., Gerothanassis, I. P., and Lauterwein, J. (1982). *Org. Magn. Reson* **18**, 120–121.

Irving, C. S., and Lapidot, A. (1971). *Nature* **230**, 224.

Irving, C. S., and Lapidot, A. (1976). *J. Chem. Soc. Chem. Commun.* 43–44.

Iwamura, H., Sugawara, T., Kawada, Y., Tori, K., Muneyuki, R., and Woyori, R. (1979). *Tetrahedron Lett.* pp. 3449–3452.

Jackson, J. A., and Taube, H. (1965). *J. Phys. Chem.* **69**, 1844–1849.

Jackson, J. A., Lemons, J. F., and Taube, H. (1960). *J. Chem. Phys.* **32**, 553–555.

Jackson, W. G., Lawrence, G. A., Lay, P. A., and Sargeson, A. M. (1982). *J. Chem. Soc. Chem. Commun.* 70–72.

Kado, R., and Takarada, Y. (1978). *J. Magn. Reson.* **32**, 89–91.

Kalabin, G. A., Kushnarev, D. F., Valeyev, R. B., Tronfinov, B. A., and Fedotov, M. A. (1982). *Org. Magn. Reson.* **18**, 1–9.

Kamper, R. A., Lea, K. R., and Lustig, C. D. (1957). *Proc. Phys. Soc.* **70B**, 897–899.

Kawada, Y., Sugawara, T., and Iwamura, H. (1979). *J. Chem. Soc. Chem. Commun.* 291–292.

Kidd, R. S. (1967). *Can. J. Chem.* **45**, 605–608.

Kintzinger, J. P. (1981). "NMR Basic Principles and Progress" (P. Diehl, E. Fluck, and R. Kosfeld, eds.), Vol. 17, pp. 1–64. Springer, Heidelberg.

Kintzinger, J. P., Delseth, C., and Nguyên, T. T. T. (1980). *Tetrahedron* **36**, 3431–3435.

Klemperer, W. G. (1978). *Angew. Chem. Int. Ed. Engl.* **17**, 246–254.

Kobayashi, K., Sugawara, T., and Iwamura, H. (1981). *J. Chem. Soc. Chem. Commun.* 479–480.

Koenig, S. H., Hallenga, K., and Shporer, M. (1975). *Proc. Natl. Acad. Sci. USA* **72**, 2667–2671.

Koenig. H. S., Bryant, R. G., Hallenga, K., and Jacob, G. S. (1978). *Biochemistry* **17**, 4348–4358.

Kump, R. L., and Todd, L. J. (1980). *J. Chem. Soc. Chem. Commun.* 292–293.

Lapidot, A., and Irving, C. S. (1972). *J. Chem. Soc. Dalton Trans.* 668–670.

Li, F., Brookeman, J. R., Rigamonti, A., and Scott, T. A. (1980). *J. Chem. Phys.* **74**, 3120.

Lippmaa, E. *et al.* (1972). *Org. Magn. Reson.* **4**, 153–160, 197–202.

Lowe, G., Potter, B. V. L., Sproat, B. S., and Hull, W. E. (1979). *J. Chem. Soc. Chem. Commun.*, 733–735.
Lumpkin, O., and Dixon, W. T. (1979a). *Chem. Phys. Lett.* **62**, 139–142.
Lumpkin, O., and Dixon, W. T. (1979b). *J. Chem. Phys.* **71**, 3550.
Lumpkin, O. Dixon, W. T., and Poser, J. (1979). *Inorg. Chem.* **18**, 982–984.
Lutz, O., and Oehler, H. (1977). *Z. Naturforsch.* **32a**, 131–133.
Lutz, O., Nepple, W., and Nolle, A. (1976a). *Z. Naturforsch.* **31a**, 978–980.
Lutz, O., Nepple, W., and Nolle, A. (1976b). *Z. Naturforsch.* **31a**, 1046–1050.
Luz, Z., and Pecht, I. (1966). *J. Am. Chem. Soc.* **88**, 1152–1154.
Luz, Z., and Silver, B. L. (1966). *J. Phys. Chem.* **70**, 1328–1331.
McKelvey, R. D., Kawada, Y., Sugawara, T., and Iwamura, H. (1981). *J. Org. Chem.* **46**, 4948–4952.
Mägi, M., Erashko, V., Shevelon, S., and Fainziberg, A. (1971). *Eesti NSV Tead. Akad. Toim. Keem. Geol.* **20**, 297–303.
Mateescu, G. D., and Benedikt, G. M. (1979). *J. Am. Chem. Soc.* **101**, 3959–3960.
Meiboom, S. (1961). *J. Chem. Phys.* **43**, 375–388.
Meirovitch, E., and Kalb, A. J. (1973). *Biochem. Biophys. Acta* **303**, 258–263.
Miller, K. F., and Wentworth, R. A. D. (1979). *Inorg. Chem.* **18**, 984–988.
Moore, R. N., Diakur, J., Nakashima, T. T., McLaren, S. L., and Vederas, J. C. (1981). *J. Chem. Soc. Chem. Commun.*, 501–502.
Nguyên, T. T. T., Delseth, C., Kintzinger, J. P., Carrupt, P. A., and Vogel, P. (1980). *Tetrahedron* **36**, 2793–2797.
Niederberger, W., and Tricot, Y. (1977). *J. Magn. Reson.* **28**, 313–316.
Olah, G. A., Berrier, A. L., and Prakash, G. K. S. (1982). *J. Am. Chem. Soc.* **104**, 2373–2376.
Pecht, I., and Luz, Z. (1965). *J. Am. Chem. Soc.* **87**, 4068–4072.
Popov, A. I., Smetana, A. J., Kintzinger, J. P., and Nguyên, T. T. T. (1980). *Helv. Chim. Acta* **63**, 668–673.
Reuben, J. (1969). *J. Am. Chem. Soc.* **91**, 5725–5729.
Reuben, J., and Brownstein, S. (1967). *J. Mol. Spectrosc.* **23**, 96–98.
Risley, J. M., and Van Etten, R. L. (1979). *J. Am. Chem. Soc.* **101**, 252–253.
Risley, J. M., and Van Etten, R. L. (1980a). *J. Am. Chem. Soc.* **102**, 4609–4614.
Risley, J. M., and Van Etten, R. L. (1980b). *J. Am. Chem. Soc.* **102**, 6699–6702.
Risley, J. M., and Van Etten, R. L. (1981). *J. Am. Chem. Soc.* **103**, 4389–4392.
Rose, K. D., and Bryant, R. G. (1980). *J. Am. Chem. Soc.* **102**, 21–24.
Rosenblum, B., and Nethercot, A. H. (1957). *J. Phys. Chem.* **27**, 828–829.
Sardella, D. J., and Stothers, J. B. (1969). *Can. J. Chem.* **47**, 3089–3092.
Schwartz, H. M., McCoss, M., and Danyluk, S. S. (1980). *Tetrahedron Lett.* pp. 3837–3840.
Shporer, M., and Achlama, A. M. (1976). *J. Chem. Phys.* **65**, 3657–3664.
Shporer, M., and Civan, M. M. (1975). *Biochim. Biophys. Acta* **385**, 81–87.
Spiess, H. W., Garret, B. B., Sheline, R. K., and Rabideau, S. W. (1969). *J. Chem. Phys.* **51**, 1201–1205.
Sugawara, T., Kawada, Y., and Iwamura, H. (1978). *Chem. Lett.* 1371–1374.
Susuki, M., and Kubo, R. (1964). *Mol. Phys.* **7**, 201–209.
Swift, T. J., and Barr, E. M. (1973). *Ann. N. Y. Acad. Sci.* 191–196.
Tiffon, B., Ancian, B., and Dubois, J. E. (1980). *Chem. Phys. Lett.* **73**, 89–93.
Townes, C. H., and Dailey, B. P. (1949). *J. Chem. Phys.* **17**, 782–796.
Tricot, Y., and Niederberger, W. (1979). *Biophys. Chem.* **9**, 195–200.
Tsai, M. D. (1979). *Biochemistry* **18**, 1468–1472.

Tsai, M. D., Huang, S. L., Kozlowski, J. F., and Chang, C. C. (1980). *Biochemistry* **19**, 3531–3536.

Valentine, B., Amour, T. E. St., Walter, R., and Fiat, D. (1980). *Org. Magn. Reson.* **13**, 232–233; *J. Magn. Reson.* **38**, 413–418.

Van Etten, R. L., and Risley, J. M. (1981). *J. Am. Chem. Soc.* **103**, 5633–5636.

Vederas, J. C. (1980). *J. Am. Chem. Soc.* **102**, 374–376.

Verhoeven, J., and Dynamus, A., and Bluyssen, H. (1969). *J. Chem. Phys.* **50**, 3330–3338.

Versmold, H., and Yoon, C. (1972). *Ber. Bunsenges. Phys. Chem.* **76**, 1164–1168.

Virlet, J., and Tantot, G. (1976). *Chem. Phys. Lett.* **44**, 296–299.

Vold, R. R., and Vold, R. L. (1974). *J. Chem. Phys.* **61**, 4360–4361.

Wallach, D., and Huntress, W. T. (1969). *J. Chem. Phys.* **50**, 1219–1227.

Weaver, M. E., Tolbert, B. M., and La Force, R. C. (1955). *J. Chem. Phys.* **23**, 1956–1957.

Wydrzynski, T. J., Marks, S. B., Schmidt, P. G., Govindjee, and Gutowsky, H. S. (1978). *Biochemistry* **17**, 2155–2162.

5 Alkali Metals

Christian Detellier

The Ottawa-Carleton Institute for Research and Graduate Studies in Chemistry
University of Ottawa
Ottawa, Canada

I. Introduction

Like some other words ("alcohol" and "algebra", for example) derived from the rich Arabian knowledge of science in the Middle Ages, the medieval Arabic language has bequeathed the term "alkali" (meaning "soda"). Sodium is one of the more abundant elements. The 16 external kilometers of the terrestrial crust and seawaters contain, respectively, 2.83 and 1.14 wt % sodium. Potassium is almost as abundant as sodium in the terrestrial crust (2.59%), and its percentage by weight is 0.04% in seawaters. A 70-kg representative of the species *Homo sapiens* contains approximately 140 g of potassium and 105 g of sodium. Needless to say, potassium and sodium play a crucial role in every kind of life. Lithium is essential for the growth of only a few plants. However, lithium carbonate has an important medical application because it has been employed, since the 1950s, in the treatment of psychiatric diseases, such as manic-depressive psychosis. Its actual role in this treatment is not yet well understood. Rubidium and cesium are poisonous to the majority of organisms, and francium has no stable isotopes.

The nuclides to be considered in this chapter are ^{6}Li, ^{7}Li, ^{23}Na, ^{39}K, ^{85}Rb, ^{87}Rb, and ^{133}Cs. Table I gives the nuclear properties of these alkali metal isotopes. Lithium-7, ^{23}Na, ^{39}K, and ^{133}Cs have been endowed with a 90% or more natural abundance and also represent the more easily observed nuclide (^{7}Li, ^{39}K) or the sole natural nuclide (^{23}Na, ^{133}Cs) of each of these elements. They are all stable isotopes, except ^{87}Rb whose half-life is 5×10^{11} yr. Among the radioactive isotopes, ^{40}K, with a natural abundance of 0.012% and a half-life of 1.28 10^9 yr, is known to have a quadrupole moment of -0.7×10^{-28} m^2 and a spin number $I = 4$. Its receptivity compared to that of ^{13}C is only 3.57×10^{-3}. Sodium-22 (half-life 2.6 yr, $I = 3$) is a potential candidate for future investigations. All these nuclei are quadrupolar, with half-integer spin numbers, except for ^{6}Li ($I = 1$) and ^{40}K ($I = 4$). The majority have a spin number of $\frac{3}{2}$, and the minority are represented by ^{85}Rb ($I = \frac{5}{2}$) and ^{133}Cs ($I = \frac{7}{2}$). Thus the predominant relaxation mechanism is quadrupolar. For ^{6}Li and ^{133}Cs, quadrupole moments are low ($Q = -8 \times 10^{-4}$ and $-3 \times 10^{-3} \times 10^{-28}$ m^2, respectively), and other mechanisms can compete or even dominate. The receptivities of ^{7}Li, ^{23}Na, ^{87}Rb, and ^{133}Cs are high compared to that of ^{13}C, and these nuclei are easily observed. Although ^{41}K receptivity is very low, this nuclide could be useful because it provides the possibility of a complementary study involving ^{39}K, at a different field strength, after acoustic ringing problems have been solved.

Excellent reviews on the subject of alkali metal NMR have appeared as such (Forsén and Lindman, 1978), incorporated into a broader topic (Dechter, 1982; Forsén and Lindman, 1981; Wehrli, 1979), or focused on a single nuclide (^{23}Na) (Laszlo, 1978). Solid state studies will not be considered in this chapter: one can, however, mention the report of the first triple-quantum ^{23}Na Fourier transform (FT) spectra (Vega and Naor, 1981) and the use of high-speed sample spinning techniques (Oldfield *et al.*, 1982) to obtain high-resolution spectra in solids.

II. Chemical Shifts

The paramagnetic term is the predominant term of the chemical shift for the alkali metals, except for ^{6}Li and ^{7}Li where diamagnetic and paramagnetic terms are on the same order of magnitude. Although correlations have been found among ^{23}Na, ^{39}K, and ^{133}Cs chemical shifts and the donicity number of the solvent (Popov, 1979), such a correlation does not exist in the case of ^{7}Li where paramagnetic and diamagnetic contributions nearly cancel out each other (Section IV,A,2). Chemical shifts relative to

TABLE I
Nuclear Properties of the Alkali Metals.[a]

Isotope	Natural abundance (%)	I	Gyromagnetic ratio γ (10^7 rad T^{-1} s^{-1})	Quadrupole moment Q (10^{-28} m^2)	Receptivity referred to ^{13}C	Resonance frequency ^{1}H at 80 MHz (MHz)	Resonance frequency ^{1}H at 400 MHz (MHz)
^{6}Li	7.42	1	3.9366	-8×10^{-4}	3.58	11.772	58.862
^{7}Li	92.58	$\frac{3}{2}$	10.3964	-4.5×10^{-2}	1540	31.091	155.454
^{23}Na	100	$\frac{3}{2}$	7.0761	0.12	525	21.161	105.805
^{39}K	93.1	$\frac{3}{2}$	1.2483	5.5×10^{-2}	2.69	3.733	18.666
^{41}K	6.88	$\frac{3}{2}$	0.6851	6.7×10^{-2}	0.0328	2.049	10.245
^{85}Rb	72.15	$\frac{5}{2}$	2.5828	0.25	43	7.724	38.620
^{87}Rb	27.85	$\frac{3}{2}$	8.7532	0.12	277	26.177	130.885
^{133}Cs	100	$\frac{7}{2}$	3.5087	-3×10^{-3}	269	10.493	52.468

[a] The majority of data are taken from Brevard and Granger, 1981.

TABLE II

Chemical Shifts for Alkali Ions[a]

	δ (ppm)		
Nuclide	M⁻ in THF	M⁺ in NM	M⁺ in DMSO
^{7}Li	—	+10.6	+10.0
^{23}Na	−2.3	+44.9	+60.4
^{39}K	—	+84.0	+112.9
^{87}Rb	+14.4	—	—
^{133}Cs	+52.3	+284.5	+412.3

[a] Expressed in parts per million relative to M(g). Data for M⁻ are from Dye (1979). Data for M⁺ are from Popov (1979), based on reference values from Forsén and Lindman (1978). THF, Tetrahydrofuran; NM, Nitromethane (donicity number 2.7); DMSO, dimethyl sulfoxide (donicity number 29.8).

Na(g) are given in Table II for cations dissolved in a solvent of low donicity number (nitromethane, DN = 2.7) and in a solvent of high donicity number [dimethyl sulfoxide (DMSO), DN = 29.8]. The difference is indicative of the chemical shift range displayed by the observed cationic nuclide. ^{23}Na⁺, ^{39}K⁺, and ^{133}Cs⁺ have, respectively, ranges of 20, 30, and 120 ppm. Since the preparation of alkali metal anion salts by Dye and his co-workers (Dye, 1979), chemical shift ranges have been greatly broadened (Table II): ^{23}Na⁻, ^{87}Rb⁻, and ^{133}Cs⁻ resonate high field of the cation resonance. The chemical shift is close to the value for Na(g) and to the calculated diamagnetic shift for Na⁻ compared to Na(g) (Dye, 1979) (see also Section IV,A,3). Sodium-23 chemical shifts have been correlated with observed line widths in amine solutions (Delville et al., 1981c; Section IV,A,2).

III. Relaxation

All alkali metal nuclides are characterized by a quadrupole moment, and consequently their relaxation behavior is influenced by the powerful quadrupolar relaxation mechanism. For nuclides with a low quadrupole moment (^{6}Li, ^{7}Li, ^{133}Cs), other relaxation mechanisms can compete with the quadrupolar mechanism, and their behavior will be considered here. Line widths at half-height for concentrated aqueous cation solutions are given in Table III. Three groups appear: non-purely-quadrupolar nuclides (^{6}Li,

TABLE III

Line Widths[a] and Sternheimer[b] Antishielding
Factors for Alkali Metal Cationic Nuclides

Nuclide	Solution	$\nu_{1/2}$ (Hz)	$1 + \gamma_\infty$
^{6}Li	LiCl–D$_2$O, 1 M	1	0.249
^{7}Li	LiCl–D$_2$O, 1 M	1	0.249
^{23}Na	NaBr–D$_2$O, 9.8 M	14	−5.26
^{39}K	KBr–D$_2$O, saturated	13	−19.96
^{41}K	KBr–D$_2$O, saturated	16	−19.96
^{85}Rb	RbCl–D$_2$O, 5.6 M	150	−47.66
^{87}Rb	RbCl–D$_2$O, 5.6 M	132	−47.66
^{133}Cs	CsNO$_3$–D$_2$O, saturated	0.6	−95.16

[a] From Brevard and Granger (1981).
[b] From Schmidt et al. (1980), for Alkali Metal Cationic Nuclides.

^{7}Li, ^{133}Cs), purely quadrupolar nuclides with narrow line widths (^{23}Na, ^{39}K, ^{41}K), and purely quadrupolar nuclides with broad line widths (^{85}Rb, ^{87}Rb). The low quadrupole moment explains the line width narrowness of the first group, and the high value of the Sternheimer antishielding factor explains the line width broadness of the third group. Sternheimer antishielding factors, which indicate that nuclei are being influenced by a magnified electrostatic field gradient due to the atomic electrons, have been calculated (Schmidt et al., 1980) and are given in Table III.

Except under extreme narrowing conditions, the relaxation of a quadrupolar nucleus is the sum of $2I$ exponentials for nuclei of integral spin I and of $2I + 1$ exponentials for nuclei with spin I, where I is half of an odd integer. The case of a spin-$\frac{3}{2}$ nucleus is particularly interesting because the decay of the transverse relaxation can be calculated (Bull, 1972; Hubbard, 1970) to be the sum of two exponentials given by

$$M_t = Mo[0.6 \exp(-t/T_2') + 0.4 \exp(-t/T_2'')] \tag{1}$$

This equation is valid for a spin-$\frac{3}{2}$ nucleus exchanging between two states, f and b, if the extreme narrowing condition is not satisfied at one of the sites (b) and if the relaxation times and exchange times at one site are much smaller than those at the other site.

T_2' is given by

$$\frac{1}{T_2'} = \frac{1}{T_{2f}} + \frac{P_b}{P_f} \frac{1}{\tau_b} \frac{(1/T_{2b}')(1/T_{2b}' + 1/\tau_b) + \Delta\omega^2}{(1/T_{2b}' + 1/\tau_b)^2 + \Delta\omega^2} \tag{2}$$

and T''_2 is given by

$$\frac{1}{T''_2} = \frac{1}{T_{2f}} + \frac{P_b}{P_f}\frac{1}{\tau_b}\frac{(1/T''_{2b})(1/T''_{2b} + 1/\tau_b) + \Delta\omega^2}{(1/T''_{2b} + 1/\tau_b)^2 + \Delta\omega^2} \tag{3}$$

where p is the population, τ_b the average lifetime at site b, $\Delta\omega$ the chemical shift difference between ω_f and ω_b, and T'_{2b} and T''_{2b} are given by

$$\frac{1}{T'_{2b}} = \frac{1}{20}\left(1 + \frac{\eta^2}{3}\right)\chi^2\left(\tau_c + \frac{\tau_c}{1 + \omega^2\tau_c^2}\right) \tag{4}$$

$$\frac{1}{T''_{2b}} = \frac{1}{20}\left(1 + \frac{\eta^2}{3}\right)\chi^2\left(\frac{\tau_c}{1 + 4\omega^2\tau_c^2} + \frac{\tau_c}{1 + \omega^2\tau_c^2}\right) \tag{5}$$

where τ_c is the correlation time for the electric field gradient fluctuation, χ the quadrupolar coupling constant (QCC) e^2qQ/h incorporating also $1 + \gamma_\infty$ (the Sternheimer antishielding factor), η the asymmetry factor (varying between 0 and 1), eq the electric field gradient at the nucleus site, and eQ the electric quadrupole moment. If T'_{2b} and T''_{2b} can be determined, their ratio will give τ_c:

$$\frac{T''_{2b}}{T'_{2b}} = \frac{1 + 1/(1 + \omega^2\tau^2)}{1/(1 + 4\omega^2\tau_c^2) + 1/(1 + \omega^2\tau_c^2)} \tag{6}$$

Equation (6) assumes that the residence time τ_b is small with respect to T'_{2b} and T''_{2b} and can safely be neglected. As shown in Fig. 1, deconvolution of the experimental non-Lorentzian line shape provides T''_2 and T'_2 and permits the calculation of τ_c and $P_b\chi^2$ (if $\Delta\omega^2$ can be neglected) (Delville et al., 1979; Gustavsson et al., 1978; Monoi and Uedaira, 1980; Mulder et al., 1980; Venkatachalam and Urry, 1980; Section IV,B). One must be aware of all the assumptions made throughout the course of this determination. Both τ_c and $P_b\chi^2$ can also be determined by comparison of T''_{2b} or T'_{2b} at two different field strengths. A simple method for determining T''_{2b} and T'_{2b} from a non-Lorentzian signal of a spin-$\frac{3}{2}$ nucleus without the necessity of a full deconvolution procedure has been proposed by Delville et al. (1979). The two components, however, can be separated because of the possibility of a second-order dynamic frequency effect (Werbelow and Marshall, 1981).

Under extreme narrowing conditions ($\omega^2\tau_c^2 \ll 1$) the decay of the relaxation becomes exponential and

$$\frac{1}{T_1} = \frac{1}{T_2} = \frac{3}{40}\frac{2I + 3}{I^2(2I - 1)}\chi^2\tau_c \tag{7}$$

Under these conditions, it is impossible to determine χ^2 and τ_c separately from the observation of only $1/T_2$.

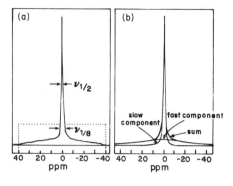

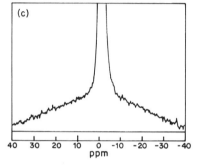

Fig. 1 Non-Lorentzian ^{23}Na line shape obtained from solutions of sodium chloride (0.35 M) and malonyl gramicidin incorporated into lysolecithin micelles (2.6 mM). (a) The experimental line shape. The dotted boundary indicates the portion shown enlarged in (c). Computer-calculated Lorentzians are shown in (b). (From Venkatachalam and Urry, 1980.)

When analytical solutions are available for spin-$\frac{3}{2}$ nuclei, equations must be solved numerically for nuclei with larger spins, because as the relaxation parameters change, both the relaxation rates and preexponential factors vary (Bull et al., 1979). Halle and Wennerstrom (1981) have described a perturbation treatment applicable to $I = \frac{5}{2}$ (^{85}Rb) and $I = \frac{7}{2}$ (^{133}Cs) nuclei, which describes the relaxation behavior as a simple biexponential decay.

Since the quadrupole moment is very low for ^{6}Li, ^{7}Li, and ^{133}Cs, one can expect the appearance of competitive relaxation mechanisms. However, a 0.2 M CsCl aqueous solution does not show signal enhancement after irradiation at the proton frequency, and the dependence of ln T_1 versus inverse absolute temperature between 30 and 100°C is linear, giving an activation energy of 1.93 kcal mol^{-1} (Wehrli, 1977). Under these conditions, no dipolar or spin–rotation relaxations have been established for ^{133}Cs, and the quadrupolar relaxation mechanism remains the only plausible one, despite the low quadrupole moment.

Individual contributions to the experimental ^{6}Li T_1 relaxation rate for aqueous lithium chloride have been determined (Wehrli, 1976, 1978a,b). The dipole–dipole (^{6}Li—^{1}H) contribution is by far the most important at 40°C (84%), spin–rotation accounts for 8%, and the quadrupolar contribution is negligible. Lithium-6 behaves like a spin-$\frac{1}{2}$ nucleus (Wehrli, 1978a,b).

The contribution of the dipolar mechanism is also important (30%) in ^{7}Li relaxation in LiI aqueous solutions (Weingärtner, 1980).

IV. Applications

A. Local Molecular Ordering around Alkali Metal Cations

1. Ion Pairing

Many types of ion pairs can coexist in a solution of at least one salt dissolved in water or in a nonaqueous solvent. Arbitrarily taking the cation as the center around which other molecular entities organize, one can consider mainly two types of molecular entities competing for occupation of the coordination shell of the cation: the conjugate anion and the solvent molecule. At a second stage, one considers the possibility of different solvent molecules competing for entry into the coordination shell, in the case of binary mixtures, for example, or competing with a ligand molecule. We shall discuss these cases in Sections IV,A,2 and IV,A,3.

In a simplified (perhaps oversimpled) way we shall call the contact (or tight) ion pair the ion pair, which consists of a cation surrounded by a number of solvent molecules with an anion in its first coordination shell. A solvent-separated (or loose) ion pair consists of a cation surrounded by a complete solvation shell associated with an anion at close proximity (in the second solvation shell), which can eventually distort the symmetry of the first shell. If the lifetime of these species is long compared to the reorientation correlation time of the complex entity (cation–solvent molecules–anion), one is dealing with complexes fully characterized by NMR spectroscopy.

Direct metal cation NMR is a powerful tool for differentiating between the types of ion pairs (and their aggregates). Quadrupolar ion NMR is particularly well adapted for this, because the electric field gradient is very sensitive to the symmetry of the charge distribution around the observed nucleus. For example, replacement of a solvent molecule in the first cation coordination shell by an anion enhances dramatically the field gradient at the cation nucleus site and lowers the quadrupolar relaxation rate.

A basic type of experiment seems obvious for determining ionic association: one based on the observation that the chemical shifts and the relaxation rates of a quadrupolar ionic nucleus depend on the salt concentration. Thus Weingärtner and Hertz (1977) have followed the variations in the ^7Li, ^{23}Na, ^{87}Rb, ^{133}Cs, ^{35}Cl, ^{81}Br, and ^{127}I nuclei relaxation rates for the corresponding salts dissolved in a series of solvents characterized by a high dipole moment (>1.5 D) and by a dielectric constant $\varepsilon > 25$ (Fig. 2). In all cases the spin–lattice relaxation rates of the cation (and the anion as well) increase with the concentration of the salt. Extrapolation to a zero salt concentration gives ionic relaxation rates characteristic of the structuring of the solution around a given ion, without any perturbation due to the presence of the associated counterion. In comparison with calculated values based on models of the structure of the solvation shell, the infinite dilution relaxation rates can give indications of the solution structure in closest proximity to the ion. Even if the concentration dependence is not discussed in detail, one can note that the extent of the ionic association is qualitatively related to the dielectric constant of the solvent. For example, in the case of RbF the extent of ionic association varies in the sequence $HCONH_2$ ($\varepsilon = 109.5$) > MeOH ($\varepsilon = 32.7$) > EtOH ($\varepsilon = 24.5$).

The variation in the chemical shift with the salt concentration has been studied more extensively. Qualitatively, one finds the same sequence of increased shielding in the case of halogen anions for $^{23}Na^+$ (Deverell and Richards, 1966; Templeman and Van Geet, 1972), $^{39}K^+$ (Bloor and Kidd, 1972), $^{87}Rb^+$ (Deverell and Richards, 1966), $^{85}Rb^+$ (Lutz and Nolle, 1972), and $^{133}Cs^+$ (Lutz, 1967): $I^- < Br^- < Cl^-$.

Another type of ionic interaction is the collision of two cations: The shielding of $^{133}Cs^+$ is sensitive to the presence of other diamagnetic cations in the solution, such as Li^+, Na^+, and K^+ (Carrington et al., 1960). In the presence of paramagnetic cations, the chemical shifts of alkali metal nuclei move to higher frequencies in aqueous solutions (Goeller et al., 1972). A relaxation study on Li^+–Ni^{2+} (Hirata et al., 1980) and Al^{3+}–Ni^{2+} (Holz et al., 1982) interactions in aqueous solutions has been undertaken by Friedman and Hertz and their co-workers. Relaxation data can be compared with models of cation collisions, probably the core of the electron exchange process in electron transfer reactions. Similarly, an ionic aggregate (Cs^+, X^-, FeX^{2+}) is formed in solutions of cesium and ferric

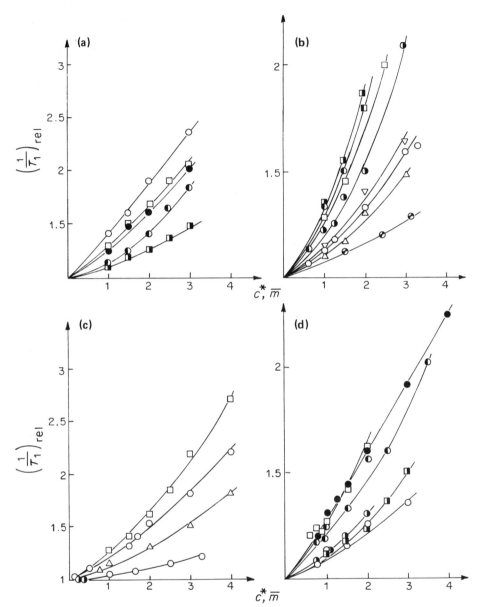

Fig. 2 (a) Relative ⁷Li relaxation rates for Li⁺ salts in nonaqueous solutions: LiCl–DMF-d_7 (●), LiClO₄–NMF-d_5 (◐), LiCl–NMF-d_5 (○), LiClO₄–MeCN-d_3 (□), LiClO₄–Me₂CO-d_6 (▣). (b) Relative ²³Na relaxation rates for Na⁺ salts in solution: NaCl–HCONH₂ (□), NaBr–HCONH₂ (◐), NaClO₄–HCONH₂ (□), NaI–HCONH₂ (◑), NaBr–NMF (○), NaI–NMF (△), NaClO₄–NMF (▽), NaBr–DMF (⊘). (c) Relative ⁸⁷Rb relaxation rates in the systems: RbI–MeOH (◑), RbF–MeOH (○), RbF–EtOH (△), RbI–HCONH₂ (◐), RbF–HCONH₂ (□). (d) Relative ¹³³Cs relaxation rates in the systems: CsF–MeOH (◐), CsF–EtOH (□), CsI–HCONH₂ (▣), CsI–HCOOH (◑), CsI–NMF (○). (From Weingärtner and Hertz, 1977.)

chlorides: The two cations Cs^+ and FeX^{2+} or FeX_2^+ share the same halogen counterion as a ligand (Shporer et al., 1972). In low-polarity media, such as nitromethane solutions, large chemical shift variations are found in the $^{23}Na^+$ spectrum in the presence of paramagnetic cations (Mn^{2+} or Ni^{2+}) (Detellier and Stöver, to be published). The problem of ion association phenomena and more precisely of triple-ion or larger cluster formation is a debated one, and it is not our purpose here to discuss it in detail. We shall only mention the findings by Chabanel and coworkers (Vaes et al., 1978) of centrosymmetric dimers of ion pairs and polymeric ionic chains in lithium thiocyanate solutions in aprotic solvents. They used ^{15}N and ^7Li NMR in parallel with IR spectroscopy. In the same way, solvated contact ion pairs and solvent-bridged ion pairs of sodium iodide or tetraphenylborate aggregate in isopropylamine solutions (Detellier and Gerstmans, 1982); previously, sodium iodide had also been shown by ^{23}Na NMR to form dimers in tetrahydrofuran (THF) (Detellier and Laszlo, 1975). Metal ion NMR is a powerful tool for investigating the formation of supramolecular entities organized around a cation. A good example of the use of this technique can be used for the system formed by alkali ethylacetoacetate ion pairs in organic solvents. The conformations of these ion pairs are closely related to their ambient reactivity, leading, on alkylation, to cis-O-, trans-O-, or C-alkylation (Bram, 1981). Lithium forms a triple anion with ethylacetoacetate that is very stable (Bram, 1981). Cambillau and Ourevith (1981) have shown with ^7Li NMR the presence of complexes (**I**) and (**II**) in dichloromethane and dimethyl sulfoxide

I

II

(DMSO) solutions (Fig. 3). Previously, ^{23}Na NMR had detected the presence of a crown-separated dimer of contact ion pairs in a THF solution of sodium ethylacetoacetate (Cornélis et al., 1978). A multinuclear approach (^1H, ^{13}C, ^7Li) was used by Fraenkel and Hallden-Abberton (1981) to exam-

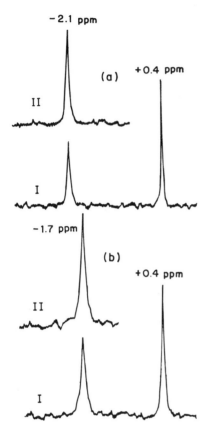

Fig. 3 Lithium-7 NMR spectra of complexes **I** and **II** [0.1 M solutions in CH_2Cl_2 (a) and Me_2SO (b)]. (From Cambillau and Ourevitch, 1981.)

ine the structure of peralkylcyclohexadienyl lithium compounds (**III**) in nonpolar media.

III

In the presence of tertiary amines these compounds form tight ion pair aggregates, whereas in THF, glymes, or Hexamethylphosphoric triamide (HMPT), loose ion pair dimers are favored. As in the case of lithium ethylacetoacetate, lithium, being sandwiched between two anions, is involved in a triple ion.

Directly bonded ^{13}C—^7Li scalar couplings have been observed for methyl lithium (McKeever *et al.*, 1969) and *n*-butyl lithium (McKeever and Waack, 1969) in ether, which provides important experimental evidence regarding the nature of alkyl–lithium bonding ($J_{^7Li^{13}C}$ values are in the range 10–15 Hz). An ionic bond should not give rise to the observation of this coupling constant. Calculations with INDO have shown that $J_{^7Li^{13}C}$ should be very large (>200 Hz) in monomeric methyl lithium, confirming the predominantly covalent nature of the C—Li bond (Clark *et al.*, 1980). However, the couplings are not observed in many cases because of averaging by ^7Li quadrupole-induced relaxation. Fraenkel *et al.* (1980) have avoided this problem by using ^{13}C, ^6Li-enriched samples of propyl lithium. The quadrupole moment of ^6Li is too low to interfere with the coupling. In this work, Fraenkel *et al.* used high-field ^{13}C and ^6Li NMR to identify the fluxional aggregates of propyl lithium in cyclopentane. Species with aggregation numbers of 6, 8, and 9, undergoing fast intraaggregate carbon—lithium bond exchange, coexist in solution. Colquhoun *et al.* (1982) observed ^{31}P—^7Li spin couplings in solutions of lithiated organophosphorus species. The ^7Li and ^{31}P spectra of 0.5 M Ph$_2$PLi in diethyl ether is shown in Fig. 4. Each lithium is coupled to two phosphorus nuclei ($J_{^{31}P^7Li}$ = 45 Hz), which is corroborated by the septet ($J = 45$ Hz) observed on the proton-decoupled ^{31}P spectrum. The major species in the solution (**IV**) is thus covalent and dimeric.

IV

2. Solvation

In solution, anion and solvent molecules compete for occupation of the coordination shells of a cation. Chemical shifts and relaxation rates, as seen previously, are very dependent on the nature of the molecular entities occupying the first coordination shell. And it is not surprising that the first studies dealing with this problem involved variations in chemical shifts with salt concentration, in water (Templeman and Van Geet, 1972; Wertz and Jardetzky, 1956; Richards and Yorke, 1963; Deverell and Richards, 1966) and in nonaqueous solvents (Bloor and Kidd, 1968; Richards and Yorke, 1963). The first studies (Richard and Yorke, 1963; Wertz and Jardetzky, 1956) failed to show any dependence of the ^{23}Na resonance position on the salt concentration. Later (Deverell and Richards, 1966) a linear dependence of the chemical shift on the concentra-

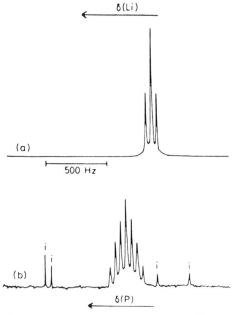

Fig. 4 (a) Lithium-7 NMR spectrum of Ph_2PLi in Et_2O at 200 K (34.8 MHz). (b) Phosphorus-31 NMR of the same solution (36.2 MHz). Peaks marked i are impurities. (From Colquhoun *et al.*, 1982.)

tions of sodium halides and nitrate in water was reported, and in 1971 Templeman and Van Geet discussed the chemical shift of the ^{23}Na signal in aqueous solutions of $NaClO_4$, NaOH, $NaBPh_4$, and NaCl. They attribute the shifts observed after increasing the salt concentration to the formation of contact ion pairs. It is very interesting to note that, when the shifts in water are extrapolated to pure salt, the shift obtained is on the order of magnitude of the chemical shift of crystalline halides (Deverell and Richards, 1966; Templeman and Van Geet, 1972). In nonaqueous solvents, the ^7Li resonances of lithium perchlorate and lithium bromide cover a range of 6 ppm, depending on the solvent (Bloor and Kidd, 1968). The range is larger for ^{23}Na (Bloor and Kidd, 1968) and much larger for ^{133}Cs (Erlich and Popov, 1971; Section II). So the different values of the chemical shift obtained in different solvents were much easier to compare in the case of $^{23}Na^+$ ions than in the case of $^7Li^+$ ions. And it was found that the ^{23}Na (Erlich *et al.*, 1970), the ^{39}K (Shih and Popov, 1977), and the ^{133}Cs (De Witte *et al.*, 1977) chemical shifts were all linearly related to a parameter characterizing the solvating power: the Gutmann donicity number (Gutmann, 1968) (the enthalpy change for the reaction with $SbCl_5$

in 1,2-dichloroethane) (Fig. 5). The ^{23}Na study (Erlich *et al.,* 1970) followed the work of Bloor and Kidd (1968) who tried to correlate the ^{23}Na chemical shift with the pK_a values of the solvent. The salt used in the latter study was sodium iodide, giving rise to ionic associations to a large extent. Popov and co-workers (De Witte *et al.,* 1977; Erlich *et al.,* 1970; Shih and Popov, 1977) used tetraphenylborate and perchlorate salts, anions much less associated with the cation in low-polarity media (Detellier and Gerstmans, 1982), and isolated the specific solvent effects on the ^{23}Na resonance. However, it must be stressed that sodium tetraphenylborate forms in some cases, such as amine solutions, ion pair aggregates (Detellier and Gerstmans, 1982). Thus the results of extrapolation of the observed linear relationship (Fig. 5) to more downfield shifts to obtain the donicity number in these very basic media must be critically examined (Herlem and Popov, 1972). A relationship to the donicity number was not found in the case of ^{7}Li (Cahen *et al.,* 1975c). This is understandable if one considers that in this case the diamagnetic and paramagnetic screening constants are on the same order of magnitude. The two effects can influence the chemical shifts in an opposite manner (Popov, 1979). By using a series of solvents varying relatively little in viscosity, one may find that the relaxation rate is affected principally by changes in the electrostatic field gradient caused by ion–solvent molecule interactions (Section III). For instance, it has been reported that the ^{23}Na$^+$ line width is approxi-

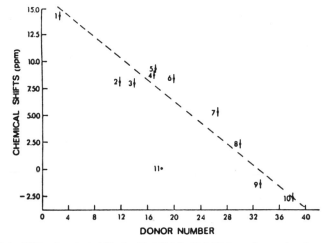

Fig. 5 Plot of ^{23}Na chemical shifts for Na$^+$(Bϕ_4)$^-$ (0.5 *M*) in various solvents versus donor number of the solvents: (1) nitromethane, (2) benzonitrile, (3) acetonitrile, (4) acetone, (5) ethylacetate, (6) THF, (7) dimethylformamide, (8) DMSO, (9) pyridine, (10) hexamethylphosphoramide, (11) water. (From Erlich *et al.,* 1970.)

mately linear with respect to the Gutmann donicity number in a series of 10 solvents (Kessler *et al.*, 1977). These variations in the relaxation rates and in the chemical shifts with the nature of the solvent raise the question of the nature of the alkali metal cation–solvent bonding. Because a down-field shift is observed as the solvent becomes a better electron donor (Fig. 5), the paramagnetic term is expected to make the most important contribution to the chemical shift. In fact, it has been calculated (Bloor and Kidd, 1968) that this paramagnetic term amounts to ~270 ppm per unit electronic charge, whereas the diamagnetic contribution amounts to only 10 ppm. But, because Na^+ has no unpaired electrons in its p, d, or f orbitals, it is necessary to assume that electrons are donated by the solvent to a $3p$ orbital of the sodium ion (Bloor and Kidd, 1968). Are we really dealing with a true electron transfer from the solvent molecules to the p orbitals of the sodium ion (Laszlo, 1978)? Refined calculations, at the ab initio level, fail to show the existence of such a transfer with the required magnitude (Kistenmacher *et al.*, 1974). Experimentally, a linear relationship has been shown between the square root of the ^{23}Na line width corrected for viscosity and the chemical shift in the case of amines (aniline, pyridine, piperidine, pyrrolidine, *n*-propylamine, isopropylamine) and THF solvates of the sodium cation (Fig. 6) by Delville *et al.* (1981c). The approach of this group to preferential solvation and to factors determining chemical shifts and relaxation rates is also described, in a somewhat different context, in Chapter 6, Vol. I. The proportionality of the QCC $\chi = e^2qQ/\hbar$ to the paramagnetic part of the shielding constant σ_p was predicted by Deverell (1969a, 1969b) based on previous work (Kondo and Yamashita, 1959):

$$(\Delta\nu^*_{1/2})^{1/2}_i = C_i \, \bar{\Delta}E_i \, (\tau_c/\eta)^{1/2}_i \, (\sigma_i - \sigma_d) \tag{8}$$

where $i = 1, 2, 3, 4$ (corresponding to the number of amine molecules in the first coordination shell of the sodium cation), $\bar{\Delta}E_i$ is the mean excitation energy, $(\tau_c)_i$ the correlation time for the fluctuating electric field gradient, and η_i the macroscopic viscosity. Invariability of τ_c/η in all the systems studied is plausible, because the ligands have similar volumes and therefore similar reorientation correlation times. The relationship should be linear if $\bar{\Delta}E_i$ depends only on the number of amines in the first coordination shell. One other point related to Fig. 6 merits mention. The four observed straight lines are extrapolated to a single focal point which gives a direct experimental determination of the value of the diamagnetic term τ_d for Na^+ ions dissolved in mixtures of THF and amines. This study (Delville *et al.*, 1981c) gives good experimental evidence of the fact that ^{23}Na shielding constants and QCCs share a common origin—transfer of electrons donated by the ligands to $3p$ orbitals of the Na^+ ion, the correla-

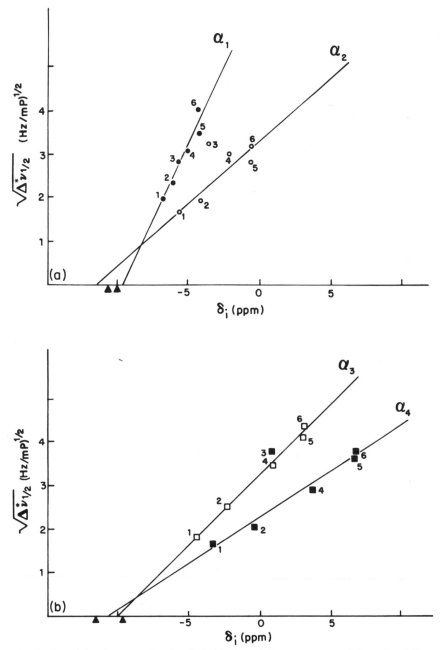

Fig. 6 Correlation between the chemical shift δ_i and the square root of the reduced line width $(\Delta\nu^*_{1/2})_i^{1/2}$ ($i = 1$ to 4): (1) aniline, (2) pyridine, (3) piperidine, (4) pyrrolidine, (5) n-propylamine, (6) isopropylamine. (a) α_1 (●) and α_2 (○); (b) α_3 (□) and α_4 (■). (From Delville *et al.*, 1981c.)

V

tion time τ_c being identified with the reorientation correlation time for the whole solvate (**V**).

The results of this study (Delville *et al.*, 1981c) are based on a detailed analysis of the ^{23}Na spectra for NaClO$_4$ dissolved in binary mixtures of THF and amines. The problem of ionic preferential solvation was of interest for a while, and NMR techniques have provided a useful tool for this kind of study. When one deals with kinetically inert solvates, it is possible to obtain precise, detailed information on the ratios of the different solvates in equilibrium. This is the case, for example, for Al^{3+} or Be^{2+} in aqueous mixtures of organophosphorus solvents. Thus ^{27}Al^{3+} or ^{9}Be^{2+} NMR has been used to determine the successive equilibrium constants for the replacement of a solvent by another one (Delpuech *et al.*, 1971, 1975). The problem is much less readily solved when all solvates are in rapid chemical exchange, which is the case for alkali cations. The observables, chemical shift and line width, vary between the two characteristic values for each solvent (Fig. 7), as exemplified by the case of the sodium cation dissolved in binary mixtures of amines and THF (Delville *et al.*, 1981a). Qualitatively, the curvature can be attributed to preferential solvation. Frankle *et al.* (1970) have proposed that the value of the molar fraction corresponding to the algebraic mean of the chemical shifts in the two pure solvents, the "isosolvation point," is a qualitative measure of this preference. This molar fraction corresponds to occupation of the solvation shell by the same number of solvent molecules, if the hypothesis of chemical shift additivity is applicable. Covington *et al.* (1973, 1974) have elaborated a treatment of the thermodynamics of preferential solvation (Covington

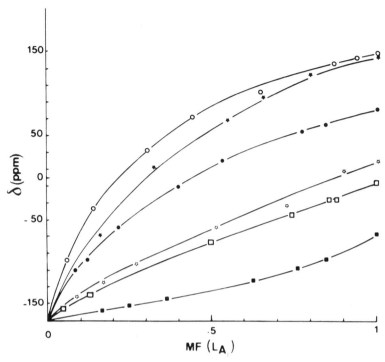

Fig. 7 Variation in the ^{23}Na chemical shift δ versus mole fraction of the amine for binary mixtures of aniline (■), pyridine (□), piperidine (☆), pyrrolidine (●), propylamine (○) and isopropylamine (★) with THF. (From Delville *et al.*, 1981a.)

and Newman, 1979), and preferential solvation data have been derived from this treatment for the sodium cation (Detellier and Laszlo, 1976a,b; Erlich *et al.*, 1973; Greenberg and Popov, 1975) and the thallium cation (Briggs and Hinton, 1979). A similar model based on application of the *Hill* formalism (Hill, 1910), widely used in biochemical studies on phenomena such as cooperative oxygen binding to hemoglobin, has been developed by Delville *et al.* (1980a, 1981a). Mathematically, preferential solvation of a tetracoordinated ion and binding of a ligand to one of four equivalent sites on a biomolecule constitute one and the same problem. This approach has permitted derivation of the intrinsic equilibrium constants for the replacement of coordinating THF by amine in the binary mixture (Table IV) and demonstration of the noncooperativity of the process. The treatment has been extended to polydentate ligands, which furnished the equilibrium constants, without any assumptions based on cooperativity (Delville *et al.*, 1980a, 1981b). As one can see from the

TABLE IV

Chemical Shifts for $^{23}NaClO_4$ (10^{-2} M) in Amines,[a]
Slope of the Hill Plot, and the Intrinsic
Equilibrium Constant[b]

Solvent	δ (ppm)	Slope of the Hill plot	Intrinsic constant K
Aniline	−3.39	1.00	0.46
Pyridine	−0.45	0.96	1.3
Piperidine	0.89	1.03	1.5
Pyrrolidine	3.67	0.96	2.8
Isopropylamine	6.80	1.12	3.0
Propylamine	6.73	0.97	4.4

[a] Reference: aqueous infinite dilution NaCl solution.
[b] From Delville et al. (1981a).

values in Table V, the entries of the first and second diamine molecules into the sodium coordination shell (Scheme 1) are independent and equally probable steps: $k_1 = k_3$ and $k_2 = k_4$. This also has permitted direct determination of the chelate effect, defined as the ratio of the number of complexes having a bidentate attachment of the ligand to the metal to the number of complexes in which only unidentate binding occurs.

A detailed study on the nuclear magnetic relaxation rates of the ionic nuclei ^{23}Na, ^{87}Rb, ^{35}Cl, ^{81}Br, and ^{127}I has been made by Holz et al. (1977). In this study the authors developed the electrostatic theory of the relaxation of quadrupolar ionic nuclei in solution (Hertz, 1973) for the case of binary mixtures. Na^+ and Rb^+ are preferentially hydrated, whereas Cl^-

TABLE V

Intrinsic Equilibrium Constants for the Successive
Replacement of THF Molecules by Amine
Solvents[a]

Solvent	k_1	$k_1 k_2$	k_3	$k_3 k_4$
1,5-Diaminopentane (cadaverin)	0.9	20	1.1	43
1,3-Diaminopropane	0.4	38	0.5	35
Ethylenediamine	1.5	110	1.1	95

[a] From Delville et al. (1980b).

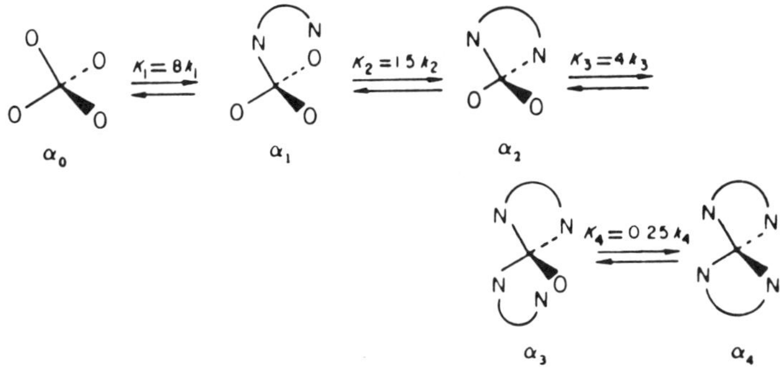

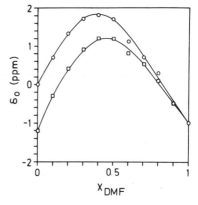

Scheme 1

and Br⁻ are preferentially solvated by methanol in the mixtures. For binary mixtures of water and amides [N-methylformamide, N,N-dimethylformamide) (DMF)], the same approach resulted in the conclusion that the solvation of Na⁺ is nonpreferential (Holz et al., 1978). The D_2O–H_2O isotope effect on the quadrupolar relaxation of an ion in binary mixtures was investigated by Holz (1978) for [23]Na, [87]Rb, and [81]Br, and by Gustavsson et al. (1978a) for [133]Cs. The D_2O–H_2O isotopic effect was proven to be a useful method for obtaining information about ionic preferential solvation and permitted calculation of the local mole fraction of water in the solvation sphere of the ions. The measurements (Holz et al., 1978) corroborated previous observations (Holz et al., 1977) that Na⁺ and Rb⁺ were preferentially hydrated in methanol–water mixtures. The [133]Cs⁺ chemical shift in H_2O–DMF and D_2O–DMF is shown in Fig. 8 (Gustavsson et al., 1978a).

Fig. 8 Chemical shift for [133]Cs⁺ in parts per million at infinite dilution in H_2O–DMF (○) and D_2O–DMF (□) solutions as a function of mole fraction of DMF. (From Gustavsson et al., 1978a.)

The chemical shift at intermediate solvent compositions extends well beyond the chemical shift range encompassed by the pure solvents, showing in this case a large deviation from the usual assumption of a linear variation in the chemical shift with the solvent sphere composition (Covington and Newman, 1979; Delville *et al.*, 1981a). Application of Covington's approach (Covington *et al.*, 1973) to this solvent isotope effect provides evidence of some preferential solvation of Cs^+ by DMF. The nature of sodium methoxide solutions in mixtures of methanol and DMSO has been studied with ^{23}Na NMR, and no strong tendency for preferential solvation was detected (Baltzer *et al.*, 1981).

3. Complexation

Having examined the situation in which two neutral solvent molecules compete for occupation of the coordination shell of a cation, one must consider the similar case of the competition between solvent and solute molecules. Solute molecules able to compete with such a concentrated species as the solvent either have a much higher donicity number than the solvent [e.g., hexamethylphosphoric triamide (HMPA) as compared to nitromethane or acetonitrile] or possess more than one site of complexation leading to a chelate effect (e.g., acyclic ligands), a macrocyclic effect (e.g., monocyclic ligands), or a cryptate effect (e.g., polycyclic ligands). We have already mentioned polyamines studied in binary mixtures with THF (Delville *et al.*, 1980, 1981b). The ligand 2,2'-bipyridine, whose coordination chemistry has been extensively studied, has received little attention concerning its role in the coordination of alkali metal cations. Using ^7Li, ^{23}Na, ^{39}K, ^{133}Cs, and ^{13}C NMR, Schmidt *et al.* (1981) studied this interaction in several nonaqueous solvents. In strongly solvating media such as DMF, DMSO, and methanol, the ^{23}Na chemical shift remains independent of the molar ratio. In solvents of lower donor ability a variation in the chemical shift and in the line width of the signal is observed, but this variation did not permit the calculation of equilibrium constants for the Na^+–bipyridine interaction. The limiting ^{23}Na chemical shift in nitromethane (5 ppm, when the shift of the solvated Na^+ is -15 ppm) suggests that bipyridine molecules insulate the sodium cation from the surrounding solvent. This is, however, in contradiction to the values for the observed line widths ($\sim$600 Hz), suggesting a very dissymmetric environment around the sodium cation. Tentatively one could propose a sodium coordination shell composed of two bipyridine and one nitromethane molecule, even though, as the authors point out, one cannot unambiguously assign a stoichiometry from these NMR measurements. The variation in the ^7Li chemical shift of $LiClO_4$ solutions in nitromethane shows a

definite break at a bipyridine/Li$^+$ molar ratio of 2.0, indicating the formation of a 2 : 1 complex. The variations of the ^{39}K or ^{133}Cs chemical shift or
line width are small compared to those of ^{23}Na or ^{7}Li. Even in a very
poorly solvating medium such as nitromethane, the interaction between
K$^+$ or Cs$^+$ and bipyridine is weak.

Tetradentate amido ethers (**VI**) possess a high alkali metal cation complexing ability (Kirsch and Simon, 1976).

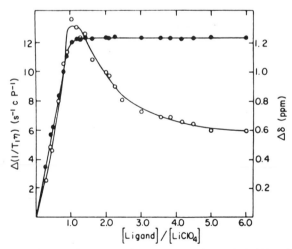

VI

Olsher *et al.* (1980) have followed the complexation of Li$^+$ with *N,N'*-
diheptyl-*N,N'*,5,5'-tetramethyl-3,7-dioxanonane diamide (NDA) (**VI**)
with ^{1}H and ^{7}Li NMR. Variations in the viscosity-normalized relaxation
rates and chemical shifts of ^{7}Li with the [NDA]/[LiClO$_4$] ratio in nitromethane are shown in Fig. 9. A similar curve was obtained in acetonitrile.

Fig. 9 Increments of normalized relaxation rates and of chemical shifts of ^{7}Li versus the
[NDA]/[LiClO$_4$] ratio. The solutions were 0.1 *M* LiClO$_4$ in nitromethane. Full lines correspond to the computed values. $\bigcirc$, $\Delta(1/T_1\eta)$; $\bullet$, $\Delta\delta$. (From Olsher *et al.*, 1980.)

The ^{7}Li relaxation rates indicate the formation of a complex having a stoichiometry greater than $1:1$, probably $2:1$, as indicated by ^{1}H NMR. This complex possesses a symmetry higher than the $1:1$ complex, because the relaxation rate decreases when it is formed. The three following equilibria account for the experimental results:

$$Li^+, ClO_4^- + NDA \overset{K_1}{\rightleftharpoons} (NDA) \, Li^+, ClO_4^-$$

$$(NDA) \, Li^+, ClO_4^- + NDA \overset{K_2}{\rightleftharpoons} (NDA)_2 \, Li^+, ClO_4^-$$

$$(NDA) \, Li^+, ClO_4^- + Li^+, ClO_4^- \overset{K_{0.5}}{\rightleftharpoons} (NDA) \, (Li^+, ClO_4^-)_2$$

In nitromethane, $K_{0.5}$, K_1, and K_2 were, respectively, found to be 50 ± 10, $10^5 < K_1 < 10^6$, and 17 ± 3.

Multidentate ligands such as noncyclic crown ethers (**VII–IX**) are good complexing agents for alkali metal and alkaline earth cations (Rasshofer *et al.*, 1978).

VII–IX

(**VII**): $n = 1$, R = —NHCOCH$_3$
(**VIII**): $n = 3$, R = —NHCOCH$_3$
(**IX**): $n = 1$, R = —CONHCH$_3$

Multinuclear magnetic resonance (^{1}H, ^{13}C, and ^{23}Na) has been applied to the study of complex formation between Na$^+$ and (**VII–IX**) (Grandjean *et al.*, 1978, 1981). Like acyclic ionophore antibiotics (monensin, lasalocid, and grisorixin, inter alia), these linear polyethers wrap themselves around the cation. The thermodynamic parameters for the complexation of Na$^+$ by (**VII–IX**) in pyridine solutions are indicated in Table VI. Small changes in the structure of the ligands lead to drastic differences in the enthalpy and entropy terms. In a ligand such as (**VIII**), in a pyridine solution, the entropy loss due to restriction of the ligand around the cation, together with the entropy loss from formation of a single complex from its two freely diffusing components, is almost fully compensated for by the gain in translational entropy on solvation of each of the two partners ($\Delta S^\circ = -11$ J K^{-1} mol^{-1}). This compensation no longer exists when at least one solvent molecule remains attached to the cation in the complex, and this

TABLE VI

Thermodynamic Parameters and Limiting Half-Height Line
Widths Obtained for the Complexation of Na^+ with
Acyclic Polyethers in a Pyridine Solution[a]

Ligand	$\Delta H°$ (kJ mol^{-1})	$\Delta S°$ (J K^{-1} mol^{-1})	$\Delta\nu_{1/2}^{B}$ at 306 K (Hz)
VII	-66 ± 10	-185 ± 45	181 ± 9
VIII	-18 ± 3	-11 ± 3	245 ± 12
IX	-45 ± 7	-115 ± 25	315 ± 15

[a] From Grandjean et al. (1981).

effect can be a major cause of the strongly negative entropy in the formation of **VII**–Na^+ and of **IX**–Na^+ (-185 and -115 J K^{-1} mol^{-1}). The
characteristic line widths of the complex do not reflect a higher symmetry
for **VIII**–Na^+ compared to **VII**–Na^+ or **IX**–Na^+. This could be due to
participation of the amido nitrogen in the complexation process.

Using ^{23}Na NMR and competition experiments Coibion and Laszlo
(1979) followed the binding of alkali cations to tetracycline. They found
the sequence of complexation at pH 8.6 to be $Li^+ > Na^+$, $Cs^+ > K^+ >$
Rb^+. The linear carboxylic polyether antibiotic ionophore lasalocid (X-
537A) transports ions and biogenic amines across membranes. Grandjean
and Laszlo (1979) have demonstrated competitive binding of Na^+ and
biogenic amines with lasalocid using ^{23}Na NMR, the equilibrium constants of complexation being, respectively, 500 ± 100 liters mol^{-1} for Na^+
and, for example, 450 ± 80 liters mol^{-1} for serotonin bimaleate.

Ab initio calculations have reproduced the trends in the variations in
the chemical shift of $^7Li^+$ complexed with small ligands such as CH_3OH,
H_2O, and NH_3 (Ribas-Prado et al., 1980). This approach is promising
because, hopefully, it will open the way to an understanding of alkali
metal cation chemical shifts.

In the complexation of a cation with an open-chain ligand, such as a
glyme, a polyamine, or an antibiotic, the ligand coils around the cation in
such a way that the cavity in which it is nested is formed during the
process of complexation.

Cyclic ligands with preformed cavities—monocyclic (crown compounds) (Pedersen, 1967), bicyclic (cryptands) (Lehn, 1973), or polycyclic
(Lehn, 1978)—have been widely used during the last 15 yr to complex
cations selectively, particularly alkali metal cations. Of course, alkali
metal ion NMR is the method of choice for studying such interactions. We

shall refer to crown compounds by using the usual nomenclature introduced by Pedersen (1967). Popov and his co-workers have been the more active group in this field (Lin and Popov, 1981; Mei *et al.*, 1977a, b, c, d; Shamsipur and Popov, 1979; Shamsipur *et al.*, 1980; Smetana and Popov, 1980), sampling interactions from lithium to cesium and from the smallest crown, 12-crown-4 (12C4), to the large dibenzo-30-crown-10 (DB30C10). When the equilibrium constant of the complex formation is large ($K >$ 10^4), the stoichiometry is very easily obtained. Thus 1 : 1 and 2 : 1 (sand-

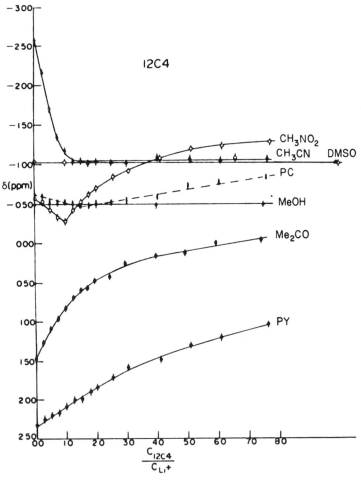

Fig. 10 Lithium-7 chemical shifts versus 12C4/Li⁺ mole ratio in various solvents. The solutions were 0.02 M in LiClO₄. (From Smetana and Popov, 1980.)

TABLE VII

Lithium-7 Chemical Shifts for Free (δ_f) and $1:1$ Complexed (δ_c) Lithium Ions in Various Solvents with Perchlorate as the Anion[a]

Solvent	δ_f (ppm)	δ_c for 12C4 (ppm)	δ_c for 15C5 (ppm)	δ_c for 18C6 (ppm)
CH_3NO_2	0.56	−0.28	−1.21	−0.80
CH_3CN	2.63	−1.04	−1.79	−1.32
Propylene carbonate	−0.64	—	−1.41	−0.99
Acetone	1.45	−0.30	−0.80	−0.52
Pyridine	2.36	−0.74	−1.15	−0.44

[a] All data taken from Smetana and Popov (1980).

wich) complexes of lithium ion and 12C4 were observed in nitromethane solutions when only a $1:1$ complex was detected in acetonitrile (Smetana and Popov, 1980) (Fig. 10). Relaxation rates can also give indications of the stoichiometry of the complexes. Cesium-133 spin–lattice relaxation rates for the Cs^+–18-crown-6 (18C6) complex show a linear increase followed by a break point at $[18C6]/[Cs^+] = 2$ (Wehrli, 1979)—a good indication of the formation of a $2:1$ complex. But no evidence is given for the formation of a $1:1$ complex, a possible species in the solution, as shown by potentiometric results (Frendsdorff, 1971) or by variation in the ^{133}Cs chemical shifts. The characteristic relaxation rate of the $1:1$ complex is probably exactly intermediate between the relaxation rates of the $2:1$ complex and the solvated cation. The presence in the solution of $2:1$ (crown–cation) complexes is corroborated by the chemical shift variations in the Cs^+–18C6 system (Mei et al., 1977a, b), but only the $1:1$ complex was shown to exist in the Cs^+–DB30C10 system. The value of the limiting chemical shift for the complex, almost constant for five different solvents, suggests folding of the large crown around the cesium ion (Shamsipur and Popov, 1979). The same trend, however, is observed in ^7Li chemical shifts for the lithium cation complexed with 12C4, 15C5, or 18C6 (Smetana and Popov, 1980) (Table VII). The ^7Li chemical shifts ranging from 2.63 ppm (acetonitrile) to −0.64 ppm (propylene carbonate) for the solvated lithium cation are tightened into a sharper domain when the cation is complexed (−0.3 to −1.0 for 12C4, −0.8 to −1.8 for 15C5, −0.4 to −1.3 ppm for 18C6). Qualitative reasoning based on the chemical shifts values could lead only to a conclusion of the close similarity of the lithium coordination shell for these three crowns in these five solvents. However, given the relative rigidity of the 12C4 ring, one cannot be sure that all solvent molecules have been extruded from the cation coordination sphere. At least in the case of ^{7}Li one has to be cautious in making a qualitative interpretation of the relative values of the chemical shifts.

The stabilities of the crown–sodium complexes decrease in the order 18C6 > 15C5 > B15C5, which does not follow the trend expected on the basis of a comparison of the dimensions of the crown cavity and the Na^+ ionic radius (Lin and Popov, 1981). Crown ethers are not conformationally rigid and can alter their conformation to accommodate the cation in a coordination shell composed of all possible crown donor atoms, which is a plausible qualitative explanation of the greater ability of 18C6 to complex sodium compared to 15C5 (Lin and Popov, 1981). Exchange processes between sodium and 18C6 in THF were shown in the same study to be anion-dependent. A solution of $NaB\phi_4$ in excess and 18C6 in THF shows two peaks: THF-solvated $^{23}Na^+$ at -7.7 ppm and crown-complexed $^{23}Na^+$ at -17.1 ppm. Only an averaged population, at the same temperature, was observed when $NaB\phi_4$ was replaced by NaI or $NaClO_4$ (Lin and Popov, 1981).

Expulsion of the solvent molecules and the counteranion from the coordination sphere of the cation was demonstrated for a complex of dibenzo-24-crown-8 (DB24C8) and Na^+ (Bisnaire *et al.*, 1982; Shamsipur *et al.*, 1980). The ^{23}Na relaxation rate of the Na^+–DB24C8 complex is linearly related to the viscosity of the solution in nitromethane, acetonitrile, pyridine, acetone, methanol, propylene carbonate, and their binary mixtures (Fig. 11) (Bisnaire *et al.*, 1982). As shown by Eq. (7), this behavior indicates (a) the invariance of the QCC in all these solvents, (b) a time-dependent fluctuation in the electric field gradient due to reorientation of the whole complex, following a Debye–Stokes–Einstein relationship, (c) the rigidity of the sodium coordination shell for a time longer than the reorientation correlation time. No conformational changes in the ligand occur in a time shorter than 60 psec. Measurement of the ^{13}C T_1 values for the complex permitted determination of the ^{23}Na QCC for the Na^+–DB24C8 complex (Table VIII), following the work of Kintzinger and Lehn (1974), who used this "double nuclear spin method" to determine the ^{23}Na QCC in sodium–cryptate complexes. This method has also been used by Drakenberg (1982) to determine ^{43}Ca QCCs. Since Live and Chan (1976) have shown that DB30C10 wraps around the sodium and the potassium cation in different ways, probably forming complexes with different geometries, sodium being much less tightly bound than potassium in the cavity of the folded crown, application of the double-nuclear-spin method to the DB30C10–Na^+ complex could lead to the determination of exchange rates for the movement of sodium to two or three different sites inside the cavity.

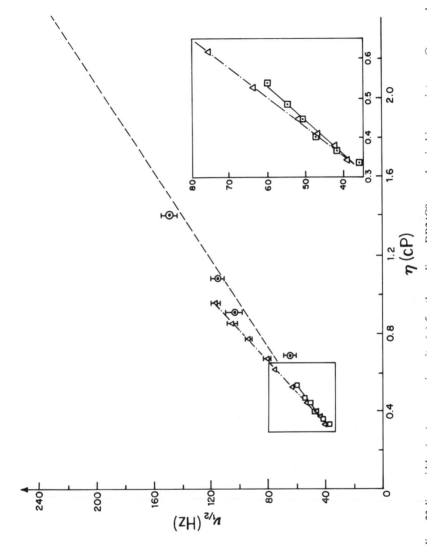

Fig. 11 Sodium-23 line widths ($\nu_{1/2}$) versus viscosity (η) for the sodium–DB24C8 complex in binary mixtures: ⊙, propylene carbonate–methanol; △, acetone–pyridine; ⊡, acetonitrile–nitromethane. (From Bisnaire et al., 1982.)

TABLE VIII

Sodium-23 QCCs of the Complex Na$^+$–DB24C8 in Solution[a]

	Solvent[b]						
QCC	NM	AN	Ac	Py	Py–THF, 60:40	PC	MeOH–PC, 15:85	CDCl$_3$
χ (MHz $\pm$ 0.2)	0.8	0.9	0.9	0.9	1.1	0.8	0.8	1.2

[a] From Bisnaire *et al.* (1982).

Thermodynamic parameters for the interaction between cation and crown can be calculated when the equilibrium constant is $<10^4$. This is the case when the solvent can compete with the crown for occupation of coordination sites. Under these conditions, the equilibrium constant is generally found to be inversely related to the solvent donicity number, as expected (Mei *et al.*, 1977a, b; Shamsipur and Popov, 1979; Shamsipur *et al.*, 1980; Smetana and Popov, 1980). In a few cases, it has been possible to derive the kinetic parameters from the quadrupolar ion spectrum of solutions containing the alkali cation and the crown. Thus a barrier of 79 kJ mol^{-1} has been measured for the decomplexation reaction of the octamethyl derivative of 16C4–Li$^+$ using ^{7}Li NMR (Aalmo and Krane, 1982). If the exchange of the cation between two sites in the solution is slow enough, one can observe two resonances for the alkali metal nucleus at low temperatures. If the characteristic relaxation rate of the cation at one of the sites is very large, the corresponding resonance line will not be observed. This is expected in the case of an alkali cation complexed with a planar crown: Its coordination sphere is very dissymmetric. Warming the solution produces, at the beginning, a narrowing of the observed line width owing to a decrease in the viscosity. Further warming produces broadening and coalescence of the lines, followed by narrowing of the single averaged resonance line at higher temperatures. Such behavior has been observed in the systems sodium–dicyclohexyl-18-crown-6 (DCC18C6) and sodium–DB18C6 in methanol and dimethoxyethane (Shchori *et al.*, 1973), Na$^+$–DB18C6 in DMF (Shchori *et al.*, 1971), K$^+$–DB18C6 in methanol (Shporer and Luz, 1975), Na$^+$–monensin (Degani, 1977), and sodium–spiro-bis-crowns (Bouquant *et al.*, 1982). The monensin– and DCC18C6–sodium complexes dissociate via a first-order dissociative mechanism (Degani, 1977; Shchori *et al.*, 1973) following Eqs. (9) and (10):

$$\text{Na}^+ + \text{Mon}^- \underset{k_\text{d}}{\overset{k_\text{a}}{\rightleftharpoons}} \text{Mon}^-, \text{Na}^+ \qquad (9)$$

$$Na^+ + crown \underset{k_d}{\overset{k_a}{\rightleftharpoons}} Na^+, crown \qquad (10)$$

The decomplexation activation parameters are similar ($\Delta H_d\ddagger$ = 10.3 kcal and 8.3 kcal mol^{-1}, respectively, for monensin and DCC18C6; $\Delta S_d\ddagger$ = -15.8 and -21 cal mol^{-1} deg^{-1}, respectively). However, a difference between the association activation parameters is noted ($\Delta S_a\ddagger$ = -3.9 cal and -23 cal mol^{-1} deg^{-1}, respectively), which was interpreted to be the consequence of conformational changes prior to complexation of the crown. These conformational changes are not observed for the monensin anion. However, the effect could also be explained by a higher degree of desolvation of the monensin anion during the complexation process.

Bicyclic and polycyclic ligands form cryptate-type cation inclusion complexes that are more stable than those formed with monocyclic ligands. In a number of cases, two different signals are observed for solvated and cryptated alkali metal cations (Fig. 12) (Ceraso and Dye, 1973; Ceraso et al., 1977). A detailed analysis of the line shapes observed when an excess of sodium cation is in the presence of the cryptand (222) has been performed (Fig. 12). The activation entropies of decomplexation were found to be negative in nonaqueous solvents, which reflects the participation of the solvent in the transition state. (Li$^+$, C211) and (Li$^+$, C221) systems are also characterized by negative activation entropies of decomplexation (Cahen et al., 1975a, b).

Popov and his co-workers have observed with ^{133}Cs NMR the formation of "inclusive" and "exclusive" conformations of the Cs$^+$–C222 complex (Mei et al., 1977a, b) [Eq. (11)]:

$$Cs^+ + C \overset{k_1}{\rightleftharpoons} (Cs^+, C)_{exc} \overset{k_2}{\rightleftharpoons} (C_s^+, C)_{inc} \qquad (11)$$

In the exclusive complex, the cation is not located at the center of the cryptand cavity and retains some residual solvent molecules in its coordination shell. In the inclusive complex, the cation is completely isolated from the bulk solvent by the bicyclic ligand.

Cryptands have permitted generation of the sodium anion in solution (Dye et al., 1975). The best evidence that the sodium anion is centrosymmetric with two electrons in the outer s orbital has been provided by the ^{23}Na$^-$-NMR spectra (Fig. 13). Sodium-23 spectra of Na$^+$–C222–Na$^-$ solutions in ethylamine, THF, or methylamine consist of two lines assignable to the cryptated cations and the anion. The solvent-induced paramagnetic shift is not observed for Na$^-$, and the resonance line is independent of the nature of the solvent, being nearly the same as that of Na$^-$ in the gas phase. This absence of a chemical shift is indicative of complete shielding of the 2p orbitals from the solvent by the presence of a filled 3s orbital. Moreover, the narrowness of the line for Na$^-$ attests to the high spherical symmetry of the anion, with a filled 3s orbital. The ^{23}Na-NMR spectrum

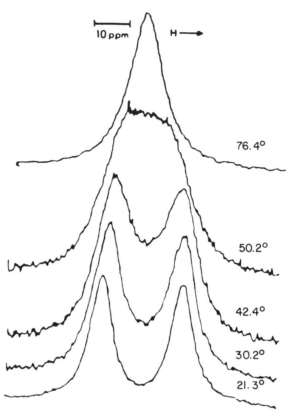

Fig. 12 Sodium-23 spectra at various temperatures for solutions of Na⁺ with cryptand C222 in ethylenediamine; $[C222] = 0.30\ M$; $[NaBr] = 0.6\ M$. (From Ceraso *et al.*, 1977.)

of the sodium anion in solutions of sodium in HMPA has been reported (Edwards *et al.*, 1981). This compound appears to fulfill many of the requirements generally sought in macrocyclic complexing agents, and Na⁻ was found to be formed in this solvent without the addition of complexing agents. The ^{23}Na⁻ chemical shift in HMPA (-62 ppm) is virtually identical to the chemical shift measured in amine solutions (-63 ppm) by Dye *et al.* (1975).

B. Fixation in Organized Systems and on Macromolecules

Fixation of ions on aggregates or on polymers, synthetic or natural, proteins or polyelectrolytes, is of great importance in diverse areas. Site binding or atmospheric binding, the exchange rate with the bulk of the solution, the exchange rate between different fixation sites, competition with other ions, the fraction of ions bound, dynamic coupling between the movement of an ion and of the macromolecule are, inter alia, problems that can be and have been approached with quadrupolar alkali metal cation NMR. A kind of interaction that can be followed is ionic interaction in amphiphilic systems (Lindman *et al.*, 1977). Because this subject is reviewed extensively in Chapter 8, Vol. I, we shall consider only the example of micellar aggregates. Micellar aggregation of sodium octanoate and sodium octyl sulfate can be described in terms of a simple two-site

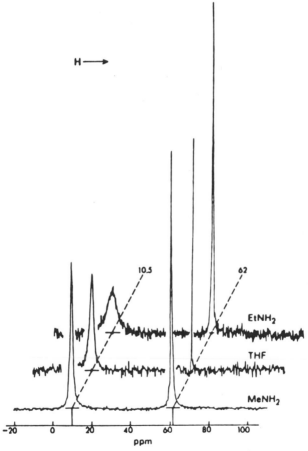

Fig. 13 Sodium-23 NMR spectra of Na^+C_{222},Na^- solutions ($\sim$0.1 M) in three solvents. All chemical shifts are referenced to aqueous Na^+ at infinite dilution. (From Dye *et al.*, 1975.)

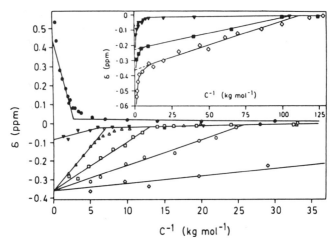

Fig. 14 Sodium-23-NMR chemical shifts (δ in parts per million) as a function of the inverse amphiphile concentration (molality) for aqueous solutions of sodium octanoate (●), octyl sulfonate (▼), octyl sulfate (△), nonyl sulfate (□), decyl sulfate (○) dodecyl sulfate (◇), and (in the inset) *p*-octylbenzene sulfonate (■). (From Gustavsson and Lindman, 1978.)

model. Sodium-23 nuclear quadrupole relaxation and shielding furnished critical micelle concentrations (CMCs) and information on premicellar aggregation and changes in the degree of counterion association (Gustavsson and Lindman, 1975). The systems sodium cholate–decanol–water and sodium deoxycholate–decanol–water display a continuous transition from normal to reversed micelles. Rubidium-85 and ^{133}Cs NMR were also used in this study (Gustavsson and Lindman, 1975), which was later extended to other systems (Gustavsson and Lindman, 1978). The phase separation model for micelle formation and the two-site model for counterion binding (free and bound counterions only) indicate that a plot of the chemical shift δ versus the inverse amphiphile concentration should consist of two straight-line segments intersecting at the critical CMC (Gustavsson and Lindman, 1978). This is experimentally observed for ^{23}Na$^+$ (Fig. 14) and for ^{133}Cs$^+$ and ^{35}Cl$^-$. The CMC decreases and the aggregation number increases with increasing alkyl chain length of the amphiphile, as expected. If the cation is necessarily associated with anionic self-aggregates such as these amphiphiles, and even if the nature of the cation has some influence on the kind of aggregate formed, the self-aggregation never shows such a dramatic dependence on the nature of the alkali metal cation as the dependence found in aggregation of the nucleotide 5'-guanosine monophosphate (5'-GMP). Alone among the nucleotides, GMP arranges itself into highly ordered aggregates in aqueous solutions (Miles and Frazier, 1972), and the alkali metal cation contributes directly to the build-up of aggregates, which are characterized by high rigidity, as shown by ^{1}H NMR (Pinnavaia *et al.*, 1975). The energy barrier of interconversion between the different species is more than 15 kcal mol^{-1}. Later, it was shown by ^{1}H NMR (Pinnavaia *et al.*, 1978) and by ^{23}Na and ^{39}K NMR (Borzo and Laszlo, 1978; Delville *et al.*, 1979; Detellier *et al.*, 1978; Laszlo and Paris, 1978) that the self-assembly process depended critically on the cation present, potassium being more able than sodium or rubidium to induce aggregation, which does not occur with cesium or lithium. A

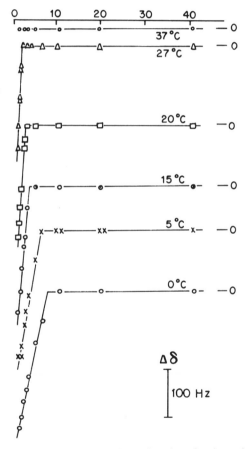

Fig. 15 Variations of the ^{23}Na chemical shift as a function of reciprocal 5'-GMP concentration in D$_2$O solution at various temperatures. (From Borzo *et al.*, 1980.)

multinuclear approach (^{1}H, ^{13}C, ^{23}Na, ^{31}P, ^{39}K, ^{87}Rb) has shown that, in the presence of sodium cations only, 5'-GMP aggregates by dimerization of the planar tetrameric arrangement of GMP molecules, which are held together by eight hydrogen bonds (Zimmerman, 1976), followed by dimerization of octamers in hexadecamers (Borzo et al., 1980). Variation in the ^{23}Na chemical shift (Fig. 15) or in the reduced line width (Fig. 16) as a function of the reciprocal 5'-GMP concentration in D$_2$O solutions consists of two straight lines with a sharp intersection, the critical concentration being identical for chemical shifts and for line widths. A maximum is reached for the line widths at high GMP concentrations where the line

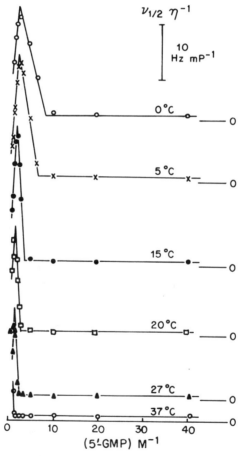

Fig. 16 ^{23}Na-reduced line widths as a function of reciprocal 5'GMP concentration in D$_2$O solution at various temperatures. (From Borzo et al., 1980.)

shape of the signal is no longer truly Lorentzian. These observations produced the following information:

(a) The ^{23}Na absorption is the superposition of two Lorentzians. Separation of these two components by a simplified procedure (Delville et al., 1979) permitted determination of the values for the product $P_B\chi^2$, where P_B is the mole fraction of sodium ions in the bound state and χ their QCC, together with τ_c, the correlation time for these bound sodium ions. Because τ_c does not follow the bulk viscosity, it is not a true reorientation correlation time but is predominantly determined by the residence time τ_B of the sodium ions on the aggregates ($\tau_c^{-1} = \tau_B^{-1} + \tau_R^{-1}$).

The observation of $\tau_B \simeq 30$ ns rules out sodium binding to a single phosphate and is consistent only with sodium binding by several groups. However, the cation release does not destroy the aggregate. Whereas sodium cations exchange with rates of 10^7–10^8 s^{-1}, the nucleotide counterions exchange with much slower rate constants of less than 10^2 s^{-1}. One sodium cation, characterized by a high QCC of 1.8 MHz (thus in a very dissymmetric environment), occupies the core position of the tetrameric assembly. Peripheral sodium cations are characterized by a QCC of 0.9 MHz, a value slightly greater than those corresponding to single-phosphate binding of 0.68 MHz (Detellier and Laszlo, 1980).

(b) The observation of critical concentrations suggests the use of a simplifying two-state phase separation model such as the one depicted previously for micelles. Application of such a model leads to the determination of thermodynamic parameters for the aggregation process: $\Delta H° = -17$ kcal mol^{-1} and $\Delta S° = -51$ cal mol^{-1} K^{-1}; the self-assembly is enthalpy-driven, which is reminiscent of the self-assembly of polynucleotides.

(c) The magnitude of the limiting chemical shift of ^{23}Na on the aggregates is unusual: One can evaluate -100 to -160 ppm for Na$^+$ in the central position of the tetramers. Such large shifts have been reported for Na$^+$ fixed on Nafion (Komorosky and Mauritz, 1978). Na$^+$ must be coordinated to atoms of very different types: OH oxygens or guanine nitrogen in addition to the four O-6 oxygens. This is consistent with the finding of a high QCC. Recent determinations of the crystallographic structures of disodium GMP heptahydrate show that the sodium coordination shells involve no direct interactions with the ionized phosphate groups, but coordination by N-7 with the guanine base or by O-2' or O-3' with the sugar moiety (Barnes and Hawkinson, 1982; Katti et al., 1981). Fisk et al. (1982) have shown with ^{1}H and ^{13}C NMR that the most stable species resulting from the self-association of Na$_2^+$–5'-GMP is a head-to-tail octamer formed by the association of two tetrameric plates. Petersen et al.

(1982), in their ^{1}H-, ^{13}C-, and ^{31}P-NMR studies on the same system (Na$_2$$^+$– 5'-GMP), found the tetramer model unable to explain some features of the ^{13}C spectrum of 5'-GMP on aggregation, such as the observed downfield shifts of the new lines relative to the parent signals. They interpret their results with a model wherein the stacking of monomers is the predominant mode of interaction at higher temperature, whereas at lower temperatures stacks of asymmetric hydrogen-bonded dimers form. This model, however, does not explain the large ^{23}Na chemical shift observed (Borzo *et al.*, 1980), because it implies sodium binding on peripheral phosphates only.

K$^+$ plays a crucial role in the GMP self-assembly process (Detellier *et al.*, 1978; Pinnavaia *et al.*, 1978). Mixed aggregates (K$^+$ and Na$^+$) have been examined using ^{23}Na, ^{39}K, ^{87}Rb, and ^{31}P NMR (Detellier and Laszlo, 1980). A remarkable duality of sites on the octameric aggregate has been found: Each type of site is selective for different cations. Solutions of 0.10 M 5'-GMP at pH 8.5 in D$_2$O give a ^{23}Na signal that is non-Lorentzian in the presence of added KCl. Some information is available from observations of such behavior:

(a)   A single correlation time τ_c extracted from the non-Lorentzian line shape (Delville *et al.*, 1979) was found for a wide variety of experimental conditions (9.4 $\pm$ 1 ns).

(b)   The values of $P_B\chi^2$ increase regularly with the concentration of K$^+$ and are linearly correlated with the fractional population of aggregates obtained directly by a measurement of relative intensities in the ^{31}P-NMR spectra.

(c)   One observes K$^+$ critical concentrations when the ^{23}Na-NMR line widths are plotted against [K$^+$]$^{-1}$. It follows from this observation that K^+ and Na$^+$ cations do not compete for the same site, but rather are together engaged, in a synergistic manner, in the formation of aggregates. This is reminiscent of the poly(I) helix formation that occurs with nucleation by alkali metal cations at two kinds of sites (Miles and Frazier, 1978).

The following equilibrium model accounts for these three observations:

$$G_4 + M^+ \overset{K_1}{\rightleftharpoons} G_4, M^+ \tag{12}$$

$$G_4, M^+ + G_4 \overset{K_2}{\rightleftharpoons} G_8, M^+ \tag{13}$$

$$G_8, M^+ + n\text{Na}^+ \overset{K_3}{\rightleftharpoons} G_8, M^+, n\text{Na}^+ \tag{14}$$

The nucleating cation, preferentially K$^+$, induces formation of the octamer. Sodium is preferentially bound at outer sites in order to minimize electrostatic repulsions. From this model, one can calculate 2.1 $< n <$ 5.6; $n = 4$ is a likely value. Spectra for ^{39}K and ^{87}Rb show that fixation of

K^+ and Rb^+ at the outer site is of the atmospheric type, a dissymmetric environment not being detected for these two cations. This is surprising because K^+ is also involved in the core of the aggregate: Two plausible interpretations are that K^+ is bound only to oxygen atoms, contrary to what was observed for Na^+, and that the exchange of K^+ between inner and outer sites is slow and the bound-$^{39}K^+$ signal is not observable; we tend to favor the latter interpretation. Studies using low-frequency Raman scattering techniques supported the conclusion that the potassium ion enhances the self-association as compared to the sodium ion (Nielsen et al., 1982).

Two kinds of sites coexist: inner sites selective for potassium ions, and outer sites selective for sodium ions, Na^+ and K^+ acting in a synergistic manner. In the presence of NH_4^+ a new type of aggregation appears, which is a result of the synergistic action of K^+ and NH_4^+ (Detellier and Laszlo, 1979). The addition of NH_4^+, in the necessary presence of K^+, although leaving virtually unchanged the proportion of phosphate-bound sodium ions in the aggregates, markedly increases the apparent correlation time for the aggregate. This is attributed to the formation of NH_4^+-bridged dimers of the $[(GMP)_8, K^+, nNa^+]$ aggregate. It is interesting to note that the K^+ octamer discriminates among NH_4^+, glycine, and alanine, which may have some prebiotic significance (Detellier and Laszlo, 1979).

In general, one can discriminate between two kinds of ionic binding in an aggregate or in a polymer: site binding and atmospheric binding. Quadrupolar nuclei NMR can be used to give useful information on this problem: If the ion, partially dehydrated, is bound at sites, the electric field gradient at the ionic nucleus and consequently the QCC will be enhanced; moreover, if the bound ion accompanies the macromolecule in its reorientation movement, the relaxation rate will no longer be exponential. On the contrary, in atmospheric binding, only a slight increase in the relaxation rate of the cationic nucleus is expected, because of the distortion of the hydration shell symmetry near the polyelectrolyte and/or because of the cylindrical potential surrounding the polyelectrolyte (Van der Klink et al., 1974, 1975).

Thus ^{23}Na NMR has been used to study the extent to which monovalent cations associate with doubled-stranded DNA (Anderson et al., 1978; Reuben et al., 1975). Anderson et al. (1978) obtained reaseonable agreement between experimental values and theoretical values based on the ion condensation model of Manning (1978). The observed line shapes of $^{23}Na^+$ in a DNA solution are Lorentzian, and the line widths vary from 6 to 20 Hz for DNA phosphate/sodium concentration ratios between 0 and 0.5. This situation is in contrast to the behavior of Mg^{2+} as shown by a ^{25}Mg-

NMR study (Rose *et al.*, 1980). The line shapes of $^{25}Mg^{2+}$ in a DNA solution are no longer Lorentzian, which is an indication of site binding. Sodium binding to chromatin has been studied with ^{23}Na NMR (Burton and Reimarsson, 1978). The Na^+-binding sites of DNA and of the DNA–histone complex are different, as indicated by different values of the ratios for $^{23}Na^+$ in the presence of histones: The histones are bound at low Na^+ concentrations and dissociated at higher Na^+ concentrations. Rubidium-85 NMR (Lindqvist and Lindman, 1970) and ^{23}Na NMR (Andrasko *et al.*, 1972) have been used to follow the fixation of these cations on humic acids.

Mucopolysaccharides are another class of biological polyelectrolytes, and chondroitin 4-sulfate, dermatan 4-sulfate (Gustavsson *et al.*, 1978b), heparin (Herwats *et al.*, 1978), and chondroitin sulfate–polypeptide solutions (Gustavsson *et al.*, 1978c); bind sodium at well-defined sites. Gramicidin A, a pentadecapeptide with a regular alternation of L and D residues, spontaneously dimerizes in protic solvents. Sodium and potassium ions diffuse across the lipid bilayer of cell and vesicle membranes doped with gramicidin A. Channels are formed, with lengths of 35 to 40 Å spanning the width of the bilayer and an ID of 5 to 6 Å sufficient for the inclusion of Na^+ or K^+ ions. Neher *et al.* (1978) have proposed the existence of two kinds of sites available to univalent cations: outer sites, at the entrance to the channel, and inner sites, in the lipophilic environment of the bilayer. Sodium-23 NMR has been applied to this problem (Cornélis and Laszlo, 1979; Venkatachalam and Urry, 1980). In ethanol–water mixtures (90:10) a correlation time for the bound Na^+ ions of $\tau_c = 1.9 \pm 0.5$ ns was found. This correlation time is a composite of the residence time τ_B at the binding sites and of the reorientation correlation time τ_R for the complex [Eq. (15)].

$$\tau_c^{-1} = \tau_B^{-1} + \tau_R^{-1} \tag{15}$$

The reorientation correlation time for the gramicidin dimers, as determined from ^{13}C longitudinal relaxation rates (Fossel *et al.*, 1974), is $\simeq 30$ ns. Hence $\tau_B \simeq 1.8$ ns (Cornélis and Laszlo, 1979). The equilibrium constant for the fixation of Na^+ on the gramicidin dimer was found to be 4.0 M^{-1} (Cornélis and Laszlo, 1979). Venkatachalam and Urry (1980) have shown multiple-ion occupancy in the channel and have estimated binding constants for two delineated sodium-binding sites.

Fixation of cations on proteins has also been studied by alkali cation NMR. A group of proteins has been quite well explored: calciproteins. Even if ^{43}Ca is the more appropriate nucleus for such studies (Andersson *et al.*, 1982), ^{23}Na can be valuable when used in competition experiments (Delville *et al.*, 1980b, c; Gerday *et al.*, 1979; Grandjean and Laszlo, 1978;

Grandjean *et al.*, 1977; Parello *et al.*, 1979). Calciproteins are reviewed in Chapter 8.

Bryant (1970) has observed the enhancement of $^{39}K^+$ relaxation rates in the presence of pyruvate kinase.

Combined use of ^{43}Ca and ^{23}Na NMR allowed two different classes of sites to be distinguished on lobster *Panulirus interruptus* hemocyanin by Andersson *et al.* (1982). Previously, the same group (Norne *et al.*, 1979) had followed fixation of Na^+ on this hemocyanin and had evaluated the number of Na^+ ions bound per oxygen carrying the hemocyanin subunit, in the absence of Ca^{2+}, to be 20. Considerable line broadening was found in the $^{23}Na^+$ spectrum in the presence of erythrocruorin, a respiratory protein from invertebrates (Chiancone *et al.*, 1976). Lithium-7 NMR has been used to characterize the binding of monovalent cations to Na^+, K^+-ATPase (Grisham and Hutton, 1978). The ^7Li longitudinal relaxation rate is increased on binding of Mn^{2+} or Cr–ATP to the enzyme, and the effects of Mn^{2+} are consistent with the existence of a Li^+-binding site at a distance 7.2 Å from the single catalytically active Mn^{2+} site on the ATPase. Hutton *et al.* (1977) had previously used the same approach to investigate the monovalent cation site on pyruvate kinase, and the same laboratory, using ^7Li NMR, has followed the interactions of gadolinium and lithium cations with calcium ion transport ATPase (Stephens and Grisham, 1979).

C. Cells and Living Systems

Nuclear magnetic resonance spectroscopy provides a useful probe for the study of intact cells and tissues and is a noninvasive technique. Phosphorus-31 has been the most employed nucleus, mainly in the study of cellular metabolism (Moon and Richards, 1973; Cohen *et al.*, 1978; Hanley, 1981), of whole organs (Hanley, 1981), and of *in vivo* tumors where response to therapeutic modalities is accompanied by distinctly different spectral changes (Ng *et al.*, 1982). Natural abundance ^{13}C NMR from human arm and rat tissues has been observed *in vivo* (Alger *et al.*, 1981).

Because sodium ions play an important role in the fundamental processes associated with cellular activity (Cameron *et al.*, 1980), and because ^{23}Na has a high receptivity, ^{23}Na NMR should be applicable to the study of whole cells and organs.

The main problem associated with this approach is that the intra- and extracellular $^{23}Na^+$ have the same chemical shift. Gupta and Gupta (1982) have shown the utility of a paramagnetic NMR shift reagent, the anionic complex of dysprosium(III) and tripolyphosphate $[Dy(PPP_i)_2^{7-}]$, to permit the observation of resolved resonances. This reagent has been applied

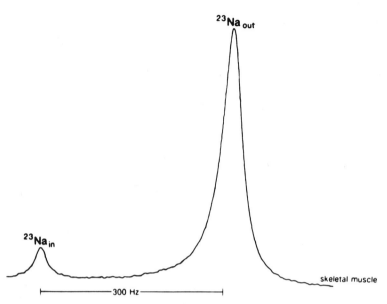

Fig. 17 Sodium-23 spectrum (53 MHz, 20°C) of a skeletal muscle from the northern frog *Rana pipiens* suspended in a physiological Ringer solution containing 104 mM Na$^+$, 2.5 mM K$^+$, 1.6 mM P$_i$, 1 mM Ca^{2+} and 2 mM Dy (PPP$_i$)$_2$ at pH 7.5. Resonances of intra- and extracellular Na$^+$ ions are labeled ^{23}Na$_{in}$ and ^{23}Na$_{out}$, respectively. (From Gupta and Gupta, 1982.)

in the study of red blood cells and frog skeletal muscle. The anionic paramagnetic shift reagent remains only on the outside of the cells and affects only the chemical shift of extracellular sodium. Figure 17 shows the ^{23}Na-NMR spectrum obtained from frog muscles suspended in a physiological solution (Gupta and Gupta, 1982). The two resonances correspond to intra- and extracellular Na$^+$ ions (^{23}Na$_{in}$ and ^{23}Na$_{out}$, respectively in Fig. 17).

Pike and Springer (1982) have developed other aqueous shift reagents for cationic NMR. Although Fe(CN)$_6^{3-}$ and Pr(EDTA)$^-$ gave disappointing results, Dy(EDTA)$^-$, Dy^{3+}–dipicolinate [Dy(DPA)$_3^{3-}$], and Dy^{3+}–nitrilotriacetate [Dy(NTA)$_2^{3-}$] complexes were found to give significant shifts. These complexes have been used to study Na$^+$ transport in model membrane vesicles (Pike *et al.*, 1982). Gated ^{23}Na-NMR images of an isolated perfused working rat heart have been obtained by Delayre *et al.* (1981). These authors gated the acquisition of signals to the heartbeat in order to overcome the problems posed by heart motion. Because the concentrations of sodium in blood and in a healthy tissue are very different, ^{23}Na NMR provided a good contrast between blood and surrounding tissues, giving negative images of the myocardium. Gated ^{23}Na-NMR images of the rat heart in diastole and in systole have been obtained by this method. This spectacular experiment demonstrates the technical potential of imaging a moving organ with NMR (Delayre *et al.*, 1982).

V. Conclusions

We have reviewed some applications of alkali metal NMR such as ion pairing and rat heart, transport across membranes and alkali metal anions, highly ordered nucleotide aggregates induced by alkali metal cation and cation–cation close encounters in solution, crown and cryptand complexation and cation–cation distances on proteins, site or atmospheric binding on polyelectrolytes.... They represent only a minute part of a world still to explore.

References

Aalmo, K. M., and Krane, J. (1982). *Acta Chem. Scand.* **36A,** 219–225.

Alger, J. R. *et al.* (1981). *Science* **214,** 660–661.

Anderson, C. F., and Record, M. T., Jr., and Hart, P. A. (1978). *Biophys. Chem.* **7,** 301–316.

Andersson, T., Chiancone, E., and Forsén, S. (1982). *Eur. J. Biochem.* **125,** 103–108.

Andrasko, J., Lindquist, I., and Bull, T. E. (1972). *Chem. Scr.* **2,** 93–95.

Baltzer, L., Bergman, N. A., and Drakenberg, T. (1981). *Acta Chem. Scand.* **A35,** 759–762.

Barnes, C. L., and Hawkinson, S. W. (1982). *Acta Crystallogr.* **B38,** 812–817.

Bisnaire, M., Detellier, C., and Nadon, D. (1982), *Can. J. Chem.* **60,** 3071–3076.

Bloor, E. G., and Kidd, R. G. (1968). *Can. J. Chem.* **46,** 3425–3430.

Bloor, E. G., and Kidd, R. G. (1972). *Can. J. Chem.* **50,** 3926–3930.

Borzo, M., and Laszlo, P. (1978). *C. R. Acad. Sci. Paris* **C287,** 475–476.

Borzo, M., Detellier, C., Laszlo, P., and Paris, A. (1980). *J. Am. Chem. Soc.* **102,** 1124–1134.

Bouquant, J., Delville, A., Grandjean, J., and Laszlo, P. (1982). *J. Am. Chem. Soc.* **104,** 686–691.

Bram, G. (1981). *J. Mol. Catal.* **10,** 223–229.

Brevard, C., and Granger, P. (1981). "Handbook of High Resolution Multinuclear NMR." Wiley, New York.

Briggs, R. W., and Hinton, J. F. (1979). *J. Solution Chem.* **8,** 519–527.

Bryant, R. G. (1970). *Biochem. Biophys. Res. Commun.* **40,** 1162–1166.

Bull, T. E. (1972). *J. Magn. Reson.* **8,** 344–348.

Bull, T. E., Forsén, S., and Turner, D. L. (1979). *J. Chem. Phys.* **70,** 3106–3111.

Burton, D. R., and Reimarsson, P. (1978). *FEBS Letters,* **89,** 183–186.

Cahen, Y. M., Dye, J. L., and Popov, A. I. (1975a). *J. Phys. Chem.* **79,** 1289–1292.

Cahen, Y. M., Dye, J. L., and Popov, A. I. (1975b). *J. Phys. Chem.* **79,** 1292–1295.

Cahen, Y. M., Handy, P. R., Roach, E. T., and Popov, A. I. (1975c). *J. Phys. Chem.* **79**, 80–85.

Cambillau, C., and Ourevitch, M. (1981). *J. Chem. Soc. Chem. Commun.* 996–997.

Cameron, I. L., Smith, N. K. R., Pool, T. B., and Sparks, R. L. (1980). *Cancer Res.* **40**, 1493–1500.

Carrington, A., Dravnicks, F., and Symons, M. C. R. (1960). *Mol. Phys.* **3**, 174–182.

Ceraso, J. M., and Dye, J. L. (1973). *J. Am. Chem. Soc.* **95**, 4432–4434.

Ceraso, J. M., Smith, P. B., Landers, J. S., and Dye, J. L. (1977). *J. Phys. Chem.* **81**, 760–766.

Chiancone, E., Bull, T. E., Norne, J. E., Forsén, S., and Antonini, E. (1976). *J. Mol. Biol.* **107**, 25–34.

Clark, T., Chandrasekhar, J., and Von Rague Schleyer, P. (1980). *J. Chem. Soc. Chem. Commun.* 672–673.

Cohen, S. M. *et al.* (1978). *Nature* **273**, 554–556.

Coibion, C., and Laszlo, P. (1979). *Biochem. Pharmacol.* **28**, 1367–1372.

Colquhoun, I. J., McFarlane, H. C. E., and McFarlane, W. (1982). *J. Chem. Soc. Chem. Commun.* 220–221.

Cornélis, A., and Laszlo, P. (1979). *Biochemistry* **10**, 2004–2007.

Cornélis, A., Laszlo, P., and Cambillau, C. (1978). *J. Chem. Res.* (S), 462–463.

Covington, A. K., and Newman, K. E. (1979). *Pure Appl. Chem.* **51**, 2041–2058.

Covington, A. K., and Thain, J. M. (1974). *J. Chem. Soc. Faraday Trans.* 1 **70**, 1879–1887.

Covington, A. K., Newman, K. E., and Lilley, T. H. (1973). *J. Chem. Soc. Faraday Trans.* 1 **69**, 973–983.

Covington, A. K., Lantzke, I. R., and Thain, J. M. (1974). *J. Chem. Soc. Faraday Trans.* 1 **70**, 1869–1878.

Dechter, J. J. (1982). "Progress Inorganic Chemistry" (S. J. Lippard ed.), Vol. 29, pp. 285–385. Wiley, New York.

Degani, H. (1977). *Biophys. Chem.* **6**, 345–349.

Delayre, J. L., Ingwall, J. S., Malloy, C., and Fossel, E. T. (1981). *Science* **212**, 935–936.

Delpuech, J. J., Peguy, A., and Khaddar, M. R. (1971). *J. Electroanal. Chem.* **29**, 31–54.

Delpuech, J. J., Khaddar, M. R., Peguy, A. A., and Rubini, P. R. (1975). *J. Am. Chem. Soc.* **97**, 3373–3379.

Delville, A., Detellier, C., and Laszlo, P. (1979). *J. Magn. Reson.* **34**, 301–315.

Delville, A., Detellier, C., Gerstmans, A., and Laszlo, P. (1980a). *J. Am. Chem. Soc.* **102**, 6559–6561.

Delville, A. *et al.* (1980b). *Eur. J. Biochem.* **105**, 289–295.

Delville, A. *et al.* (1980c). *Eur. J. Biochem.* **109**, 575–522.

Delville, A., Detellier, C., Gerstmans, A., and Laszlo, P. (1981a). *Helv. Chim. Acta* **64**, 547–555.

Delville, A., Detellier, C., Gerstmans, A., and Laszlo, P. (1981b). *Helv. Chim. Acta* **64**, 556–567.

Delville, A., Detellier, C., Gerstmans, A., and Laszlo, P. (1981c). *J. Magn. Reson.* **42**, 14–27.

Detellier, C., and Gerstmans, A. (1982). *Inorg. Chim. Acta* **65**, L157–L158.

Detellier, C., and Laszlo, P. (1975). *Bull. Soc. Chim. Belg.* **84**, 1081–1086.

Detellier, C., and Laszlo, P. (1976a). *Helv. Chim. Acta* **59**, 1333–1345.

Detellier, C., and Laszlo, P. (1976b). *Helv. Chim. Acta* **59**, 1346–1351.

Detellier, C., and Laszlo, P. (1979). *Helv. Chim. Acta* **62**, 1559–1565.

Detellier, C., and Laszlo, P. (1980). *J. Am. Chem. Soc.* **102**, 1135–1141.

Detellier, C., and Stöver, H. To be published.

Detellier, C., Paris, A., and Laszlo, P. (1978). *C. R. Acad. Sci. Paris* **286**, 781–783.
Detellier, C., Gerstmans, A., and Laszlo, P. (1979). *Inorg. Nucl. Chem. Letters* **15**, 93–97.
Deverell, C. (1969a). *Prog. Nucl. Magn. Reson. Spectrosc.* **4**, 235–334.
Deverell, C. (1969b). *Mol. Phys.* **16**, 491–500.
Deverell, C., and Richards, R. E. (1966). *Mol. Phys.* **10**, 551–564.
DeWitte, W. J., Liu, L., Mei, E., Dye, J. L., and Popov, A. I. (1977). *J. Solution Chem.* **6**, 336–347.
Drakenberg, T. (1982). *Acta Chem. Scand.* **A36**, 79–82.
Dye, J. L. (1979). *Angew. Chem. Int. Ed. Engl.* **18**, 587–598.
Dye, J. L., Andrews, C. W., and Ceraso, J. M. (1975). *J. Phys. Chem.* **79**, 3076–3079.
Edwards, P. P., Guy, S. C., Holton, D. M., and McFarlane, W. (1981). *J. Chem. Soc. Chem. Commun.*, 1185–1186.
Erlich, R. H., and Popov, A. I. (1971). *J. Am. Chem. Soc.* **93**, 5620–5623.
Erlich, R. H., Roach, E., and Popov, A. I. (1970). *J. Am. Chem. Soc.* **92**, 4989–4990.
Erlich, R. H., Greenberg, M. S., and Popov, A. I. (1973). *Spectrochim. Acta* **29A**, 543–549.
Fisk, C. L., Becker, E. D., Miles, H. T., and Pinnavaia, T. J. (1982). *J. Am. Chem. Soc.* **104**, 3307–3314.
Forsén, S., and Lindman, B. (1978). "NMR and the Periodic Table" (R. K. Harris and B. E. Mann eds.), Ch. 6. Academic Press, London.
Forsén, S., and Lindman, B. (1981). *Methods Biochem. Anal.* **27**, 289–486.
Fossel, E. T., Veatch, W. R., Ovchinnikov, Y. A., and Blout, E. R. (1974). Biochemistry **13**, 5264–5275.
Fraenkel, G., and Hallden-Abberton, M. P. (1981). *J. Am. Chem. Soc.* **103**, 5567–5664.
Fraenkel, G., Henrichs, M., Hewitt, J. H., Su, B. M., and Geckle, M. J. (1980). *J. Am. Chem. Soc.* **102**, 3345–3350.
Frankle, L. S., Langford, C. H., and Stengle, T. R. (1970). *J. Phys. Chem.* **74**, 1376–1381.
Frendsdorff, H. K. (1971). *J. Am. Chem. Soc.* **93**, 600–606.
Gerday, C., Grandjean, J., and Laszlo, P. (1979). *FEBS Letters* **105**, 384–385.
Goeller, R., Hertz, H. G., and Tutsch, R. (1972). *Pure Appl. Chem.* **32**, 149–170.
Grandjean, J., and Laszlo, P. (1978). "Protons and Ions Involved in Fast Dynamic Phenomena" (P. Laszlo, ed.), pp. 373–380. Elsevier, Amsterdam.
Grandjean, J., and Laszlo, P. (1979). *Angew. Chem. Int. Ed. Engl.* **18**, 153–154.
Grandjean, J., Laszlo, P., and Gerday, C. (1977). *FEBS Lett.* **81**, 376–380.
Grandjean, J., Laszlo, P., Vögtle, F., and Sieger, H. (1978). *Angew. Chem. Int. Ed. Engl.* **17**, 856–857.
Grandjean, J., Laszlo, P., Offermann, W., and Rinaldi, P. L. (1981). *J. Am. Chem. Soc.* **103**, 1380–1383.
Greenberg, M. S., and Popov, A. I. (1975). *Spectrochim. Acta* **31A**, 697–705.
Grisham, C. M., and Hutton, W. C. (1978). *Biochem. Biophys. Res. Commun.* **81**, 1406–1411.
Gupta, R. K., and Gupta, P. (1982). *J. Magn. Reson.* **47**, 344–350.
Gustavsson, H., and Lindman, B. (1975). *J. Am. Chem. Soc.* **97**, 3923–3930.
Gustavsson, H., and Lindman, B. (1978). *J. Am. Chem. Soc.* **100**, 4647–4654.
Gustavsson, H., Lindman, B., Bull, T., (1978a). *J. Am. Chem. Soc.* **100**, 4655–4661.
Gustavsson, H., Siegel, G., Lindman, B., Fransson, L.-A. (1978b), *FEBS Lett.* **86**, 127–130.
Gustavsson, H., Ericsson, T., Lindman, B. (1978c). *Inorg. Nucl. Chem. Lett.* **14**, 37–43.
Gutmann, V. (1968). "Coordination Chemistry in Non-aqueous Solutions." Springer-Verlag, Vienna.
Halle, B., Wennerstrom, H. (1981). *J. Magn. Reson.* **44**, 89–96.
Hanley, P. (1981). *Chem. Br.* **177**, 374–376.

Herlem, M., Popov, A. I. (1972). *J. Am. Chem. Soc.* **94,** 1431–1434.
Hertz, H. G. (1973). *Ber. Bunsenges. Phys. Chem.* **77,** 531–540, 688–697.
Herwats, L., Laszlo, P., and Genard, P. (1978). *Nouv. J. Chim.* **1,** 173–176.
Hill, A. V. (1910). *J. Physiol.* **40,** iv–vii.
Hirata, F., Friedman, H. L., Holz, M., and Hertz, H. G. (1980). *J. Chem. Phys.* **73,** 6031–6038.
Holz, M., Weingartner, H., and Hertz, H. G. (1977). *J. Chem. Soc. Faraday Trans.* **73,** 71–83.
Holz, M., Weingartner, H., and Hertz, H. G. (1978). *J. Solution Chem.* **7,** 705–720.
Holz, M., Friedman, H. L., and Tembe, B. L. (1982). *J. Magn. Reson.* **47,** 454–461.
Hubbard, P. S. (1970). *J. Chem. Phys.* **53,** 985–991.
Hutton, W. C., Stephens, E. M., and Grisham, C. M. (1977). *Arch. Biochem. Biophys.* **184,** 166–171.
Katti, S. K., Seshadri, T. P., and Viswamitra, M. A. (1981). *Acta Cryst.* **B37,** 1825–1831.
Kessler, Y. M., Mishustin, A. I., and Podkovyrin, A. I. (1977). *J. Solution Chem.* **6,** 111–115.
Kintzinger, J. P., and Lehn, J. M. (1974). *J. Am. Chem. Soc.* **96,** 3313–3314.
Kirsch, N. N. L., and Simon, W. (1976). *Helv. Chim. Acta* **59,** 357–363.
Kistenmacher, H., Popkie, H., and Clementi, E. (1974). *J. Chem. Phys.* **61,** 799–815.
Komoroski, R. A., and Mauritz, K. A. (1978). *J. Am. Chem. Soc.* **100,** 7487–7489.
Kondo, J., and Yamashita, J. (1959). *Phys. Chem. Solids* **10,** 245–253.
Laszlo, P. (1978). *Angew. Chem. Int. Ed. Engl.* **17,** 254–266.
Laszlo, P., and Paris, A (1978). *C. R. Acad. Sci. Paris,* **286D,** 717–719.
Lehn, J. M. (1973). *Struct. Bonding Berlin* **16,** 1–69.
Lehn, J. M. (1978). *Acc. Chem. Res.* **11,** 49–57.
Lin, J. D., and Popov, A. I. (1981). *J. Am. Chem. Soc.* **103,** 3773–3777.
Lindman, B., Lindblom, G., Wennerström, H., and Gustavsson, H. (1977). "Micellization, Solubilization and Microemulsions" (K. L. Mittal, ed.), pp. 195–227. Plenum, New York.
Lindquist, I., and Lindman, B. (1970). *Acta Chem. Scand.* **24,** 1097–1098.
Live, D., and Chan, S. I. (1976). *J. Am. Chem. Soc.* **98,** 3769–3778.
Lutz, O. (1967). *Z. Naturforsch.* **22A,** 286–288.
Lutz, O., and Nolle, A. (1972). *Z. Naturforsch.* **27A,** 1577–1581.
Manning, G. S. (1978). *Q. Rev. Biophys.* **11,** 179–246.
McKeever, L. D., and Waack, R. (1969). *J. Chem. Soc. Chem. Commun.* 750–751.
McKeever, L. D., Waack, R., Doran, M. A., and Baker, E. B. (1969). *J. Am. Chem. Soc.* **91,** 1057–1061.
Mei, E., Popov, A. I., and Dye, J. L. (1977a). *J. Phys. Chem.* **81,** 1877–1881.
Mei, E., Dye, J. L., and Popov, A. I. (1977b). *J. Am. Chem. Soc.* **99,** 5308–5311.
Mei, E., Liu, L., Dye, J. L., and Popov, A. I. (1977c). *J. Solution Chem.* **6,** 771–778.
Mei, E., Popov, A. I., and Dye, J. L. (1977d). *J. Am. Chem. Soc.* **99,** 6532–6536.
Miles, H. T., and Frazier, J. (1972). *Biochem. Biophys. Res. Commun.* **49,** 199–204.
Miles, H. T., and Frazier, J. (1978). *J. Am. Chem. Soc.* **100,** 8037–8038.
Monoi, H., and Uedaira, H. (1980). *J. Magn. Reson.* **38,** 119–129.
Moon, R. M., and Richards, J. H. (1973). *J. Biol. Chem.* **248,** 7276–7278.
Mulder, C. W. R., De Bleijs, J., and Leyte, J. C. (1980). *Chem. Phys. Lett.* **69,** 354–358.
Neher, E., Sandblom, J., and Eisenman, G., (1978). *J. Membr. Biol.* **40,** 97–116.
Ng, T. C. *et al.* (1982). *J. Magn. Reson.* **49,** 271–286.
Nielsen, O. F., Lund, P. A., and Petersen, S. B. (1982). *J. Am. Chem. Soc.* **104,** 1991–1995.
Norne, J. E., Gustavsson, H., Forsen, S., Chiancone, E., Kuiper, H. A., and Antonini, E. (1979) *Eur. J. Biochem.* **98,** 591–600.

Oldfield, E. *et al.* (1982). *J. Am. Chem. Soc.* **104**, 919–920.

Olsher, U., Elgavish, G. A., and Jagur-Grodzinski, J. (1980). *J. Am. Chem. Soc.* **102**, 3338–3345.

Parello, J., Reimarsson, P., Thulin, E., and Lindman, B. (1979). *FEBS Lett.* **100**, 153–156.

Pedersen, C. J. (1967). *J. Am. Chem. Soc.* **89**, 7017–7036.

Petersen, S. B., Led, J. J., Johnston, E. R., and Grant, D. M. (1982). *J. Am. Chem. Soc.* **104**, 5007–5015.

Pike, M. M., and Springer, C. S. (1982). *J. Magn. Reson.* **46**, 348–353.

Pike, M. M., Simon, S. R., Balschi, J. A., and Springer, C. S. (1982). *Proc. Natl. Acad. Sci. USA* **79**, 810–814.

Pinnavaia, T. J., Miles, H. T., and Becker, E. D. (1975). *J. Am. Chem. Soc.* **97**, 7198–7200.

Pinnavaia, T. J. *et al.* (1978). *J. Am. Chem. Soc.* **100**, 3625–3627.

Popov, A. I. (1979). *Pure Appl. Chem.* **51**, 101–110.

Rasshofer, W., Oepen, G., and Vogtle, F. (1978). *Chem. Ber.* **111**, 419–430.

Reuben, J., Shporer, M., Gabbay, E. J. (1975). *Proc. Natl. Acad. Sci. USA* **72**, 245–249.

Ribas-Prado, F., Giessner-Prettre, C., Daudey, J. P., and Pullman, A. (1980). *J. Magn. Reson.* **37**, 431–440.

Richards, R. E., and Yorke, B. A. (1963). *Mol. Phys.* **6**, 289–300.

Rose, D. M., Bleam, M. L., Record, M. T. Jr., and Bryant, R. G. (1980). *Proc. Natl. Acad. Sci. USA* **77**, 6289–6292.

Schmidt, E., Hourdakis, A., and Popov, A. I. (1981). *Inorg. Chim. Acta* **52**, 91–95.

Schmidt, P. C., Sen, K. D., Das, T. P., and Weiss, A. (1980). *Phys. Rev. B* **22**, 4167–4179.

Shamsipur, M., and Popov, A. I. (1979). *J. Am. Chem. Soc.* **101**, 4051–4055.

Shamsipur, M., Rounaghi, G., and Popov, A. I. (1980). *J. Solution Chem.* **9**, 701–714.

Shchori, E., Jagur-Grodzinski, J., Luz, Z., and Shporer, M. (1971). *J. Am. Chem. Soc.* **93**, 7133–7138.

Shchori, E., Jagur-Grodzinski, J., and Shporer, M. (1973). *J. Am. Chem. Soc.* **95**, 3842–3846.

Shih, J. S., and Popov, A. I. (1977). *Inorg. Nucl. Chem. Lett.* **13**, 105–110.

Shporer, M., and Luz, Z. (1975). *J. Am. Chem. Soc.* **97**, 665–666.

Shporer, M., Poupko, R., and Luz, Z. (1972). *Inorg. Chem.* **11**, 2441–2443.

Smetana, A. J., and Popov, A. I. (1980). *J. Solution Chem.* **9**, 183–196.

Stephens, E. M., and Grisham, C. M. (1979). *Biochemistry* **18**, 4876–4885.

Templeman, G. J., and Van Geet, A. L. (1972). *J. Am. Chem. Soc.* **94**, 5578–5582.

Vaes, J., Chabanel, M., and Martin, M. L. (1978). *J. Phys. Chem.* **82**, 2420–2423.

Van der Klink, J. J., Zuiderweg, L. H., and Leyte, J. C. (1974). *J. Chem. Phys.* **60**, 2391–2399.

Van der Klink, J. J., Prins, D. Y. H., Zwolle, S., and Van der Touw, F., Leyte, J. C., (1975). *Chem. Phys. Lett.* **32**, 287–289.

Vega, S., and Naor, Y. (1981). *J. Chem. Phys.* **75**, 75–86.

Venkatachalam, C. M., and Urry, D. W. (1980). *J. Magn. Reson.* **41**, 313–335.

Wehrli, F. W. (1976). *J. Magn. Reson.* **23**, 527–533.

Wehrli, F. W. (1977). *J. Magn. Reson.* **25**, 575–580.

Wehrli, F. W. (1978a). *J. Magn. Reson.* **30**, 193–209.

Wehrli, F. W. (1978b). *Org. Magn. Reson.* **11**, 106–108.

Wehrli, F. W. (1979). "Annual Reports on NMR Spectroscopy" (G. A. Webb, ed.), Vol. 9, pp. 125–219. Academic Press, New York.

Weingärtner, H. (1980). *J. Magn. Reson.* **41**, 74–87.

Weingärtner, H., and Hertz, H. G. (1977). *Ber. Bunsenges. Phys. Chem.* **81**, 1204–1221.

6 Aluminum-27

J. J. Delpuech

Laboratoire de Chimie Physique Organique
University of Nancy I
Vandoeuvre-les-Nancy Cedex, France

I. Introduction

Group III of the periodic table contains three typical metallic elements—aluminum, gallium, and indium—closely related to each other as shown by similarities in their metallurgy and in the chemical properties of the elements and their derivatives (Banister and Wade, 1975). However, the lightest element of this triad is by far the most important one in practice. This is due to (a) the high natural abundance of aluminum in the earth's crust (8.8 mass %) in the form of aluminates and aluminosilicates, (b) the outstanding properties of the metal and its alloys or intermetallic compounds, and (c) the exceptional role played by aluminum compounds in some major fields of industrial chemistry.

153

The importance of this element accounts for the continued interest in its NMR properties. Aluminum-27 has been studied extensively since the discovery of NMR in the mid-1940s in the metallic or intermetallic state by both wideline NMR and nuclear quadrupole resonance (NQR). Many publications are presently appearing in this area, and important reviews exist on this topic (Carter *et al.*, 1977). We have, however, restricted the scope of this chapter to the *high-resolution NMR of nonmetallic aluminum compounds,* mostly involving (but not exclusively) molecular species in the liquid state. This field of investigation has been covered in the past by two reviews, the first by Akitt (1972) and the second by Hinton and Briggs (1978). However, significant results have appeared in the literature since the time of these publications. Progress in this area has been magnified by the advent of high-frequency Fourier transform (FT) spectrometers allowing higher resolution and sensitivity in the liquid state, and also in the solid state through the use of magic angle spinning (MAS). Most of the pioneering work described in the previously mentioned reviews has been thoroughly reexamined in recent years, leading to new conclusions, sometimes in contrast to erroneous statements from previous investigations. It seems to be the appropriate time to present a report on the results presently obtained and an overall view of the potential of this spectroscopy with its difficulties and limitations.

All three elements of the group possess magnetically active isotopes: ^{27}Al, ^{69}Ga, ^{71}Ga, ^{113}In, and ^{115}In. They are characterized by a high sensitivity to detection and large chemical shift ranges, two factors that make their study relatively easy. Fortunately, aluminum has proved to be still more amenable to NMR studies than its companion elements because of a higher intrinsic sensitivity and the existence of only one isotope, which results in high receptivity of the ^{27}Al nucleus as shown in Table I. Another attractive feature of this nucleus is the relatively small value of its quadrupole moment eQ associated with a high spin number, $I = \frac{5}{2}$. These two favorable factors result in much higher relative peak heights—by one order of magnitude or more (Table I)—than for its companion elements. This means that sufficient signal/noise ratios can be obtained even with dilute ($\sim$0.01 M) solutions of aluminum compounds.

However, even under these favorable conditions, ^{27}Al line widths may vary from 3 Hz to several kilohertz, and the signal may even completely vanish into the baseline noise in some instances. This negative aspect of quadrupolar relaxation is counterbalanced by the additional information obtained about molecular symmetries from the magnitude of quadrupolar line broadenings. Sharper lines reveal a more-or-less perfect cubic symmetric (O_h or T_d) arrangement of substituents about the observed nucleus.

TABLE I

Nuclear Magnetic Resonance Properties of Aluminum, Gallium, and
Indium Isotopes

Isotope X	Spin, I_X	Isotopic abundance, a_X (%)[a]	NMR frequency $\nu_0{}^{a,b}$	Quadrupole moment $Q(10^{-28} \text{ m}^2)^a$	Relative receptivity $R_X{}^{c,d}$	Relative line width, $W_X{}^e$	Relative peak height, $H_X{}^f$
^{27}Al	5/2	100	26.057	0.149	0.206, 1170	1.00	100.0
^{69}Ga	3/2	60.4	24.001	0.178	0.042, 237	5.94	3.4
^{71}Ga	3/2	39.6	30.496	0.112	0.056, 319	2.35	11.6
^{113}In	9/2	4.28	21.865	1.14	0.015, 84	13.55	0.54
^{115}In	9/2	95.72	21.912	0.83	0.332, 1890	7.18	22.4

[a] Quoted from BRUKER NMR Tables and from Brevard and Granger (1981).
[b] In megaherts for an induction of 2.348 T (^{1}H at 100 MHz).
[c] With respect to ^{1}H or ^{13}C nuclei (first and second numbers, respectively).
[d] Computed as the ratio R of the receptivities $a_X\gamma_X I_X(I_X + 1)$ at constant field of the mentioned isotope and of the reference nucleus.
[e] Computed as the ratio W_X of the values taken by the function $(2I + 3)Q^2/I^2(2I - 1)$ for the mentioned isotope and for ^{27}Al nuclei (assuming identical EFG tensors and correlation times).
[f] Computed as 100 times the ratio of the values taken by the function R_X/W_X for the mentioned isotope and for ^{27}Al nuclei.

An important consequence is that tetra- and hexacoordinate aluminum(III) compounds (taking the element in its main oxidation state only) may be differentiated from tri- and pentasubstituted compounds on the basis of line widths smaller or larger than an ill-defined limiting value (100–300 Hz) for each group, respectively.

Quadrupolar relaxation effects are examined with some detail in Section II which is entirely devoted to relaxation mechanisms and dynamic and chemical exchange processes. The general features of ^{27}Al spectroscopy are discussed in Section III which may serve as a key to the tables of ^{27}Al parameters given as a conclusive section.

II. Relaxation Mechanisms and Dynamic Processes

A. Measurement of T_1 and T_2 Relaxation Times

Most relaxation data for ^{27}Al NMR are based on the line widths W of the absorption signals or on the peak-to-peak interval $\Delta\nu_{1/2}$ (in hertz) or

ΔB_0 (in gauss) of their first derivatives, according to the classic formulas

$$W = \frac{1}{\pi T_2} \quad \text{or} \quad \Delta\nu_{1/2} = \frac{1}{\pi T_2 \sqrt{3}} \quad \text{and} \quad \Delta B_0 = \frac{2}{\gamma_{Al} T_2 \sqrt{3}} \tag{1}$$

Overlapping lines require prior deconvolution of the spectrum. This is the case for (a) a series of mixed Al^{3+} solvates with closely spaced resonances, $AlX_n Y_{6-n}{}^{3+}$ ($n = 0$ to 6), where X = TMPA (see notations in Table VIII), Y = H_2O (Canet et al., 1973; Delpuech et al., 1974) or X = Br or Cl, Y = MeCN (Dalibart et al., 1981, 1982a,b); (b) the tris-chelate Al-(NIPA)$_3{}^{3+}$, in which the ^{27}Al nucleus is bonded to 6 equivalent ^{31}P nuclei (Rubini et al., 1979) (the observed broad ^{27}Al singlet is consequently assumed actually to result from seven broad overlapping Lorentzian curves shifted from each other by $^2J_{AlP}$ of relative intensities $1:6:15:20:15:6:1$ and with a common line width $W = 1/\pi T_2$—the unknown parameters J and T_2 are then adjusted by trial and error so as to obtain the best fit between experimental and theoretical curves); (c) the tetramethylaluminate anion $AlMe_4{}^-$ in which the ^{27}Al nucleus is coupled to 12 equivalent protons and the observed ^{27}Al broad singlet is similarly analyzed as a 13-line multiplet (Gore and Gutowsky, 1969).

The direct measurement of T_1 by the usual $180°\text{-}\tau\text{-}90°$ pulse method has been carried out in three main studies, on the $Al(H_2O)_6{}^{3+}$ cation in aqueous solutions (Hertz et al., 1971; Holtz et al., 1982; Takahashi, 1977, 1978, 1980; Vashman et al., 1979; Zaripov and Nikiforov, 1979), on the $AlCl_4{}^-$ anion in nonaqueous solvents (Nöth et al., 1982), and on the trimethylaluminum dimer and its derivatives (Yamamoto, 1975a).

Indirect measurements of ^{27}Al relaxation times result from the analysis of coupled spin-$\frac{1}{2}$ nuclei (^{13}C, ^1H, ^{31}P), provided that the ^{27}Al relaxation rate is not so fast as completely to average out spin coupling patterns (Section II,D).

B. Quadrupolar Relaxation in Al(III) Compounds

The main contribution to ^{27}Al relaxation rates arises from quadrupolar interactions. If one assumes (a) a time-independent arrangement of substituents about the aluminum atom, (b) extreme narrowing conditions, and (c) isotropic molecular reorientation with a single correlation time τ_c, then the quadrupolar relaxation rate is given by the equation (Abragam, 1961)

$$\frac{1}{T_{1Q}} = \frac{1}{T_{2Q}} = \frac{3\pi^2}{10} \frac{2I + 3}{I^2(2I - 1)} \left(1 + \frac{\eta^2}{3}\right) \left(\frac{e^2 qQ}{h}\right)^2 \tau_c \tag{2}$$

where eq is the main component of the electric field gradient (EFG) tensor at the ^{27}Al nucleus, $e^2 qQ/h$ the quadrupolar coupling constant, and η the asymmetry parameter. In a few instances, as in the case of dimeric triethyl aluminum in toluene solution, values of $\eta = 0.87$ and $e^2qQ/h = 23.23$ MHz are known from NQR measurements in the solid state (Dewar *et al.*, 1973) and thus allow one to compute correlation times for ^{27}Al nuclei (Vestin *et al.*, 1981). An interesting point of this study was to show that these values were clearly larger than those obtained from the methylene and methyl ^{13}C nuclei, thus demonstrating the existence of rapid internal motions of CH_2 and CH_3 groups superimposed on the overall tumbling of the dimer.

Ions with tetrahedral (T_d) or octahedral (O_h) symmetry, in which $q = 0$, have relaxation rates as small as a few reciprocal seconds. This is the case for $Al(H_2O)_6^{3+}$ [5.7 s^{-1} in an acidic solution of 0.2 M $Al(ClO_4)_3$, Holtz *et al.*, 1982], $AlCl_4^-$ in MeCN ($W = 6$ Hz, Dalibart *et al.*, 1982), $AlMe_4^-$ in monoglyme (5 Hz, Oliver and Wilkie, 1967), and $Al(TMPA)_6^{3+}$ in MeNO$_2$ (3 Hz, Delpuech *et al.*, 1974). In such cases, line widths are determined by other relaxation mechanisms: (a) scalar relaxation induced by the rapid quadrupolar relaxation of coupled ^{35}Cl or ^{37}Cl nuclei in $AlCl_4^-$, and (b) dipolar Al–H or Al–P interactions in the other examples, with the exception of $Al(H_2O)_6^{3+}$, in which case it has been shown that quadrupolar interactions remain the dominant relaxation mechanism (Hertz *et al.*, 1971).

The O_h and T_d symmetry requirements should be consequently analyzed more closely as follows:

(a) The first coordination sphere should contain four (or six) substituents.

(b) These substituents should be identical.

(c) If X denotes the coordinating atom in the ligand molecule RX, the four (or six) σ-type Al—X bonds should lie in a nondistorted T_d or O_h arrangement.

(d) Each coordinating Al—X bond is a revolution axis (or at least a C_n rotation axis) for the remainder R of the ligand molecule.

(e) Solvent and solute molecules in the second coordination sphere should maintain an isotropic continuum around the aluminum-containing species.

Deviations from requirements (a) to (e) result in more-or-less intense line broadenings that chemists can use to advantage in structure determination and line assignment problems. As a result of requirement a, the resonances of tri- or pentasubstituted species are expected to be clearly broader than those of tetra- or hexacoordinated species. The signal of monomeric triisobutyl aluminum is thus much broader (6000 Hz, O'Reilly, 1960) than those of the dimeric trimethyl or triethyl aluminum (400 and 1000 Hz) or of their adducts with Lewis bases (1000–2000 Hz) (Table IV). The spectrum of a tetrahydrofuran (THF) solution of $AlCl_3$ is

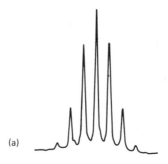

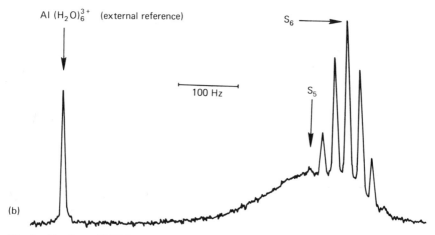

Fig. 1 Aluminum-27 NMR spectra of nitromethane solutions of $Al(TMPA)_6^{3+}$, $3ClO_4^-$ at 25°C. (a) Anhydrous solution, showing the sharp coupling pattern of the octahedral Al-$(TMPA)_6^{3+}$ ion (S_6). (b) Slightly aqueous solution showing the broad signal of the mixed solvate $Al(TMPA)_5(H_2O)^{3+}$ superimposed on the S_6 multiplet. (From Delpuech *et al.*, 1975. Copyright 1975 American Chemical Society.)

composed of several lines which have been accounted for by the formation of four species (Derouault *et al.*, 1977): $AlCl_3(THF)$, $AlCl_3(THF)_2$, $AlCl_2(THF)_4^+$, $AlCl_4^-$. The observed line widths are in the expected sequence: $AlCl_3(THF)_2$ (500 Hz) > $AlCl_3(THF)$ and $AlCl_2(THF)_4^+$ (120 and 250 Hz) >> $AlCl_4^-$ (6 Hz).

As a consequence of requirement b, replacing ligand molecule X in AlX_4 or AlX_6 by a different ligand Y so as to obtain mixed species $S_n = AlX_n Y_{4(or\ 6)-n}$ ($n = 0$ to 4 or 6, respectively) usually results into spectacular

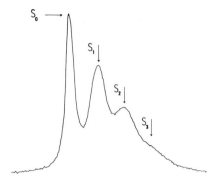

Fig. 2 Aluminum-27 NMR spectrum of a 0.5:8:1 (molar ratio) $Al(ClO_4)_3$–TMPA–H_2O mixture at 65°C, showing the simultaneous presence of the solvates $Al(TMPA)_n(H_2O)_{6-n}$ ($n = 0$ to 3) in a molar ratio $S_0 : S_1 : S_2 : S_3 = 31:39:26:4$. (From Delpuech *et al.*, 1975. Copyright 1975 American Chemical Society.)

line broadenings. This is demonstrated by the mixed solvates $Al(TMPA)_n(H_2O)_{6-n}$ shown in Fig. 1 ($n = 5$ and 6) and Fig. 2 ($n = 0, 1, 2, 3$) (Delpuech *et al.*, 1975). However, it should be realized that a more-or-less complete cancellation of the EFG tensor may occur even in the absence of O_h or T_d symmetry. This has been demonstrated theoretically using a point-charge model in which ligands X and Y are, respectively, assumed to bear an electronic charge e_1 and e_2 at distances r_1 and r_2 from the metal atom (Valiev and Zaripov, 1966). Computed values of the EFG tensor invariant $G^2 = e^2q^2$ as a function of the quantity $\delta q = e_2/r_2^3 - e_1/r_1^3$ are displayed in Table II. These values show that (a) a fortuitous cancellation

TABLE II

Electric Field Gradient Tensor Invariant $G^2 = e^2q^2$
Calculated for the Central Atom in Mixed
Tetrahedral or Octahedral Complexes

Complex	Symmetry	G^{2a}	G^2/G^2_{max}
AlX_4^b or AlY_4	T_d	0	0.00
AlX_3Y or $AlXY_3$	C_{3v}	$\frac{9}{27}\delta q^2$	0.75
AlX_2Y_2	C_{2v}	$\frac{12}{27}\delta q^2$	1.00
AlX_6^c or AlY_6	O_h	0	0.00
AlX_5Y or $AlXY_5$	C_{4v}	$9\,\delta q^2$	0.25
trans-AlX_4Y_2 (or -AlX_2Y_4)	D_{4h}	$36\,\delta q^2$	1.00
cis-AlX_4Y_2 (or -AlX_2Y_4)	C_{2v}	$16\,\delta q^2$	0.44
trans (mer)-AlX_3Y_3	C_{2v}	$27\,\delta q^2$	0.75
cis (fac)-AlX_3Y_3	C_{3v}	0	0.00

[a] Expressed as a multiple of the quantity $\delta q^2 = (e_2/r_2^3 - e_1/r_1^3)^2$.
[b] From Tarasov *et al.*, 1978.
[c] From Valiev and Zaripov, 1966.

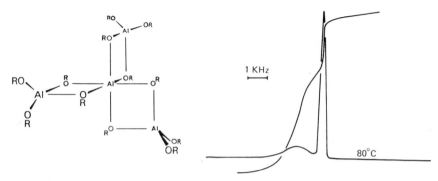

Fig. 3 Structural formula and ^{27}Al spectrum of the aluminum isopropoxide tetramer (R = CHMe$_2$) in toluene at 80°C. (From Akitt and Duncan, 1974, by permission.)

of G^2 (and therefore of quadrupolar effects) may occur within the whole series of mixed solvates if $\delta q = 0$, i.e., if ligands X and Y create the same individual EFG, $e_1/r_1^3 = e_2/r_2^3$; this is presumably the case for the series AlCl$_n$(SCN)$_{4-n}$ (Tarasov et al., 1980) where unexpectedly sharp coupling patterns for ^{27}Al—^{14}N are observed (Section III,C); (b) pairs of cis–trans isomers are endowed with largely different line widths; and (c) the resonance of the cis (or fac) AlX$_3$Y$_3$ isomer is as sharp as that of the O_h species AlX$_6$ (or AlY$_6$). The two latter points have been nicely illustrated in an investigation of AlCl$_3$ solutions in acetonitrile (Wehrli and Wehrli, 1981), where the whole series of solvates AlCl$_n$(MeCN)$_{6-n}$ was observed (Table X).

Further symmetry requirements are that individual chemical bonds themselves have an O_h or T_d arrangement about the aluminum atom. These requirements are not strictly obeyed in most compounds for the whole set of bonds in each ligand molecule. Various levels of approximation are therefore considered depending on the bond number to the metal center. Requirement c states that coordinative Al—X bonds must lie in a strict O_h or T_d arrangement. The tetramer of aluminum isopropoxide, for example, is described as the assembly of a nondistorted octahedral AlO$_6$ unit and three distorted tetrahedral AlO$_4$ units (Fig. 3), as revealed by the presence of a sharp and a broad line (50 and 500 Hz) with an intensity ratio of 1 : 3 (Akitt and Duncan, 1974).

Requirement d is more or less obeyed in practice. An important concept is that quadrupolar relaxation actually results from both the overall molecular reorientation (correlation time τ_c) and the random distortions of the EFG tensor due to internal motion. The latter effect is not included in Eq. (2). No general theory seems to exist to date dealing with the combined effects of overall and local molecular motions on quadrupolar relaxation. Thus, again in reference to the example of $Al(H_2O)_6^{3+}$, internal rotations of coordinated water molecules about their C_2 axes have been tentatively treated using a jump model involving two overall configurations of T_d and C_{2v} symmetries, for which $q = 0$ and $q \neq 0$, respectively (Takahashi, 1977, 1978, 1980). This example is of great importance in practice, because $Al(H_2O)_6^{3+}$ is the reference sample used in ^{27}Al NMR (Section IV,B). Large deviations from cubic symmetry are obtained when polymerized aquohydroxo derivatives of $Al(H_2O)_6^3$ are formed in solutions with relatively high pH values. Large polynuclear species even disappear from the total aluminum content as computed by integration over the whole ^{27}Al spectrum. Only three species have been clearly identified: (a) The aluminate ion $Al(OH)_4^-$ at high pH (one sharp line) (Moolenaar et al., 1970); (b) the bridged dimer $(H_2O)_4Al(\mu\text{-}OH)_2Al(H_2O)_4^{4+}$ in which the two AlO_6 units give rise to a broad signal located quite close to that of $Al(H_2O)_6^{3+}$ (Akitt et al., 1972); and (c) the polymeric cation $AlO_4Al_{12}(OH)_{24}(H_2O)_{12}^{7+}$ which yields one characteristic sharp, intense resonance at 62.5 ppm in the liquid (Akitt et al., 1972b) and solid Müller et al., 1981a,c) states. This line has been assigned to a central highly symmetric AlO_4 tetrahedron surrounded by 12 AlO_6 octahedra with shared edges. The spectrum of the 12 AlO_6 units has been observed as a broad resonance on increasing the temperature to 80°C so as to decrease the ^{27}Al relaxation rate (Akitt and Mann, 1981).

Deviations from O_h or T_d symmetry are nearly absent in the tetracoordinated ions AlX_4^-, where X = H, D, Cl, or Br (Table V); they are weak when X = Me (each Al—C bond is a C_3 axis for the attached methyl) and moderately strong when R = Et or Bu. A similar degradation of cubic symmetry is also observed for dimeric alkyl aluminums because the alkyl group is more bulky. This results in broader line widths (Petrakis and Dickson, 1972) and strong curvatures in Arrhenius plots.

Unusually sharp lines (3–5 Hz) are observed in certain organophosphorus solvates, such as $Al(HMPA)_4^{3+}$ and $Al(TMPA)_6^{3+}$ (see previous discussion). This shows that the solvating phosphoryl bond P=O of the ligand molecule should lie on a straight line with the cation Al^{3+}. This is somewhat unexpected because limited crystallographic data concerning solid adducts such as $VOCl_2$, 2HMPA and $SbCl_5$, $O=PMe_3$ show that the bond angle V—O—P or Sb—O—P is different from 180° (154 and 145°, respectively). The bulky organophosphorus ligands presumably cause enlargement of the bond angle up to 180° so as to minimize the repulsions between the peripheral methyl group of the four (or six) ligands. This steric hindrance, however, is decreased when the ionic radius is increased if one considers the Ga(III) and In(III) homologous complexes for the sake of comparison: The bond angle is expected to become smaller than 180° and the cubic symmetry to be progressively degraded. In effect, line widths of ^{71}Ga and ^{115}In resonances in $Ga(TMPA)_6^{3+}$ and $In(TMPA)_6^{3+}$ are multiplied by factors of 28 and 85 (relative to ^{27}Al resonance in $Al(TMPA)_6^{3+}$), which are clearly larger than the expected factors of 2.35 and 7.18 on the basis of identical chemical environments (Table I) (Rodehüser et al., 1977). The high symmetry of Al^{3+} complexes with monophosphorylated ligands is greatly destroyed with analogous diphosphorylated centrosymmetric ligands such as nonamethylimidodiphosphoramide, $(Me_2N)_2P(O)NMeP(O)(NMe_2)_2$ (NIPA). The resulting complex, Al-$(NIPA)_3^{3+}$ has D_3 symmetry only, and its line width amounts to 150 Hz (Rubini et al., 1982).

The lack of symmetry in multidentate complexes can also be accounted for by distortions in the cubic arrangement of coordinative Al—X bonds (requirement c). This is assumed to be the case for the AlO_6 octahedron in alumichrome peptides, where the line width amounts to several kilohertz, in sharp contrast to the analogous tris(acetyl)hydroxamate complex $[MeC(O)NH(O^-)]_3Al$ ($W = 370$ Hz). This effect was tentatively traced to distortions stemming from structural constraints imposed by the peptide conformation, i.e., from entasis (Llinás and De Marco, 1980). This unique example illustrates the potential of using ^{27}Al as a biological quadrupolar probe—in the present case as a model of biological iron chromophores—ferrichromes—in natural siderophores.

The presence of ions in the second coordination sphere may cancel out the effect of a satisfactory cubic symmetry in the first coordination sphere (requirement f). This is clearly illustrated by obtaining broad resonances for the previously mentioned AlR_4^- anions in poorly solvating media (diethyl ether), in contrast with the sharp coupling patterns observed in 1,1-dimethoxyethane (DME or glyme) (Gore and Gutowsky, 1969; Hermanek et al., 1975; Oliver and Wilkie, 1967). This indicates the formation of contact ion pairs or of multiple ions (Nöth, 1979) in ether solutions. Aluminum-27 NMR can thus be used as an efficient tool to distinguish Al^{3+} free ions (or solvent-separated ion pairs) from intimate ion pairs, and consequently to classify solvents according to their coordinating properties (Oliver and Wilkie, 1967; Westmoreland et al., 1972, 1973).

C. Aluminum-27 Dynamic NMR

Chemical exchange can contribute significantly to ^{27}Al line broadening in either of two ways:

(a) *Two (or more) species A, B, ..., each endowed with its own EFG tensor* $(q_A, q_B, ...)$ *have a fast exchange rate on the NMR time scale*. The resulting line width is the weighted mean over exchanging sites. This is equivalent to considering an averaged EFG tensor invariant $\bar{q} = p_A q_A + p_B q_B + ...$, where p_A, p_B, ..., are the mole fractions of sites A, B, ... Examples of such a situation have been studied in the case of (i) dimeric aluminum alkyls $Al_2R_6(A)$ in fast equilibrium with the monomer $AlR_3(B)$ and a neutral adduct $AlR_3 ... D(C)$ (in the presence of a donor Lewis acid, which may be the solvent itself), with q_A and $q_C < q_B$ (Petrakis and Dickson, 1972) and (ii) intimate ion pairs (A) in fast equilibrium with solvent-separated ion pairs (B), with $q_A \gg q_B$ (see Sections II,B and II,D).

(b) *The exchange rate is within the NMR time scale window*. This has been reported to be the case (i) for the previously mentioned mixed ^{3+}Al solvates S_n as a consequence of the following equilibria

$$S_n + Y \underset{K_n}{\rightleftharpoons} S_{n-1} + X, \qquad n = 1 \text{ to } 6 \text{ (or 4)} \tag{3}$$

(ii) for chlorosolvato complexes such as $AlCl_4^-$ and $AlCl_2(THF)_4^+$ in THF solutions of $AlCl_3$, the line width of $AlCl_4^-$ passing from 6 to 90 Hz on raising the temperature from 0 to 27°C as a result of the equilibrium $2(AlCl_3, 2THF) \rightleftharpoons AlCl_4^- + AlCl_2(THF)_4^+$ (Derouault *et al.*, 1977); and (iii) for bulk and bound ligand molecules in an organophosphorus Al^{3+} solvate, such as $Al(TMPA)^{3+}$, dissolved in an inert cosolvent, $Al(TMPA)_6^{3+} + TMPA^*(\text{free}) \overset{k}{\to} Al(TMPA)_5(TMPA^*)^{3+} + TMPA$, in which case the chemical exchange induces a NMR site exchange between the components of the sharp ^{27}Al multiplet due to six equivalent $^{27}Al-^{31}P$ couplings (Section III,C) (Delpuech *et al.*, 1975).

In the latter example, precise kinetic measurements were performed at variable concentrations and temperatures using a series of hexacoordinated complexes and a tetracoordinated complex $Al(HMPA)_4^{3+}$. A positive cross-checking was possible by comparing the ^{27}Al results with those obtained from ^1H dynamic NMR of the ligand molecules. The rate laws are, respectively, zero- and first-order in free ligand for octahedral and tetrahedral solvates. These data are consistent with a dissociative and an associative substitution mechanism, respectively. This mechanistic change is accompanied by a strong decrease in the activation enthalpies and entropies. Typical values are $k_{25°C} = 5.1$ s^{-1} and 4.8×10^3 M^{-1} s^{-1}, $\Delta H\ddagger = 79.5$ and 32.2 kJ mol^{-1}, and $\Delta S\ddagger = 33.0$ and -42.7 J mol^{-1} for $Al(\text{dimethylmethylphosphonate})_6^{3+}$ and $Al(HMPA)_4^{3+}$, respectively. These investigations illustrate the potential of ^{27}Al NMR for study of the solution chemistry and electrochemistry of this element.

D. Line Broadening Processes in the Spectra of
Coupled Nuclei

Quadrupolar relaxation also affects the spectrum of spin-$\frac{1}{2}$ nuclei (X) coupled to a relaxing ^{27}Al nucleus. For a set of equivalent nuclei X, the ^{27}Al coupling pattern is composed of six equidistant resonances (by J_{AlX}) of equal intensity but unequal line width (and therefore unequal peak height) in a ratio of $15 : 23 : 18 : 18 : 23 : 15$. This results from unequal transition probabilities for the six magnetic states of the ^{27}Al nuclei (Abragam, 1961), themselves inducing unequal exchange probabilities for the six corresponding components of the multiplet. These probabilities are displayed here in a so-called exchange matrix (Martin *et al.*, 1980).

$$
\frac{1}{8T_{1Q}} \times
\begin{vmatrix}
-15 & 10 & 5 & 0 & 0 & 0 \\
10 & -23 & 4 & 9 & 0 & 0 \\
5 & 4 & -18 & 0 & 9 & 0 \\
0 & 9 & 0 & -18 & 4 & 5 \\
0 & 0 & 9 & 4 & -23 & 10 \\
0 & 0 & 0 & 5 & 10 & -15
\end{vmatrix}
$$

The sextet is observed only in the slow-exchange region when the ^{27}Al relaxation rate is slow ($1/T_{1Q} \ll J$). Its components are averaged out into a sharp singlet when the quadrupolar relaxation is fast ($1/T_{1Q} \gg J$). Coupling constants J_{AlX} and relaxation rates $1/T_{1Q}$ may be extracted through complete line shape analysis when $JT_{1Q} \sim 1$. The ^{27}Al relaxation rates in the solvate Al(DMMP)(H$_2$O)$_5^{3+}$ between -30 and $+50°$C have been measured from both ^{27}Al and ^{31}P spectra, and they have been effectively found to be equal within experimental error (Delpuech *et al.*, 1975). Chemical exchange effects may interfere with purely quadrupolar effects. One should note that the influence of temperature is opposite for these two processes; i.e., quadrupolar effects decrease (and coupling patterns are sharper) as the temperature increases. In the example, ligand exchange between bulk and bound DMMP molecules takes place on the NMR time scale at temperatures above 60°C. This results in an opposite line broadening of the ^{31}P sextet as the temperature is raised above 60°C (Fig. 4).

The determination of quadrupolar relaxation rates and coupling constants therefore requires a lowering of the temperature until the undesired chemical exchange is stopped, provided that, simultaneously, the quadrupolar relaxation rate is not beyond the fast-exchange limit. This is the case

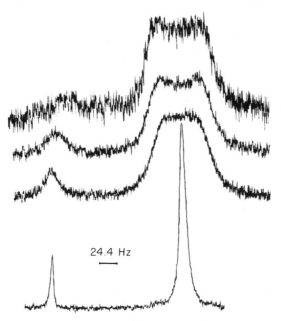

24.4 Hz

Fig. 4 Phosphorus-31 spectrum of an aqueous solution of the solvate $Al(DMMP)(H_2O)_5^{3+}$ (S_1) at 77, 54, 30, and $-30°C$ (from top to bottom, upfield signal). The signal to low field refers to bulk DMMP molecules. Note the progressive appearance of the expected sextet from -30 to $54°C$ (as a result of decreased quadrupolar interaction) and the opposite degradation of signals from 54 to $77°C$ (as the result of a predominant chemical exchange rate). (From Delpuech *et al.*, 1975. Copyright 1975 American Chemical Society.)

for the dimer Al_2Me_6 in toluene solution, which is brought to $-60°C$ where the methyl exchange between terminal and bridging positions becomes negligible, whereas the ^{27}Al relaxation rate still corresponds to the coalescence of the observed ^{13}C spectrum (Yamamoto, 1975a).

Another possibility is raising the temperature so as to go beyond the limits of fast exchange and consequently averaging out the effect of chemical exchange. This is the case for lithium tetramethylaluminate in THF solution where intimate ion pairs (A) are quickly converted to dissociated ions (B), and vice versa, according to the chemical equilibrium

$$Li^+ \,||\, AlMe_4^-(A) \overset{K}{\rightleftharpoons} Li^+ + AlMe_4^-(B)$$

The relaxation rate $R_1 = 1/T_{1Q}$ to be introduced into the exchange matrix to simulate the ^1H spectrum is the weighted mean of the ^{27}Al relaxation rates R_{1A} and R_{1B} at sites A and B of mole fractions p_A, p_B: $R_1 = R_{1A}p_A + R_{1B}p_B$ (Gore and Gutowsky, 1969). Besides ^{27}Al relaxation rates and coupling constants, variable-temperature experiments yield, through the appropriate line shape analysis, the thermodynamic parameters for the equilibrium $(K, \Delta H, \Delta S)$.

III. Main Features of ^{27}Al-NMR Spectroscopy

A. Instrumentation

The first spectra of ^{27}Al samples were obtained from wide-line spectrometers recording the first derivative of resonance absorption (O'Reilly, 1960). *Liquid samples* are presently studied using high-resolution spectrometers and recording the absorption signal. Single-scan continuous-wave (CW) NMR may be conveniently used for concentrated solutions $(> \sim 0.5$ M). Fourier transform NMR is necessary in most cases where sensitivity requirements necessitate the fast accumulation of a large number of free induction decays (FIDs). Resolution and sensitivity have been greatly improved in recent years by the advent of multinuclei high-field superconducting spectrometers operating at frequencies of up to 104 MHz.

However, some care should be exercised to obtain reliable and reproducible results. Even with the minimum delay time permissible (10–100 μs), the intensity of broad lines may be more or less suppressed. In order to make sure that no broad line has escaped observation, a comparison of the total aluminum concentration obtained from integration over the entire ^{27}Al spectrum with the aluminum analytical concentration in the sample is recommended whenever possible. This requires a comparison of the total integrated ^{27}Al intensity of the investigated sample with the intensity of a titrated reference sample, generally $Al(H_2O)_6^{3+}$ or $Al(D_2O)_6^{3+}$. This can be done by using successively two NMR tubes of the same diameter (Akitt and Farthing, 1978, 1981a) or by placing the two solutions in concentric tubes of calibrated diameters (Dalibart *et al.*, 1981). A fast accumulation rate is generally recommended because of short quadrupolar relaxation times T_1. However, it should be noted that accurate intensity measurements of sharp lines, if any, necessitate a long waiting time after each acquisition (Dalibart *et al.*, 1981).

The detection of sharp weak lines buried in broad resonances can be made easier by using a long delay time (Gray and Maciel, 1981). The opposite situation of a weak broad line superimposed on an intense sharp line may be resolved by a simple inversion recovery experiment in which the period between the inversion and detection pulses is short relative to T_1 for ^{27}Al in the sharp signal but long relative to T_1 for the broad signal (Wehrli and Wehrli, 1981).

These aids to signal detection should not be used in quantitative measurements of line intensities. Overlapping lines require in this case the production of simulated spectra (Section II,A) to be compared to the experimental profile. Even in the event of well-separated resonances, the integration of a broad line remains a difficult problem because of the necessarily limited integration domain. For a measured area A between $-L$ and $+L$, the error $\Delta A/A$ amounts to $(200/\pi) \times \arctan(2L/W)$ (as a percentage), and corrections of up to 10% can be made in this way (Dalibart et al., 1982a).

The poor sensitivity obtained in conventional FT NMR for fast-decaying FIDs (broad lines) seems to be improved by using rapid-scan FT spectroscopy which affords the advantage of efficiently recording the spectrum while the excitation is on (Llinás and De Marco, 1980).

A major event in the history of ^{27}Al spectroscopy was the extension of high-resolution NMR to the study of polycrystalline and amorphous solid samples. This involved the combination of two techniques: (a) using MAS to suppress dipolar interactions, chemical shift anisotropies, and quadrupolar interactions in first order; and (b) using high magnetic field strengths to decrease quadrupolar interactions to second order in the central transition $(-\frac{1}{2} \leftrightarrow +\frac{1}{2})$, which is the only transition observed. The residual line width $w = 25\nu_Q^2/18\nu_L$ (where ν_L is the Larmor frequency and ν_Q the fixed quadrupole frequency) is important at relatively low field strengths: 900 ppm for a compound with a quadrupolar coupling constant of 1 MHz at a 35-MHz Larmor frequency, and reasonably small at higher frequencies—3 ppm at 100 MHz (Fyfe et al., 1982). This method has been applied in the study of important classes of chemicals—aluminum oxides, aluminates, and silicoaluminates (Mastikhin et al., 1981; Müller et al., 1981a–c)—where the fundamental new information is the distinction between octahedral and tetrahedral Al—O coordination (Tables V and XI, Fig. 5).

B. Chemical Shifts

An exhaustive compilation of ^{27}Al chemical shifts presently available in the literature is given at the end of this chapter. An inspection of Tables III to XI shows that ^{27}Al chemical shifts may be categorized according to the degree of substitution at the aluminum atom. Hexacoordinate Al(III) compounds afford signals covering approximately 60 ppm, from -40 to $+20$ ppm (region I, Fig. 6), most of which are located upfield of the $Al(H_2O)_6^{3+}$ reference signal. Exceptions to this rule are the octahedral tris hydroxamate complexes (36.5 ppm) and the related alumichrome cyclohexapeptides (41.5 ppm) (Table XI). Tetracoordinate species appear

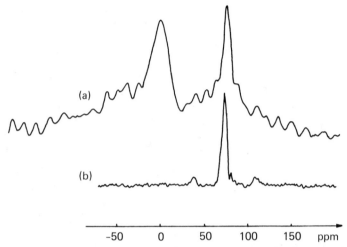

Fig. 5 Solid-state ^{27}Al-NMR spectra of an alkaline-rich sodium aluminate sample (a) and β-NaAlO$_2$ (b), showing the presence of AlO tetrahedra only (at 77 ppm) in the reference sample (b), and of both AlO tetrahedra and octahedra (at 77 and 0 ppm, respectively) in the test sample (a). Note that downfield shifts are directed to the right of the figure. (From Müller *et al.*, 1981 (1981c), by permission.)

within two separate regions, region II (60–110 ppm) and region III (140–180 ppm), depending on whether alkyl substitution is absent or present, respectively. Exceptions to this rule are shown by iodine-containing neutral and anionic species, which cover a wide range of frequencies from −28(AlI$_4^-$) to 86 ppm (AlCl$_3$I$^-$). Tri- and pentasubstituted compounds are too exceptional to allow us to draw a general conclusion. Let us say that the pentacoordinate adducts of AlCl$_3$ and AlBr$_3$ presently known appear between 35 and 63 ppm (Table VII) and that at very low field lies the resonance of triisobutylalane which seems to be the only aluminum alkyl investigated to date that may be safely assumed to be monomeric (Table III). Thus it may be said that most resonances are grouped within three separate windows, and that regions I to III each span a well-defined class of molecular structures, if we ignore the previously mentioned exceptions. This feature is consequently a useful guide in chemical investigations of Al(III) compounds. For example, ^{27}Al NMR allows a clear distinction between AlO$_6$ octahedral and AlO$_4$ tetrahedral sites in complex aluminum–oxygen compounds (Tables V.D and XI), such as the aluminum isopropoxide tetramer in a toluene solution (Fig. 3) and aluminum oxides and alkali and alkali earth aluminates in the solid state (Fig. 5).

One may wonder whether the ^{27}Al upfield shift in compounds of increasing coordination number is due to tetrahedral or octahedral hybridization of aluminum atomic orbitals or simply to the subsequent increased

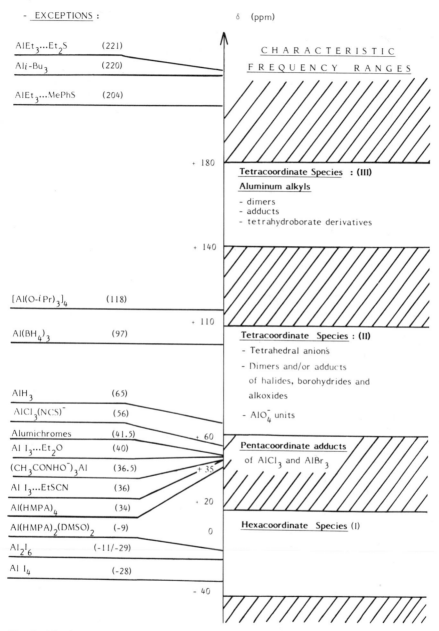

Fig. 6 Aluminum-27 NMR correlation chart. Shifts are referred to external $Al(H_2O)_6^{3+}$, and a positive shift is downfield.

number of substituents and therefore to the increased overall electron donation to the metal center. The existence of additivity laws governing structurally related series of compounds seems to support the latter assumption and, more generally, the contention that the electron-donating properties of the substituents are a major factor in determining chemical shifts. Thus, the order of appearance of the successive signals in the series AlX_nY_{4-n} ($n = 0$ to 4) and AlX_nY_{6-n} ($n = 0$ to 6) is strictly monotonic. In some instances, e.g., X = H_2O, Y = TMPA, TEPA, DMMP, DEEP, and DMHP (Table IX), the successive chemical shifts δ_n are simply a linear function of n, $(\delta_n - \delta_0)/(\delta_6 - \delta_0) = n/6$, with a constant frequency interval between two consecutive signals. The same is true for tetramethyl aluminosilicate aqueous solutions (Table V.D) which give rise to four equally spaced signals assigned to AlO_4 tetrahedrons bonded to zero, one, two, or three silicon atoms, respectively. In many cases, however, the second-order approach introduced by Vladimiroff and Malinowski (1967) is necessary to account for the observed chemical shifts in the series involving the following pairs of ligands: (Cl^-, Br^-), (Cl^-, I^-), (Br^-, I^-), (Cl^-, NCS^- or NCO^-) with $n = 4$ (Table V.C); (MeCN, H_2O), (DMF, DMSO), (TMPA, DMSO) (Table IX), and (Cl^-, MeCN) (Table X), with $n = 6$. This approach uses a pairwise additivity model (Malinowski, 1969)

$$\delta_n = \sum C_{jk}\eta_{jk} \tag{4}$$

where C_{jk} is the number of possible interactions η_{jk} (in parts per million) between ligands j and k (η_{XX}, η_{XY}, and η_{YY} in the previous series). Unequal frequency intervals are observed between successive signals, and distinct shifts are obtained for pairs of cis–trans isomers of the species AlX_2Y_4, AlX_3Y_3, and AlX_4Y_2, if $\eta_{XY} \neq (\eta_{XX} + \eta_{YY})/2$. Only three terms of each series are necessary to compute the set of increments η. This allows control of the position of the remaining species, if their identity is known, or assignment of unknown resonances in the series. If fewer than three resonances have been identified, assignments of unknown resonances remain feasible on the basis of Eq. (4) and of trial-and-error procedures. The sets of increments thus obtained may be further combined to yield the chemical shifts of species containing ligands of any type, e.g., $AlClBrI_2^-$ and $AlClBr_2I^-$, using the series $AlCl_nBr_{4-n}^-$, $AlCl_nI_{4-n}^-$, and $AlBr_nI_{4-n}^-$ Table V.C).

C. Coupling Constants and Coupling Patterns

Well-resolved coupling patterns are of fundamental importance to both the chemist, as an aid in determining molecular structures, and the NMR spectroscopist, for extracting coupling constant values. The occurrence of fine structures is exceptional in ^{27}Al-NMR spectroscopy. Known examples involve ^{27}Al coupling to ^{1}H, ^{2}D, ^{31}P, or ^{14}N nuclei. Couplings between ^{27}Al and ^{1}H have been observed in two classes of compounds: (a) a 13-line multiplet in amine adducts of aluminum borohydride (Table IV.D), confirming the existence of 12 coupled equivalent ^{1}H nuclei, and (b) a quintuplet for alkali tetrahydroaluminates in DME (Table V.A). Couplings between ^{27}Al and ^{2}D have been observed with the homologous alkali tetradeuteroaluminates in the form of a nonet, with line intensities in the ratio $1:4:10:16:19:16:10:4:1$ (Table V.A). ^{27}Al–^{31}P coupling patterns have been obtained with three classes of compounds: (a) phosphine adducts of AlCl$_3$ and AlBr$_3$ (Table IV.B), (b) octahedral or tetrahedral solvates of the Al^{3+} ion (Tables VIII and VI), and (c) mixed-solvation complexes Al(DMMP)(H$_2$O)$_5$$^{3+}$ and Al(DMMP)$_2$(H$_2$O)$_4$$^{3+}$ (Table IX). Unexpectedly sharp ^{27}Al–^{14}N coupling patterns are obtained with the anions AlCl$_3$(XCN)$^-$ and AlCl$_2$(XCN)$_2$$^-$, where X = S or O (Table V.C), whose existence is thus unequivocally established.

Coupling constants J_{AlX} may also be obtained from well-resolved coupling patterns in the spectrum of *coupled nuclei X*. A *sextet* is typically expected for a set of equivalent X nuclei coupled to one ^{27}Al nucleus. In fact, it has been observed in a very few examples: (a) the ^{1}H and ^{13}C spectra of alkali tetramethylaluminates in DME (Table V.B), (b) the ^{13}C spectrum of NaAlEt$_4$ and NaAlBu$_4$ solvent-separated ion pairs in DME or C$_6$H$_6$ (Table V.B), and (c) the ^{31}P spectra of Al(HMPA)$_4$$^{3+}$ (Table VI), Al(TMPA)$_6$$^{3+}$, Al(DMMP)$_6$$^{3+}$, Al(TEPA)$_6$$^{3+}$, etc. (Table VIII). The latter example is the only one, to our knowledge, in which J values have been obtained from both spectroscopies (^{27}Al and X) and shown to be equal.

The last source of ^{27}Al coupling constants arises from the line shape analysis of partially averaged out multiplets (Section II,D). Four examples of such calculations have been described in the literature: (a) The ^{1}H spectrum of LiAlMe$_4$ or NaAlMe$_4$ contact ion pairs in ether and THF (Gore and Gutowsky, 1969), (b) the ^{13}C spectrum of the terminal and bridging methyls in Al$_2$Me$_6$, which contains the only example of a spin $\frac{1}{2}$ coupled to two equivalent spins $\frac{5}{2}$ (Yamamoto, 1975a), (c) the ^{31}P spectrum of the solvate Al(DMMP)(H$_2$O)$_3$$^{3+}$ (Fig. 4)—again an example where J values have been obtained from both ^{27}Al and ^{31}P NMR (Delpuech *et al.*, 1975), and (d) the tris-chelate Al(NIPA)$_3$$^{3+}$ (Rubini *et al.*, 1979).

The J values presently known deserve some comment. Couplings between ^{27}Al and ^{1}H through one, two, and three bonds have an order of magnitude of ~ 170, 7, and 6 Hz, respectively. The exceptional value of 44 Hz in Al(BH$_4$)$_3$ reveals a very rapid intramolecular exchange of terminal and bridging protons, so that the observed coupling is the average of a large and a small value for the bridge and terminal positions, respectively

(Table III). There is a general increase in J on passing from aluminum to ^{71}Ga and eventually to ^{115}In: $J = 170$, ~500, and 970 Hz (estimated) in tetrahydrometallates (Tarasov and Bakum, 1975); $J = 19.9$, 33, and 48 Hz, respectively, in TMPA complexes (Rodehüser et al., 1977); $J = 40$ and 95 Hz in cholorothiocyanato anions (Tarasov et al., 1980). Consideration of the corresponding reduced coupling constants, however, shows that these sequences constitute evidence for an increase in the contribution of the valence s electrons to the chemical bond in AlH_4^- relative to GaH_4^- (Tarasov et al., 1981). Couplings between ^{27}Al and ^{31}P through one or two bonds are in the range 250–290 and 12–30 Hz, respectively. The value of $^2J_{PAl}$ increases on passing from a hexacoordinate solvate, e.g., $Al(TMPA)_6^{3+}$, to the tetracoordinate complex $Al(HMPA)_4^{3+}$ ($J = 19.9$ and 30 Hz, respectively) in accord with the classic dependency of J on the proportion of s-character in the aluminum atom orbitals, namely, $\frac{1}{6}$ and $\frac{1}{4}$ for d^2sp^3 and sp^3 hybridizations, respectively (Delpuech et al., 1975). The magnitudes of ^{27}Al—^{13}C couplings in $AlMe_4^-$ again point to a predominant Fermi contact interaction (Yamamoto, 1975b) on the basis of the correlations proposed in the literature of J_{XC} with J_{CC} and of J_{XCH} with J_{HCH} (Smith, 1963; Weigert et al., 1968). The exceptionally small coupling $^1J_{AlC}$ $= 19$ Hz found for the bridging methyl in Al_2Me_6 indicates the formation of a fairly strong Al—Al bond, mainly σ in character, and a consequent reduction in the s-electron contribution to the bridge bond.

D. Line Intensities

Line intensities can be used for either of two purposes: (a) as an aid in line assignment and structure determination, and (b) in aluminum quantitative analysis. In any case, accuracy is a limiting factor (Section III,A). Relative intensity measurements may help in formulating unknown molecular structures, and this was the case for the isopropoxide tetramer (Fig. 3) where a ratio of 1:3 for hexa- to tetracoordinate aluminum is the principal argument in favor of the proposed structure (Akitt and Duncan, 1974). More generally, quantitative ^{27}Al analysis is used to determine the percentages of tetrahedral and octahedral sites of coordination in aluminum–oxygen compounds, especially in the solid state (Fig. 5, Müller et al., 1981c). Intensity measurements are especially useful in solution chemistry, where unexpected solvates or solvent–counterion complexes may be formed. Typical examples are solutions of aluminum halides in nonaqueous solvents or of aluminum perchlorate in solvent mixtures. Assumptions on the species being formed can be checked by verifying that the corresponding line intensities obey (a) the law of conservation of mass (this helps to make the distinction between mono and polynuclear complexes), (b) the law of conservation of electric charge (this helps to differentiate mono from di and tri positive or negative ions, Dalibart *et al.*, 1982a), and (c) mass action laws [Eq. (3)], assuming unknown activity coefficients equal to unity (this assumption was shown to be invalid in the case of the mixed-solvation complexes of Table IX, Delpuech *et al.*, 1975). Absolute quantitative measurements using a reference sample are seldom performed, although this should be the case in all instances where broad lines may have escaped detection (Section III,A).

E. Line Assignments

Let us briefly summarize the various methods mentioned in the previous paragraphs for solving the related problems of structure determination and line assignment:

(a) The chemical shift correlation chart (Fig. 6)
(b) Pairwise additivity correlations (Section III,B)
(c) The magnitude of quadrupolar line widths (Section II,B)
(d) Coupling patterns (Section III,C)
(e) Line intensities (Section III,D).

In fact all these possibilities are seldom available to the experimentalist, and a number of errors have been made in line assignments. Supplementary evidence from solid state X-ray diffraction, IR, and Raman spectroscopy, and from the NMR spectrum of all types of nuclei in the ligand molecules, is usually necessary to establish molecular formulas on a firm basis.

IV. Main Fields of Application

A. Typical Examples

Let us mention some well-known items of interest in aluminum chemistry that have stimulated important investigations using ^{27}Al NMR:

(a) *Aluminum alkyls* which constitute a chemically interesting and industrially significant class of compounds because Al—H bonds can add to olefins (especially in the presence of transition-metal complexes in the

Ziegler–Natta low-pressure stereospecific polymerization of olefins) and metallic aluminum can fix molecular hydrogen in the presence of aluminum alkyls to yield hydridoaluminum alkyls

(b)   Aluminum oxides as essential constituents of many solid materials (cements, bricks, adsorbents, zeolites, etc.) and as a matrix for metallic impurities (ruby, sapphire, etc.) or metal ions in heterogeneous supported catalysis

(c)   Aluminum halides in isomerization reactions of alkanes (reforming of petroleum) and in Friedel–Crafts reactions

(d)   Aluminum tri-*tert*-butoxide in the Oppenauer oxidation of secondary alcohols

(e)   Aluminum triisopropoxide, aluminum hydride and lithium aluminum hydride as reagents in the reduction of carbonyl compounds

(f)   Aqueous solutions of partially hydrolyzed aluminum salts in the hydrometallurgy of aluminum, in the tanning of collagen, etc.

These fields of application can be found as subsections of Tables III to XI, with the relevant references, under the appropriate headings: aluminum alkyls and their adducts (Table IV.A); hydrides, borohydrides, and adducts (Tables III and IV.D); alkoxides (Tables IV.C and XI.C); haloalkyls (Table IV.E) and halides and their adducts (Tables IV.B and VII) in nonpolar solvents; halides in nonaqueous polar solvents—the adducts (Table IV.B) and anionic and cationic (Tables V.C, VI and X) species formed; solvates of the fully dissociated Al^{3+} ion in organic solvents and solvent mixtures (Tables VIII and IX); tetrahydro (Table V.A) and tetraalkyl and alkoxyaluminates (Table V.B); aquohydroxocomplexes (Tables V.D and XI); solid state NMR of aluminum oxides and aluminates (Tables V.D and XI); multidentate complexes of Al^{3+} (Table XI.E).

B. Tables of NMR Parameters of Tri-, Tetra-, Penta-, and Hexacoordinate Al(III) Compounds

Chemical shifts, line widths, and coupling constants presently known from the literature are displayed in Tables III to XI. These tables are arranged according to the coordination number of the aluminum atom, for the NMR data given in Sections II and III. Each of the chemical classes thus defined (namely, tri-, tetra-, penta-, and hexacoordinated compounds) has been further divided into subsections, when necessary, for the chemical reasons explained in the previous paragraph.

It should be emphasized that most data refer to tetra- or hexacoordinate compounds because (a) pentacoordination is relatively exceptional in aluminum chemistry, and (b) trisubstituted aluminum compounds are strong Lewis acids which give either self-adducts (dimers or polymers) or heteroadducts with any base present in the solution, including polar solvents themselves. Genuine trivalent and pentavalent derivatives are therefore exceptional, and the relevant NMR data often unreliable because of the variety of species in equilibrium, depending on the solvent, the temperature, the concentration, and the nature of the other substrates present in the solution.

These facts account for some discrepancies among data from different reference sources. In this respect the most puzzling discrepancies are those relative to chemical shifts, where systematic differences of up to 5 ppm are noted for the data from experienced laboratories. A solution to this problem could be improving the definition of the standard reference, $Al(H_2O)_6^{3+}$. As shown in Section II,B, the reference signal is sharp at low concentrations only (3 Hz in a 0.2 M solution). Moreover, aluminum chloride, which is generally used to prepare the standard solution, gives rise to signals that are systematically broader than those obtained from the perchlorate salt (Holtz $et\ al.$, 1982). This may be due to contamination of the chloride by Fe^{3+} impurities and to the existence of first-sphere or second-sphere chloro complexes at higher concentrations, two causes which could also account for erratic chemical shift differences. It may be usefully suggested here that all ^{27}Al chemical shifts should be referred to a common standard aqueous solution of 0.2 M aluminum perchlorate and 0.1 M perchloric acid.

Finally, it should be mentioned for the sake of comparison that gallium and indium isotopes have chemical shift ranges larger than those of their companion element: ~1400 and 1100 ppm instead of ~450 ppm, respectively. The chemical shifts of ^{69}Ga and ^{71}Ga nuclei seem to follow the same trends as those found for the analogous aluminum compounds (Akitt, 1972). Spin coupling data are almost nonexistent with the exception of $^1J_{GaH}$ in the GaH_4^- anion (cf. Section III,C) and $^1J_{GaN}$ in the $Cl_nGa(NCS)_{4-n}$ ($n = 2$ or 3) anion (Tarasov $et\ al.$, 1980). Nuclear magnetic resonance studies on ^{113}In and ^{115}In isotopes are relatively rare, on account of large quadrupolar moments and low receptivities (cf. Table I) (Hinton and Briggs, 1978).

TABLE III

Aluminum-27 NMR Parameters of Al(III) Trisubstituted Compounds

Compound	Solvent	δ (ppm)	W (Hz)	J (Hz)	Reference
$AlH_3{}^a$	THF	65	3000	—	Huet *et al.* (1976)
Al-i-Bu$_3$	C_6H_6	220	6000	—	O'Reilly (1960)
Al(BH$_4$)$_3$	C_6H_6	97.4	180	—	Lauterbur *et al.* (1968)
			275		Oddy and Wallbridge (1976)
				$^2J_{AlH} = 44{}^b$	Ogg and Ray (1955)
Al(BH$_4$)$_2$Et	C_6H_6	152	830	—	Oddy and Wallbridge (1978)
Al(BH$_4$)Et$_2$	C_6H_6	162	2400	—	Oddy and Wallbridge (1978)
Al(BH$_4$)$_2$-n-Pr	C_6H_6	154	1000	—	Oddy and Wallbridge (1978)
Al(BH$_4$)$_2$-i-Bu	C_6H_6	155	1000	—	Oddy and Wallbridge (1978)

^{a} Prepared from LiAH$_4$ + H$_2$SO$_4$ in THF.
^{b} From ^{1}H NMR.

TABLE IV

Aluminum-27 NMR Parameters of Al(III) Neutral Tetracoordinate Compounds

Compound	Solvent	δ (ppm)	W (Hz)	J (Hz)	Reference
A. Aluminum alkyls					
Al_2Me_6	Pure	156	780	—	O'Reilly (1960)
	C_5H_{10} or PhMe	—	400	$^1J_{AlC} = 110$ and 19[a]	Yamamoto (1975a)
Al_2Et_6	Pure	171	780	—	O'Reilly (1960)
	C_6H_{12}	174	1,000	—	Di Carlo and Swift (1964)
	C_6H_{12}	142	1,000	—	Swift et al. (1964)
	CH_2Cl_2	139	—	—	Glavincevski and Brownstein (1981)
$(HAl\text{-}i\text{-}Bu_2)_n$					
$AlEt_3 \ldots D$ (adducts)					
D = Et_2O	Pure	162	10,000	—	O'Reilly (1960)
	C_6H_{12}	178	1,160	$^1J_{AlC} = 91$[a]	Yamamoto (1975a)
D = THF	C_6H_{12}	176	1,300	—	Swift et al. (1964)
D = anisole	C_6H_{12}	176	2,100	—	Swift et al. (1964)
D = Et_2S	C_6H_{12}	221	1,900	—	Swift et al. (1964)
D = thiophene	C_6H_{12}	154	1,000	—	Swift et al. (1964)
D = MePhS	C_6H_{12}	204	2,200	—	Swift et al. (1964)
D = Et_3N	C_6H_{12}	165	1,100	—	Swift et al. (1964)
D = pyrrolidine	C_6H_{12}	176	1,200	—	Swift et al. (1964)
D = pyridine	C_6H_{12}	167	1,400	—	Swift et al. (1964)
D = quinoline	C_6H_{12}	164	2,500	—	Swift et al. (1964)
B. Aluminum halides					
Al_2Cl_6	PhMe	91	570	—	O'Reilly (1960)
	C_6H_6	98	185	—	Nöth et al. (1982)
	SO_2	83	—	—	Glavincevski and Brownstein (1981)

TABLE IV (Continued)

Compound	Solvent	δ (ppm)	W (Hz)	J (Hz)	Reference
Al₂Br₆	C₆H₆	75	510	—	Haraguchi and Fujiwara (1969)
	C₆H₆	85	132	—	Nöth et al. (1982)
	Br₂	101	300	—	O'Reilly (1960)
Al₂I₆	PhMe	−11	600	—	O'Reilly et al. (1964)
	C₆H₆	−25	800	—	Haraguchi and Fujiwara (1969)
AlX₃ ... D (adducts)					
X = Cl, D = Me₂O	SO₂	103.9	—	—	Glavincevski and Brownstein (1982)
X = Cl, D = Et₂O	C₆H₆	98	182	—	Nöth et al. (1982)
	Et₂O	105	145ᵇ	—	Nöth et al. (1982)
	Et₂O	105	190	—	O'Reilly (1960)
	Et₂O	105	126	—	Haraguchi and Fujiwara (1969)
	Et₂O	100.1	158	—	Kidd and Truax (1969a)
X = Cl, D = Bu₂O	Bu₂O	115	250	—	Nöth et al. (1982)
X = Cl, D = THF	CH₂Cl₂	94	120	—	Derouault et al. (1977)
X = Br, D = Et₂O	C₆H₆	93	118	—	Nöth et al. (1982)
	Et₂O	96	170	—	O'Reilly (1960)
	Et₂O	91.7	130	—	Kidd and Truax (1969a)
	Et₂O	95	104	—	Haraguchi and Fujiwara (1969)
X = Br, D = Bu₂O	Bu₂O	94	200	—	Nöth et al. (1982)
X = I, D = Et₂O	Et₂O	40	81	—	Haraguchi and Fujiwara (1969)
	Et₂O	39	154	—	O'Reilly (1960)
	Et₂O	34	93	—	Kidd and Truax (1969a)
X = Cl, D = PMe₃	PhMe	108.2	—	$^1J_{AlP} = 263$	Vriezen and Jellinek (1967)
X = Cl, D = PMe₂Ph	CH₂Cl₂	—	—	$^1J_{AlP} = 280^c$	Laussac et al. (1975)
	CH₂Cl₂	—	—	$^1J_{AlP} = 290^c$	Laussac et al. (1975)
X = Br, D = PMe₃	PhMe	100.8	—	$^1J_{AlP} = 248$	Vriezen and Jellinek (1970)
X = Cl, D = PEt₃	PhMe	109.1	—	$^1J_{AlP} = 263$	Vriezen and Jellinek (1970)

178

X, D	Solvent	δ			Reference
X = Br, D = PEt₃	PhMe	97.8	—	$^1J_{AlP} = 239$	Vriezen and Jellinek (1970)
X = Cl, D = PPh₃	PhMe	104	—	$^1J_{AlP} = 190$	Vriezen and Jellinek (1970)
X = Br, D = PPh₃	PhMe	98	—	$^1J_{AlP} = 190$	Vriezen and Jellinek (1970)
X = Cl, D = PPhMe₂	CH₂Cl₂	—	—	$^1J_{AlP} = 290^c$	Laussac et al. (1975)
X = Cl, D = Ph₂P(CH₂)PPh₂	CH₂Cl₂	109.3	500	—	Laussac (1979)
X = Cl, D = Ph₂P(CH₂)₂Ph₂	CH₂Cl₂	109.2	600	—	Laussac (1979)
X = Cl, D = cis-Ph₂PCH=CHPPh₂	CH₂Cl₂	103.1	400	—	Laussac (1979)
X = Cl, D = trans-Ph₂PCH=CHPPh₂	CH₂Cl₂	103.3	500	—	Laussac (1979)
X = Cl, D = Ph₂P(CH₂)₂PPh₂	CH₂Cl₂	108.2	450	—	Laussac (1979)
	SO₂	95.1	—		Glavincevski and Brownstein (1981)
X = Cl, D = Me₂S	MeNO₂	98.8	29		Schippert (1976)
X = Cl, D = DMF	EtNCS	105	100		Haraguchi and Fujiwara (1969)
X = Cl, D = EtNCS	EtNCS	36	83		Haraguchi and Fujiwara (1969)
X = I, D = EtNCS	MeCN	87	270		Wehrli and Hoerd (1981)
X = Cl, D = MeCN	MeCN	96	45		Dalibart et al. (1982a)
	SO₂	95.5	—		Glavincevski and Brownstein (1981)
X = Cl, D = PhCN	PhCN	98	<100^d		Wehrli and Hoerd (1981)
X = Cl, D = MeNO₂	MeNO₂	96	700		Dalibart et al. (1982b)
	MeNO₂	95.2	208		Kidd and Truax (1969a)
X = Cl, D = MeCOCl	CH₂Cl₂	92.4	500		Wilinski and Kurland (1978)
	SO₂	105	—		Glavincevski and Brownstein (1982)
X = Cl, D = 2MeCOCl	SO₂	105^e	—		Glavincevski and Brownstein (1982)
	SO₂	95	—		Glavincevski and Brownstein (1982)
X = Cl, D = MeCOCl	MeCOCl	97^f	166		Haraguchi and Fujiwara (1969)
X = Br, D = MeNO₂	MeNO₂	82.9	186		Kidd and Truax (1969a)
X = I, D = MeNO₂	MeNO₂	34.0	93		Kidd and Truax (1969a)

179

TABLE IV (Continued)

Compound	Solvent	δ (ppm)	W (Hz)	J (Hz)	Reference
X = Cl, D = CoCl$_2$	C$_6$H$_6$	−225[g]	800	—	O'Reilly et al. (1964)
X = Br, D = CoCl$_2$	C$_6$H$_6$	−144[g]	700	—	O'Reilly et al. (1964)
C. Alkoxides					
[Al(O-i-Pr)$_3$]$_4$	C$_6$H$_6$	118[h]	~500[d]	—	Akitt and Duncan (1974)
Al(OR)$_3$ ... D (adducts)					
R = i-Pr, D = H$_2$N(CH$_2$)$_2$NH$_2$	C$_6$H$_6$	~110[d,i]	~100[d,i]	—	Akitt and Duncan (1974)
R = (CF$_3$)$_2$CH, D = Ph$_2$P(CH$_2$)$_2$PPh$_2$	CH$_2$Cl$_2$	60	180	—	Laussac (1979)
D = Ph$_2$P(CH$_2$)$_2$AsPh$_2$	CH$_2$Cl$_2$	59.8	110	—	Laussac (1979)
D. Adducts of aluminum borohydrides					
Al(BH$_4$)$_3$ ... D					
D = NH$_3$	C$_6$H$_6$	80.1[j]	—	$^2J_{AlH}$ = 44[k]	Boiko et al. (1976)
D = MeNH$_2$	C$_6$H$_6$	76.9[j]	—	$^2J_{AlH}$ = 44[k]	Boiko et al. (1976)
D = Me$_2$NH	C$_6$H$_6$	73.3[j]	—	$^2J_{AlH}$ = 44[k]	Boiko et al. (1976)
D = Me$_3$N	C$_6$H$_6$	71.2[j]	—	$^2J_{AlH}$ = 44[k]	Boiko et al. (1976)
D = Me$_3$N	C$_6$H$_6$	71.1	<20	$^2J_{AlH}$ = 45	Lauterbur et al. (1968)
D = Me$_3$P	C$_6$H$_6$	63.2	—	$^1J_{AlP}$ = 265	Oddy and Wallbridge (1976)
D = Et$_2$O	C$_6$H$_6$	83.1	>45	—	Boiko et al. (1974)
Al(BH$_4$)$_{3−x}$R$_x$ ... D					
x = 1, R = Cl, D = Et$_2$O	C$_6$H$_6$	90.4	>45	—	Boiko et al. (1974)
x = 2, R = Cl, D = Et$_2$O	C$_6$H$_6$	94.5	>45	—	Boiko et al. (1974)
x = 2, R = Et, D = Et$_2$O	C$_6$H$_6$	142	2,900	—	Oddy and Wallbridge (1976)
x = 1, R = Et, D = Et$_2$O	C$_6$H$_6$	102	1,700	—	Oddy and Wallbridge (1976)
x = 1, R = i-Bu, D = Et$_2$O	C$_6$H$_6$	102	2,300	—	Oddy and Wallbridge (1976)

180

$x = 1$, R = H, D = Et$_2$O	C$_6$D$_6$	84	256	—	Oddy and Wallbridge (1978)
$x = 1$, R = H, D = Me$_2$S	C$_7$D$_8$	70–86	—	—	Oddy and Wallbridge (1978)
E. Miscellaneous					
(AlH-i-Bu$_2$)$_n$	Pure	162	1,000	—	O'Reilly (1960)
Al$_2$Me$_3$Cl$_3$	Pure	93	Broad	—	O'Reilly (1960)
Al$_2$(OMe)$_2$Cl$_4$	Pure	90	Broad	—	O'Reilly (1960)
Al$_2$Cl$_7$	Melt	115	650	—	Gray and Maciel (1981)
					Anders and Plambeck (1978)
(AlMe$_2$Cl)$_2$	PhMe	—	—	$^1J_{AlC} = 105^a$	Yamamoto (1975a)
(AlMe$_2$Br)$_2$	PhMe	—	—	$^1J_{AlC} = 105^a$	Yamamoto (1975a)
(AlEt$_2$Cl)$_2$	PhMe	—	—	$^1J_{AlC} = 105^a$	Yamamoto (1975a)

[a] From ^{13}C NMR.
[b] At −60°C three signals are observed at 104.2, 98, and 95.8 ppm; they are assigned to AlCl$_4^-$, AlCl$_3$ ... Et$_2$O, and Cl$_2$Al(Et$_2$O)$_2^+$, respectively.
[c] From ^{31}P NMR.
[d] Estimated.
[e] Two complexes are observed: 1:1, MeClC=O : AlCl$_3$ (105 ppm) and 2:1, Me(AlCl$_3$; Cl)C=O ... AlCl$_3$ (Al : O, 105 ppm and Al : Cl, 95 ppm).
[f] Assigned to AlCl$_4^-$ ... $^+$OCMe (Haraguchi and Fujiwara, 1969) or to (MeCO)$_2$CHCO$^+$ ... $^-$AlCl$_4$ (Wilinski and Kurland, 1978).
[g] Contact shifts in the paramagnetic complexes CoX$_2$, 2AlX$_3$.
[h] Tetrahedral site only (cf. Table XI).
[i] Main line only, assigned to (RO)$_3$AlenAl(OR)$_3$.
[j] From ^{1}H{^{27}Al} NMR.
[k] From ^{1}H NMR.

TABLE V
Aluminum-27 NMR Parameters of Al(III) Tetracoordinate Anionic Complexes

Compound	Solvent	δ (ppm)	W (Hz)	J (Hz)	Reference
A. Tetrahydroaluminates					
LiAlH₄	Et₂O	103	121	$^1J_{AlH} = 110$	O'Reilly (1960)
	Et₂O	100	420	—	Haraguchi and Fujiwara (1969)
	Et₂O	101	—	—	Hermanek et al. (1975)
	Et₂O	103	140–700[a]	—	Nöth (1979)
	THF	98	460	$^1J_{AlH} = 155$	Huet et al. (1976)
	Et₂O	97.7	—	$^1J_{AlH} = 170$	Hermanek et al. (1975)
	Et₂O	97.7	—	$^1J_{AlH} = 172$	Tarasov et al. (1981)
	Et₂O	99.6	88[a,b]	$^1J_{AlH} = 170$	Hermanek et al. (1975)
	Glyme[c]	101.2	25	—	Nöth et al. (1980)
	Diglyme	98.8	—	$^1J_{AlH} = 172$	Tarasov et al. (1981)
	Diglyme	101.5	80	—	Nöth et al. (1980)
LiAlD₄	Et₂O	102.1	—	—	Hermanek et al. (1975)
	THF	97.8	—	—	Hermanek et al. (1975)
	Diglyme	98.8	—	$^1J_{AlD} = 26$	Tarasov et al. (1981)
NaAlH₄	THF	96.7	—	$^1J_{AlH} = 171$	Hermanek et al. (1975)
	THF	95.5	—	$^1J_{AlH} = 164$	Tarasov et al. (1981)
	THF	97	220	$^1J_{AlH} = 175$	Nöth et al. (1980)
	Glyme	101	195	$^1J_{AlH} = 170$	Nöth et al. (1980)
	Diglyme	103	205	$^1J_{AlH} = 171$	Nöth et al. (1980)
	Diglyme	98.5	—	$^1J_{AlH} = 172$	Tarasov et al. (1981)
NaAlD₄	THF	98	—	—	Hermanek et al. (1975)
	THF	95.7	—	—	Tarasov et al. (1981)
	Glyme	98.2	—	—	Hermanek et al. (1975)
	Diglyme	98.3	—	—	Tarasov et al. (1981)
CsAlH₄	Diglyme	101.6	—	$^1J_{AlH} = 174$	Tarasov et al. (1981)

Compound	Solvent	δ	J		Reference
Bu₄NAlH₄	Diglyme	99.4	—	$^1J_{AlH} = 174$	Tarasov et al. (1981)
	Diglyme	100	—	$^1J_{AlH} = 172$	Nöth et al. (1980)
Bu₄NAlD₄	C₆H₆	106	210	—	Nöth et al. (1980)
	Diglyme	99.4	—	$^1J_{AlH} = 26$	Tarasov et al. (1981)
B. Tetraalkyl and tetraalkoxy aluminates					
Li(t-BuO)₄Al	THF	51.0	257		Horne (1980)
Li(t-BuO)₃AlH	THF	75.8	800		Horne (1980)
Li(t-BuO)₂AlH₂	THF	~100	~800		Horne (1980)
LiAlMe₄	Glyme	—	—	$^2J_{AlH} = 6.3^d$	{ Oliver and Wilkie (1967) / Ross and Oliver (1970)
	Glyme	—	24	$^2J_{AlH} = 6.3^d$	Gore and Gutowsky (1969)
	Glyme	—	—	$^1J_{AlC} = 71.2^e$	Yamamoto (1975b)
	Et₂O	—	182	$^2J_{AlH} = 7.1^d$	Gore and Gutowsky (1969)
	THF	—	35	$^2J_{AlH} = 6.4^d$	Gore and Gutowsky (1969)
NaAlMe₄	Glyme	—	24	$^2J_{AlH} = 6.2^d$	Gore and Gutowsky (1969)
	Et₂O	—	102	$^2J_{AlH} = 9.1^d$	Gore and Gutowsky (1969)
	THF	—	63	$^2J_{AlH} = 6.4^d$	Gore and Gutowsky (1969)
NaAlEt₄	DMSO, glyme, or THF	—	—	$^2J_{AlH} = 7.3^d$	Westmoreland et al. (1972)
	DMSO, glyme, or THF	—	—	$^3J_{AlH} = 5.8^d$	Westmoreland et al. (1972)
	DMSO, glyme, or THF	—	—	$^1J_{AlC} = 73^e$	Westmoreland et al. (1972)
	DMSO, glyme, or THF	—	—	$^2J_{AlC} \sim 1^e$	Westmoreland et al. (1972)
NaAlBu₄	DMSO	—	—	$^1J_{AlC} = 71.6^e$	Westmoreland et al. (1973)
	DMSO	—	—	$^2J_{AlC} \sim 4^e$	Westmoreland et al. (1973)
C. Haloaluminates					
AlCl₄⁻	MeCN	102	33		Haraguchi and Fujiwara (1969)
	MeCN	103	4		Jones (1972)
	Et₂O	104.1	—		Nöth et al. (1982)
	THF	102	6		Nöth et al. (1982)
	CH₂Cl₂	102.4	15		{ Derouault et al. (1977) / Kidd and Truax (1968)
	MeNO₂	102.3^f	3		Schippert (1976)
	POCl₃	102	10–30		Kidd and Truax (1969b)

TABLE V (*Continued*)

Compound	Solvent	δ (ppm)	W (Hz)	J (Hz)	Reference
AlBr$_4^-$	MeCN	80	35	—	Haraguchi and Fujiwara (1969)
	MeCN	80	16	—	Dalibart et al. (1981)
	CH$_2$Cl$_2$	79.6	—	—	Kidd and Truax (1968)
AlI$_4^-$	MeCN	−28	58	—	Haraguchi and Fujiwara (1969)
	CH$_2$Cl$_2$	−27	24	—	Kidd and Truax (1968)
AlCl$_{4-n}$Br$_n^-$					
$n = 1$	CH$_2$Cl$_2$, MeCN	99	—	—	Jones (1972); Kidd and Truax (1968)
$n = 2$	CH$_2$Cl$_2$, MeCN	94	—	—	
$n = 3$	CH$_2$Cl$_2$, MeCN	87.2	—	—	
AlCl$_{4-n}$I$_n^-$					
$n = 1$	CH$_2$Cl$_2$	86.2	42	—	Kidd and Truax (1968)
$n = 2$	CH$_2$Cl$_2$	39.4	57	—	
$n = 3$	CH$_2$Cl$_2$	21.7	46	—	
AlBr$_{4-n}$I$_n^-$					
$n = 1$	CH$_2$Cl$_2$	60.6^g	—	—	Kidd and Truax (1968)
$n = 2$	CH$_2$Cl$_2$	37	—	—	Kidd and Truax (1968)
$n = 3$	CH$_2$Cl$_2$	8	—	—	Kidd and Truax (1968)
AlClBrI$_2^-$	CH$_2$Cl$_2$	47.7	—	—	Kidd and Truax (1968)
AlClBr$_2$I$^-$	CH$_2$Cl$_2$	69.3	—	—	Kidd and Truax (1968)
AlCl$_2$BrI$^-$	CH$_2$Cl$_2$	79.0	—	—	Kidd and Truax (1968)
AlCl$_{4-n}$(NCX)$_n^-$					
X = S or (O) $n = 1$	MeCN	88.7	—	$^1J_{AlN} = 40$	Tarasov et al. (1980)
	MeCN	(89.8)	—	$^1J_{AlN} = (40)$	Tarasov et al. (1980)
X = S or (O) $n = 2$	MeCN	73.3	—	$^1J_{AlN} = 43$	Tarasov et al. (1980)
	MeCN	(78.7)	—	—	Tarasov et al. (1980)
X = S or (O) $n = 3$	MeCN	56.0	—	—	Tarasov et al. (1980)
	MeCN	(66.7)	—	—	Tarasov et al. (1980)
AlBr$_3$(NCS)$^-$	MeCN	76.7	—	—	Tarasov et al. (1980)

184

D. The AlO_4 unit

	Medium	δ	W		Reference
$Al(OH)_4^-$	H_2O–KOH	80	155	—	O'Reilly (1960)
	H_2O–KOH	80	100	—	Haraguchi and Fujiwara (1969)
	H_2O–KOH	80	23	—	Moolenaar et al. (1970)
$AlO_4Al_{12}(OH)_{24}(H_2O)_{12}^{7+}$	H_2O	62.5[h]		—	Akitt et al. (1972b)
	H_2O	62.5[h]		—	Bottero et al. (1977)
	H_2O	62.5[h]		—	Fedotov et al. (1978)
	Solid	62.5	Broad	—	Müller et al. (1981a)
Tetrahedral unidentified aquo complexes	H_2O	71.2	Broad	—	Akitt and Mann (1981)
	H_2O	77	Broad	—	Akitt and Farthing (1981a)
	H_2O	75.3	Broad	—	Akitt and Farthing (1981b)
Tetrahedral sites in Heteropolyanion $AlW_{12}O_{40}^{5-}$	H_2O	71.2	8	—	Akitt and Farthing (1981c)
		71.7	2.5	—	
Aluminosilicates $(-SiO)_n\text{-}AlO_{4-n}$					
$n = 0$ (Q^0)	H_2O	79.5		—	Müller et al. (1981b)
$n = 1$ (Q^1)	H_2O	74.3		—	Müller et al. (1981b)
$n = 2$ (Q^2)	H_2O	69.5		—	Müller et al. (1981b)
$n = 3$ (Q^3)	H_2O	64.2		—	Müller et al. (1981b)
Aluminates	Solid	60–77	~1000	—	Müller et al. (1981c)
				—	Schiller and Müller (1982)
$\gamma, \eta, \chi\text{-}Al_2O_3$	Solid	62–66	~1000	—	Mastikhin et al. (1981)
Zeolites	Solid	54–65	400–1000	—	Fyfe et al. (1982)
				—	Ameriev et al. (1978)

[a] Values depending on the concentration (W is minimum for 0.06 M solutions) and the temperature.

[b] At 243 K.

[c] 1,1-Dimethoxyethane.

[d] From ^1H NMR at −60°C.

[e] From ^{13}C NMR.

[f] At 105.6 ppm from $Al(DMF)_6^{3+}$, itself taken at −3.3 ppm from $Al(H_2O)_6^{3+}$ (Gudlin and Schneider, 1974).

[g] Estimated from the graphs published in the mentioned reference.

[h] Aluminum-27 nucleus in the central AlO_4 tetrahedron (italicized).

TABLE VI

Aluminum-27 Parameters of Al(III) Tetracoordinate Cationic Complexes

Compound	Solvent	δ (ppm)	W (Hz)	J (Hz)	Reference
Al^{3+} solvates					
$Al(HMPA)_4^{3+}$	$MeNO_2$	34	5	$^2J_{AlP} = 30$	Delpuech et al. (1975)
$Al(HMPA)_3(DMSO)^{3+}$	$MeNO_2$	−3.3	32	—	Benter (1975)
$Al(HMPA)_2(DMSO)_2^{3+}$	$MeNO_2$	−9.6	88	—	Benter (1975)
$AlCl_2^+$ solvates					
$AlCl_2(DMF)_2^+$	$MeNO_2$	79.4	63	—	Schippert (1976)
$AlCl_2(Et_2O)_2^+$	Et_2O	95.8	—	—	Nöth et al. (1982)

TABLE VII

Aluminum-27 Parameters of Al(III) Pentaccordinate Derivatives

Compound	Solvent	δ (ppm)	W (Hz)	J (Hz)	Reference
$Cl_3Al \ldots 2PEt_3$	PhMe	55.0^a	—	—	Vriezen and Jellinek (1970)
	C_6H_6	55.9^a	—	—	Vriezen and Jellinek (1970)
$Br_3Al \ldots 2PEt_3$	PhMe	34.7^a	—	—	Vriezen and Jellinek (1970)
	C_6H_6	35.0^a	—	—	Vriezen and Jellinek (1970)
$Cl_3Al \ldots 2PMe_3^b$	CH_2Cl_2	57.4^a	—	—	Laussac (1979)
$Cl_3Al \ldots 2PEt_3$	CH_2Cl_2	59.3^a	—	—	Laussac (1979)
$Cl_3Al \ldots 2Me_2PPh$	CH_2Cl_2	58.5^a	—	—	Laussac (1979)
$Cl_3Al \ldots 2THF^c$	CH_2Cl_2–THF	63.0	240 to 500	—	Derouault et al. (1977)
$Br_3Al \ldots 2THF$	CH_2Cl_2–THF	49	~500	—	Derouault et al. (1977)
$Al(MeCN)_5^{3+e}$	MeCN	−12 to −10^d	1130 to 180^d	—	Dalibart et al. (1982a)

[a] Assuming the equilibrium $AlX_3 \ldots PR_3$ (I) + $PR_3 \rightleftharpoons AlX_3 \ldots 2PR_3$ (II) and fitting the chemical shift of the averaged NMR signal of (I) and (II) as a function of added phosphine.

[b] The crystal $AlCl_3$, $2NMe_3$ shows trigonal-bipyramidal coordination, and powder X-ray diffraction patterns of $AlCl_3$, $2PMe_3$ show this compound to be isomorphous with $AlCl_3$, $2NMe_3$ (Beattie et al., 1969).

[c] The crystal structure of $AlCl_3$, $2THF$ shows a geometry around aluminum very close to that of an idealized trigonal bipyramid (Cowley et al., 1981).

[d] The NMR parameters vary as the concentration of $AlCl_3$ is increased.

[e] Nuclear magnetic resonance studies at higher frequencies (93.8 MHz) and lower concentrations show the splitting of the broad, unresolved band mentioned in the reference into a series of sharper lines which were assigned to octahedral mixed species $AlCl_n(MeCN)_{6-n}$ (Wehrli and Wehrli, 1981).

TABLE VIII

Aluminum-27 NMR Parameters of Octahedrally Solvated Al^{3+} Ions

Ion	Solvent	δ (ppm)	W (Hz)	J (Hz)	Reference
In a pure solvent					
$Al(H_2O)_6^{3+}$	H_2O	0 (ref.)	2.5–40 (pH < 1.5)	—	Epperlein and Lutz (1968) Hertz *et al.* (1971) Takahashi (1977, 1978, 1980)
$Al(DMF)_6^{3+}$	DMF	−1.6	39	—	Movius and Matwiyoff (1967)
$Al(POCl_3)_6^{3+}$	$POCl_3$	−21.2	55–77	—	Kidd and Truax (1969b)
$Al(EtOH)_6^{3+}$	EtOH[a]	~0[b]	70	—	Haraguchi and Fujiwara
$Al(PrOH)_6^{3+}$	PrOH[a]	~0[b]	70	—	(1969) Haraguchi and Fujiwara (1969)
$Al(MeCN)_6^{3+}$	MeCN	−33	—	—	Akitt and Duncan (1979)
	MeCN	−32	20	—	Wehrli and Hoerd (1981)
	MeCN	−34	15	—	Dalibart *et al.* (1982a)
	MeCN	−34	25	—	Dalibart *et al.* (1981)
	MeCN	−32.6	15	—	Wehrli and Wehrli (1981)
$Al(PhCN)_6^{3+}$	PhCN	−29	20	—	Wehrli and Hoerd (1981)
$Al(EtNCS)_6^{3+}$	EtNCS	20	60	—	Haraguchi and Fujiwara (1969)
$Al(MeNO_2)_6^{3+}$	$MeNO_2$	−13	20	—	Dalibart *et al.* (1982b)
In an inert cosolvent					
$Al(TMPA)_6^{3+}$ [b]	$MeNO_2$	−20.5	3	$^2J_{AIP} = 19.5$	Delpuech *et al.* (1974)
$Al(TEPA)_6^{3+}$ [b]	$MeNO_2$	−21.9	3	$^2J_{AIP} = 19$	Delpuech *et al.* (1975)
$Al(DMMP)_6^{3+}$ [b]	$MeNO_2$	−20.2	—	$^2J_{AIP} = 15$	Delpuech *et al.* (1975)
$Al(DEEP)_6^{3+}$ [b]	$MeNO_2$	−20.2	—	$^2J_{AIP} = 12$	Delpuech *et al.* (1975)
$Al(DMHP)_6^{3+}$ [b]	$MeNO_2$	−17.7	—	$^2J_{AIP} = 13.4$	Delpuech *et al.* (1975)
$Al(NCS)_6^{3+}$	$MeNO_2$	−33.7	40	—	Tarasov *et al.* (1980)
$Al(DMSO)_6^{3+}$	$MeNO_2$	1.69	5.5	—	Benter (1975)
$Al(DMF)_6^{3+}$	$MeNO_2$	−3.6[c]	5.6	—	Schippert (1976)

[a] A mean solvation number of 4 has been found by ^1H NMR in alcoholic solutions of $AlCl_3$ (Grasdalen, 1971, 1972), showing the existence of polynuclear or uncompletely dissociated species in which hexacoordination is maintained (Buslaev *et al.*, 1977; Martin and Stockton, 1973).

[b] TMPA, Trimethylphosphate; TEPA, triethylphosphate; DMMP, dimethylmethylphosphonate; DEEP, diethylethylphosphonate; DMHP, dimethylhydrogen phosphite.

[c] Estimated from the graphs published by Gudlin and Schneider (1974) and taking δ for $Al(DMSO)_6^{3+}$ to be 1.69 ppm (Benter, 1975).

TABLE IX
Aluminum-27 NMR Chemical Shifts (ppm) of Mixed Solvation Complexes $AlX_n Y_{6-n}^{3+}$

X =	TMPA[a,b]	TEPA[c]	DMMP[a]	DEEP[c]	DMHP[a]	MeCN[d,e]	DMF[f,g]	TMPA[f,h]	MeCONH$_2$[i,j]
Y =	H$_2$O	H$_2$O	H$_2$O	H$_2$O	H$_2$O	H$_2$O	DMSO	DMSO	H$_2$O
$n = 0$	0[i]	0[i]	0[i]	0	0[k]	0	1.7	1.7	0
$n = 1$	−3.7[i]	−3.6[i]	−3.5[i]	−3.75	−3.3[k]	−4.3	0.6[i]	−2.1	−3.0
$n = 2$	−6.7[i]	−7.0[i]	−6.8[i]	−7.5	−6.6[k]	−9.1	−0.45[i]	−5.9	−5.3
$n = 3$	−10.0[i]	−11.0[i]	−10.1[i]	−11.0	−9.1[k]	−14.2 (cis), −15.6 (trans)	−1.30[i]	−9.5	—
$n = 4$	−14.0[k]	−14.7	−14.8[k]	−14.8	−14.0[k]	−22.3 (cis), −19.9 (trans)	−2.1[i]	−13.9	—
$n = 5$	−17.5[k]	−17.5	−17.5[k]	−17.5	−15.9[k]	−26.1	−2.8	−17.5	—
$n = 6$	−20.5[k]	−21.9	−20.2[k]	−20.2	−17.7[k]	−32.7	−3.3[i]	−20.5	—

[a] Delpuech et al., 1974.
[b] Canet et al., 1973.
[c] Delpuech et al., 1975.
[d] In slightly aqueous MeCN.
[e] Wehrli and Wehrli, 1981.
[f] In MeNO$_2$ as an inert cosolvent.
[g] Gudlin and Schneider (1974).
[h] Benter, 1975.
[i] In aqueous solutions.
[j] Rubini, 1978.
[k] In slightly aqueous MeNO$_2$.
[l] Estimated from the graphs published by Gudlin and Schneider, 1974.

TABLE X

Aluminum-27 NMR Parameters of Mixed Solvent
(L)–counterion (X) Complexes $AlX_nL_{6-n}^{3-n}$

Complex	δ (ppm)	W (Hz)
X = Cl, L = MeCN		
$n = 0$	-32.6^a, -34^k, -33^l	$15^{a,k}$
$n = 1$	-21.6^a, -24^k, -23^l	160^a
$n = 2$, cis	-14.3^a, $-14^{k,l}$	—
$n = 2$, trans	-12.3^b	—
$n = 3$, fac	-7.5^a	30^a
$n = 3$, mer	-6.2^a	$\sim170^a$
$n = 4$, cis	-1.2^a	—
$n = 4$, trans	-1.2^a	—
$n = 5$	2.8^a	—
$n = 6$	7.4^a	—
X = Br, L = MeCN		
$n = 1$	-30.1^a–31^m	80^m
$n = 2$	-32^a	—
X = Cl, L = THF, $n = 2$	9^b	250^b
X = Cl, L = PhCN, $n = 2$	-18^c	1900^c
X = HSO_4, L = H_2O, $n = 2$	-6.8^d	80^d
X = SO_4, L = H_2O, $n = 1$	-3.3^e	5^e
X = F, L = H_2O, $n = 1$ to 3	2 to $15^{f,g,h,i}$	$550^{f,g,h}$
H_3PO_4–Al complexes	~7 (unresolved)^h	Broad^h
X = Cl, L = EtOH or MeOH, $n = ?$	-5 to 10^j	Broad^j

[a] Wehrli and Wehrli, 1981.
[b] Derouault et al., 1977.
[c] Wehrli and Hoerd, 1981.
[d] Epperlein and Lutz, 1968.
[e] Akitt et al., 1972a.
[f] O'Reilly, 1960.
[g] Matwiyoff and Wageman, 1970.
[h] Akitt et al., 1971.
[i] Buslaev and Petrosyants, 1979.
[j] Buslaev et al., 1977, 1978.
[k] Dalibart et al., 1982a.
[l] Akitt and Duncan, 1979.
[m] Dalibart et al., 1981.

TABLE XI

Aluminum-27 NMR Parameters for Octahedral Sites in Complex Structures

Compound	Solvent	δ (ppm)	W (Hz)	J (Hz)	Reference
A. Polynuclear aquohydroxo complexes					
$(H_2O)_4Al(\mu\text{-}OH)_2Al(H_2O)_4^{4+}$	H_2O	4.2	450	—	Akitt et al. (1972b)
$AlO_4Al_{12}(OH)_{24}(H_2O)_{12}^{7+}$	H_2O	0^a	$\sim 200^b$	—	Akitt and Mann (1981)
					Akitt and Mann (1981)
B. Al(III) oxides and aluminates					
$\alpha\text{-}Al_2O_3$	Solid	5	1200	—	Müller et al. (1981c)
	Solid	8	—	—	Mastikhin et al. (1981)
$\beta\text{-}Al_2O_3$	Solid	9	—	—	Müller et al. (1981c)
$\gamma\text{-}Al_2O_3$	Solid	4–9	1800	—	Mastikhin et al. (1981)
$\eta\text{-}Al_2O_3$	Solid	1–2	3000	—	Mastikhin et al. (1981)
$\chi\text{-}Al_2O_3$	Solid	3	2000	—	Mastikhin et al. (1981)
$Mg(Al_2O_2)_2$	Solid	22	—	—	Müller et al. (1981c)
$BaO \cdot 6Al_2O_3$	Solid	5	—	—	Müller et al. (1981c)
$2CaO \cdot Al_2O_3 \cdot 8H_2O$	Solid	9	—	—	Müller et al. (1981c)
Gibbsite $Al(OH)_3$	Solid	0	—	—	Müller et al. (1977)
$(NH_4)_3AlF_6$	Solid	10	—	—	Müller et al. (1977)
Na_3AlF_6	Solid	0	—	—	Grimmer et al. (1982)
K_3AlF_6	Solid	0	—	—	Grimmer et al. (1982)
C. Alkoxides					
$[Al(OiPr)_3]_4$	PhMe	0^a	50^b	—	Akitt and Duncan (1974)

	Solvent	δ	Linewidth	Coupling	Reference
D. Heteropolyanions					
$AlMo_6O_{21}^{3-}$	H_2O	15.4	50	—	Akitt and Farthing (1981c)
E. Complexes with multidentate ligands					
$Al(C_2O_4)_3^{3-}$	H_2O	16.1	140[b]	—	Jaber et al. (1977)
$Al(C_2O_4)_2(H_2O)_2^{-}$	H_2O	13.1	—	—	Jaber et al. (1977)
$Al(C_2O_4)(H_2O)_4^{+}$	H_2O	6.4	450	—	Jaber et al. (1977)
$Al(OMPA)_3^{3+c}$	$MeNO_2$	−13.2	95	—	Rubini et al. (1982)
$Al(NIPA)_3^{3+c}$	H_2O	−13.7	150	$^2J_{AlP} = 19$	Rubini et al. (1979)
$Al(OHIPA)_3^{3+c}$	H_2O	−14.3	240	—	Rubini et al. (1982)
$Al(ODIPA)_3^{3+c}$	H_2O	−13.7	405	—	Rubini et al. (1982)
Hydrocarboxylic acid chelates[d]	H_2O, pH = 2.5	−5–6	—	—	Toy et al. (1973)
Biological phospho ligands[e]	H_2O, pH = 1.5–2	3–7	—	—	Karlik et al. (1982)
$Al(acac)_3^{f}$	C_6H_6	0	93	—	Haraguchi and Fujiwara (1969)
$Al(acac)_3$-$Ln(fod)^g$ (1:1 complex)	C_6H_6	2[h] to −81.7	160–230	—	Hirayama and Kitami (1980)
$(MeCONHCO^-)_3Al$	H_2O	36.5	370	—	Llinás and De Marco (1980)
Alumichromes	H_2O	~41.5	1400–6600	—	Llinás and De Marco (1980)

[a] For octahedral sites only.
[b] Estimated value.
[c] Biphosphorylated neutral ligands of general formula $(Me_2N)_2P(O)XP(O)(NMe_2)_2$, with X = O(OMPA), NMe(NIPA), NC_6H_{12}(OHIPA), $NC_{12}H_{25}$(ODIPA).
[d] Tartaric, mandelic, malic, citric, etc., acids (no structural identification).
[e] Orthophosphate, diphosphate, triphosphate, AMP, ADP, ATP, NADP, etc. (no structural identification).
[f] acac, Acetylacetonato.
[g] fod (1,1,1,2,2,3,3-heptafluoro-7,7-dimethyloctane-4,6-dionato; Ln, La, Pr, Nd, Sm, and Eu (δ = 2, −70.2, −44.5, −2, and 81.7 ppm, respectively).
[h] Contact and pseudocontact shifts.

Acknowledgments

The author gratefully acknowledges the assistance of his co-workers, Drs. P. R. Rubini and L. Rodehüser, in collecting the ^{27}Al NMR data.

References

Abragam, A. (1961). "The Principles of Nuclear Magnetism." Clarendon, Oxford.
Akitt, J. W. (1972). *Ann. Reports NMR Spectry* **5A,** 465.
Akitt, J. W., and Duncan, R. H. (1974). *J. Magn. Reson.* 15, 162.
Akitt, J. W., and Duncan, R. H. (1979). *J. Magn. Reson.* **34,** 435.
Akitt, J. W., and Farthing, A. (1978). *J. Magn. Reson.* **32,** 345.
Akitt, J. W., and Farthing, A. (1981a). *J. Chem. Soc. Dalton Trans.,* 1606.
Akitt, J. W., and Farthing, A. (1981b). *J. Chem. Soc. Dalton Trans.,* 1624.
Akitt, J. W., and Farthing, A. (1981c). *J. Chem. Soc. Dalton Trans.,* 1615.
Akitt, J. W., and Mann, B. E. (1981). *J. Magn. Reson.* **44,** 584
Akitt, J. W., Greenwood, N. N., and Lester, G. D. (1971). *J. Chem. Soc. A,* 2450.
Akitt, J. W., Greenwood, N. N., and Khandelwal, B. (1972a). *J. Chem. Soc. Dalton Trans.,* 1226.
Akitt, J. W., Greenwood, N. N., Khandelwal, B., and Lester, G. D. (1972b). *J. Chem. Soc. Dalton,* 604.
Ameriev, D. M., Mastikhin, V. M., Ione, K. G., and Lapina, O. B. (1978). *Dokl. Akad. Nauk. SSSR* **243,** 945.
Anders, U., and Plambeck, J. A. (1978). *J. Inorg. Nucl. Chem.* **40,** 387.
Banister, A. J., and Wade, K. (1975). "The Chemistry of Aluminum, Gallium, Indium, and Thallium." Pergamon, Oxford.
Beattie, I. R., Ozin, G. A., and Glayden, H. E. (1969). *J. Chem. Soc. A,* 2535.
Benter, G. (1975). Ph.D. Thesis, Göttingen.
Boiko, G. N., Fokin, V. N., Malov, Y. I., and Semenenko, K. N. (1974). *Zh. Neorgan. Khim.* **19,** 3006.
Boiko, G. N, Malov, Y. U., Semenenko, K. N., and Shilkin, S. P. (1976). *Koord. Khim.* **2,** 686.
Bottero, J. Y., Cases, J. M., Rubini, P. R., and Fiessinger, F. (1977). *C. R. Acad. Sci.* **284D,** 1033.
Brevard, C., and Granger, P. (1981). "Handbook of High Resolution Multinuclear NMR", pp. 98–99. Wiley, New York.
Buslaev, Y. A., and Petrosyants, S. P. (1979). *Koord. Khim.* **5,** 163.
Buslaev, Y. A., Tarasov, V. P., Petrosyants, S. P., and Kirakosyan, G. A. (1977). *Koord. Khim.* **3,** 1316.
Buslaev, Y. A., Tarasov, V. P., Petrosyants, S. P., and Kirakosyan, G. A. (1978). *Dokl. Akad. Nauk SSSR* **241,** 838.
Canet, D., Delpuech, J.-J., Khaddar, M. R., and Rubini, P. R. (1973). *J. Magn. Reson.* **9,** 329.
Carter, G. C., Bennett, L. H., and Kahan, D. J. (1977). "Metallic Shifts in NMR", (B. Chalmers, J. W. Christian, and T. B. Massalski, eds.), Progress in Materials Science, Vol. 20. Pergamon, Oxford.

Cowley, A. H., Cushner, M. C., Davis, R. E., and Riley, P. E. (1981). *Inorg. Chem.* **20,** 1179.

Dalibart, M., Derouault, J., and Granger, P. (1981). *Inorg. Chem.* **20,** 3975.

Dalibart, M., Derouault, J., Granger, P., and Chapelle, S. (1982a). *Inorg. Chem.* **21,** 1040.

Dalibart, M., Derouault, J., and Granger, P. (1982b). *Inorg. Chem.* **21,** 2241.

Delpuech, J.-J., Khaddar, M. R., Peguy, A., and Rubini, P. R. (1974). *J. Chem. Soc. Chem. Commun.* 154.

Delpuech, J.-J., Khaddar, M. R., Peguy, A., and Rubini, P. R. (1975). *J. Am. Chem. Soc.* **97,** 3373.

Derouault, J., Granger, P., and Forel, M. T. (1977). *Inorg. Chem.* **16,** 3214.

Dewar, M. J. S., Patterson, D. B., and Simpson, W. I. (1973). *J. Chem. Soc. Dalton Trans.,* 2381.

Di Carlo, E. N., and Swift, H. E. (1964). *J. Phys. Chem.* **68,** 551.

Epperlein, B. W., and Lutz, O. (1968). *Z. Naturforsch.* **23A,** 1413.

Fedotov, M. A., Krivoruchko, O. P., and Buyanov, R. A. (1978). *Zh. Neorgan. Khim.* **23, 2326.**

Fyfe, C. A., Gobbi, G. C., Hartman, J. S., Klinowski, J., and Thomas, J. M. (1982). *J. Phys. Chem.* **86,** 1247.

Glavincevski, B., and Brownstein, S. K. (1981). *Can. J. Chem.* **59,** 3012.

Glavincevski, B., and Brownstein, S. K. (1982). *J. Org. Chem.* **47,** 1005.

Gore, E. S., and Gutowsky, H. S. (1969). *J. Phys. Chem.* **73,** 2515.

Grasdalen, H. (1971). *J. Magn. Reson.* **5,** 86.

Grasdalen, H. (1972). *J. Magn. Reson.* **8,** 336.

Gray, J. L., and Maciel, G. E. (1981). *J. Am. Chem. Soc.* **103,** 7147.

Grimmer, A. R., Müller, D., Bentrup, U., and Kolditz, L. (1982). *Z. Chem.* **22,** 43.

Gudlin, D., and Schneider, H. (1974). *J. Magn. Reson.* **16,** 362.

Haraguchi, H., and Fujiwara, S. (1969). *J. Phys. Chem.* **73,** 3467.

Hermanek, S., Kriz, O., Plesek, J., and Hanslik, T. (1975). *Chem. Ind. London* **72,** 42.

Hertz, H. G., Tusch, R., and Versmold, H. (1971). *Ber. Bunsenges. Phys. Chem.* **75,** 1177.

Hinton, J. F., and Briggs, R. W. (1978). "NMR and the Periodic Table" (R. K. Harris and B. E. Mann, eds.), pp. 279–308. Academic Press, New York.

Hiramaya, M., and Kitami, K. (1980). *J. Chem. Soc. Chem. Commun.* 1030.

Holtz, M., Friedman, H. L., and Tembe, B. L. (1982). *J. Magn. Reson.* **47,** 454.

Horne, D. A. (1980). *J. Am. Chem. Soc.* **102,** 6011.

Huet, J., Durand, J., and Infarnet, Y. (1976). *Org. Magn. Reson.* **8,** 382.

Jaber, M., Bertin, F., and Thomas-David, G. (1977). *Can. J. Chem.* **55,** 3689.

Jones, D. E. H. (1972). *J. Chem. Soc. Dalton Trans.,* 567.

Karlik, S. J., Elgavish, G. A., Pillai, R. P., and Eichhorn, G. L. (1982). *J. Magn. Reson.* **49,** 164.

Kidd, R. G., and Truax, D. R. (1968). *J. Am. Chem. Soc.* **90,** 6867.

Kidd, R. G., and Truax, D. R. (1969a). *Can. Spectrosc.* **14,** 1.

Kidd, R. G., and Truax, D. R. (1969b). *J. Chem. Soc. Chem. Commun.,* 160.

Laussac, J. P. (1979). *Org. Magn. Reson.* **12,** 237.

Laussac, J. P., Laurent, J. P., and Commenges, G. (1975). *Org. Magn. Reson.* **7,** 72.

Lauterbur, P. C., Hopkins, R. C., King, R. W., Ziebarth, O. V., and Heitsch, C. W. (1968). *Inorg. Chem.* **7,** 1025.

Llinás, M., and De Marco, A. (1980). *J. Am. Chem. Soc.* **102,** 2226.

Malinowski, E. R. (1969). *J. Am. Chem. Soc.* **93,** 4418.

Martin, J. S., and Stockton, G. W. (1973). *J. Magn. Reson.* **12,** 218.

Martin, M. L., Delpuech, J.-J., Martin, G. J. (1980). "Practical NMR Spectroscopy", Ch. 8. Heyden-Wiley, London.

Mastikhin, V. M., Krivoruchko, O. P., Zolotovskii, B. P., and Buyanov, R. A. (1981). *React. Kinet. Catal. Lett.* **18**, 117.

Matwiyoff, N. A., and Wageman, W. E. (1970). *Inorg. Chem.* **9**, 1031.

Moolenaar, R. J., Evans, J. C., and McKeever, L. D. (1970). *J. Phys. Chem.* **74**, 3629.

Movius, W. G., and Matwiyoff, N. A. (1967). *Inorg. Chem.* **6**, 847.

Müller, D., Gessner, W., and Grimmer, A. R. (1977). *Z. Chem.* **17**, 453.

Müller, D., Gessner, W., Schönherr, S., and Görz, H. (1981a). *Z. Anorg. Allg. Chem.* **483**, 153.

Müller, D., Hoebbel, D., and Gessner, W. (1981b). *Chem. Phys. Lett.* **84**, 25.

Müller, D., Gessner, W., Behrens, H. J., and Scheler, G. (1981c). *Chem. Phys. Lett.* **79**, 59.

Nöth, H. (1979). *Z. Naturforsch.* **35B**, 119.

Nöth, H., Rurländer, R., and Wolfgardt, P. (1980). *Z. Naturforsch.* **36B**, 31.

Nöth, H., Rurländer, R., and Wolfgardt, P. (1982). *Z. Naturforsch.* **37B**, 29.

Oddy, P. R., and Wallbridge, M. G. H. (1976). *J. Chem. Soc. Dalton Trans.,* 2076.

Oddy, P. R., and Wallbridge, M. G. H. (1978). *J. Chem. Soc. Dalton Trans.,* 572.

Ogg, R. A., and Ray, J. D. (1955). *Discuss. Faraday Soc.* **19**, 239.

Oliver, J. P., and Wilkie, C. A. (1967). *J. Am. Chem. Soc.* **89**, 163.

O'Reilly, D. E. (1960). *J. Chem. Phys.* **32**, 1007.

O'Reilly, D. E., Poole, C. P., Belt, R. F., and Scott, H. (1964). *J. Polym. Sci.* **A2**, 3257.

Petrakis, L., and Dickson, F. E. (1972). *Appl. Spect. Rev.* **4**, 1.

Rodehüser, L., Rubini, P R., and Delpuech, J.-J. (1977). *Inorg. Chem.* **16**, 2837.

Ross, J. F., and Oliver, J. P. (1970). *J. Organ. Chem.* **22**, 503.

Rubini, P. R. (1978). Ph.D. Thesis, Nancy.

Rubini, P. R., Rodehüser, L., and Delpuech, J.-J. (1979). *Inorg. Chem.* **18**, 2962.

Rubini, P. R., Rodehüser, L., Doucet-Ladevèze, G., and Delpuech, J.-J. (1982). Unpublished results.

Schiller, W., and Müller, D. (1982). *Z. Chem.* **22**, 44.

Schippert, E. (1976). *Adv. Mol. Relaxation Processes* **9**, 167.

Smith, G. W. (1963). *J. Chem. Phys.* **39**, 2031.

Swift, H. E., Poole, C. P., and Itzel, J. F. (1964). *J. Phys. Chem.* **68**, 2509.

Takahashi, A. (1977). *J. Phys. Soc. Japan* **43**, 968, 976.

Takahashi, A. (1978). *J. Phys. Soc. Japan* **44**, 1946.

Takahashi, A. (1980). *J. Phys. Soc. Japan* **48**, 1049.

Tarasov, V. P., and Bakum, S. I. (1975). *J. Magn. Reson.* **18**, 64.

Tarasov, V. P., Privalov, V. I., and Buslaev, Y. U. (1978). *Zh. Strukt. Khim.* **19**, 1012.

Tarasov, V. P., Petrosyants, S. P., Kirakosyan, G. A., and Buslaev, Y. A. (1980). *Koord. Khim.* **6**, 52.

Tarasov, V. P., Bakum, S. I., Privalov, V. I., and Buslaev, Y. A. (1981). *Koord. Khim.* **7**, 674.

Toy, A. D., Smith, T. D., and Pilbrow, J. R. (1973). *Aust. J. Chem.* **26**, 1889.

Valiev, K. A., and Zaripov, M. M. (1966). *Zh. Strukt. Khim.* **7**, 494.

Vashman, A. A., Pronin, I. S., Brylkina, T. V., and Samsonov, V. E. (1979). *Russ. J. Inorg. Chem.* **24**, 1001.

Vestin, U., Kowalewski, J., and Henriksson, U. (1981). *Org. Magn. Reson.* **16**, 119.

Vladimiroff, T., and Malinowski, E. R. (1967). *J. Chem. Phys.* **46**, 1830.

Vriezen, W. H. N., and Jellinek, F. (1967). *Chem. Phys. Lett.* **1**, 284.

Vriezen, W. H. N., and Jellinek, F. (1970). *Rec. Trav. Chim.* **89**, 1306.

Wehrli, F. W., and Hoerd, R. (1981). *J. Magn. Reson.* **42**, 334.

Wehrli, F. W., and Wehrli, S. (1981). *J. Magn. Reson.* **44**, 197.

Weigert, F. J., Winokur, M., and Roberts, J. D. (1968). *J. Am. Chem. Soc.* **90**, 1566.

Westmoreland, T. D., Bhacca, N. S., Wander, J. D., and Day, M. C. (1972). *J. Organ. Chem.* **38,** 1.

Westmoreland, T. D., Bhacca, N. S., Wander, J. D., and Day, M. C. (1973). *J. Am. Chem. Soc.* **95,** 2019.

Wilinski, J., and Kurland, R. J. (1978). *J. Am. Chem. Soc.* **100,** 2233.

Yamamoto, O. (1975a). *J. Chem. Phys.* **63,** 2988.

Yamamoto, O. (1975b). *Chem. Lett.* **6,** 511.

Zaripov, M. M., and Nikiforov, E. A. (1979). *Russ. J. Phys. Chem.* **53,** 39.

7 Applications of Silicon-29 NMR Spectroscopy

Brian Coleman

Koninklijke/Shell-Laboratorium, Amsterdam
(Shell Research B.V.)

I. Introduction

The basic principles of and general problems associated with ^{29}Si NMR have been extensively discussed (Harris *et al.*, 1978; Marsmann, 1981a; Williams and Cargioli, 1979), and there will be no repetition in detail here. The low natural abundance of ^{29}Si (4.7%), the low negative magnetogyric ratio, and the fractional nuclear Overhauser enhancement (NOE) effects on proton irradiation combine to make the accumulation of many transients almost a necessity. Because ^{29}Si relaxation times are comparatively long, large delays between pulses are also required. There are, however, ways partially to avoid the low sensitivity (see Vol. I, Chapter 2, for details). The INEPT pulse sequence (Morris and Freeman, 1979) has been applied in the observation of ^{29}Si (Doddrell *et al.*, 1981; Kowalewski and Morris, 1982).

TABLE I
Chemical Shift Increments
for Permethylpolysilanes[a]

Position 1	Increment A_1 (ppm)
α	-28.5 ± 0.10
β	3.9 ± 0.05
γ	1.2 ± 0.09
δ	0.2 ± 0.01

[a] Stanislawski and West, 1981a.

The enhancement factors can be high: 9.2 for tetramethylsilane (TMS) (85% of theoretical) and 8.5 for trimethylsilyl chloride (90% of theoretical).

With the use of the INEPT method the T_1 of 1,1,3,3-tetramethyldisilazane was measured in one-eighth of the time required for the conventional method (Kowalewski and Morris, 1982). It is quite likely that the INEPT sequence will also become popular for ^{29}Si relaxation time measurements.

II. Solution Studies

A. Polysilanes

A general equation for the calculation of chemical shifts has been derived from linear regression analysis of the ^{29}Si chemical shifts of 11 linear permethylpolysilanes (Stanislawski and West, 1981a).

The chemical shift of the kth silicon is given (in parts per million) by

$$\delta_{Si(k)} = 8.5 + \sum A_1 n_{k1}$$

where the increments A_1 for n neighbors α to δ from the kth nucleus are listed in Table I.

The equation gives a good approximation to the experimental values of the chemical shifts for linear permethylpolysilanes (within ~ 0.2 ppm), except for the tetrasiline in which the end silicon is calculated to be at -14.9 versus -16.1 ppm experimentally.

This equation does not apply to cyclic and branched permethylpolysilanes, presumably because of steric interaction between methyl groups.

TABLE II

Chemical Shift Ranges in Permethylpolysilanes[a]

Type of silicon	Chemical shift range (ppm)
Primary bonded to secondary	−15.0 to −16.8
Primary bonded to tertiary	−10.1 to −12.5
Primary bonded to quaternary	−5.1 to −9.8
Secondary	−25 to −49
Tertiary	−68 to −89
Quaternary	−126 to −136

[a] Ishikawa et al., 1981.

This had led other authors to use the characteristic shift ranges given in Table II for the types of silicon in permethylpolysilanes (Ishikawa et al., 1981). These ranges hold for 21 widely varying polysilanes. There is one exception: The quaternary silicon in $(Me_3Si)_3SiSiMe_2Si(SiMe_3)_3$ resonates at −118.2 ppm.

Stanislawski and West (1981b) have also derived additivity rules for halogen-substituted permethylpolysilanes. Because the effect of halogen substitution is to produce a systematic trend in chemical shift, there appears to be no need to invoke increased back-donation of electrons from chlorine to silicon with increasing chain length, as was claimed by Boberski and Allred (1974).

Newman et al. (1980) have reported ^{29}Si-NMR data for phosphapermethylcyclopolysilanes (Table III). Surprising are the changes in $^2J_{SiP}$ on

TABLE III

Nuclear Magnetic Resonance Data for Phosphapermethylcyclopolysilanes[a]

Compound	SiMe₂ group	δ_{Si} (ppm)	J_{SiP} (Hz)
MeP(SiMe₂)₄	α	−8.24	35.4
	β	−41.67	1.0
MeP(SiMe₂)₅	α	−16.67	43.0
	β	−43.37	14.5
	γ	−42.87	2.5
MeP(SiMe₂)₆	α	−15.97	50.0
	β	−42.80	19.1
	γ	−42.16	4.5

[a] Newman et al., 1980.

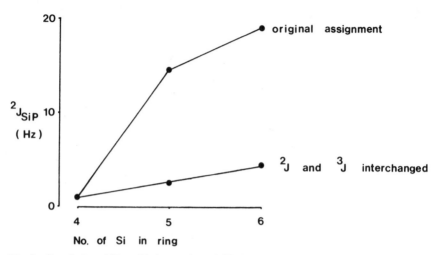

Fig. 1 Correlation of $^2J_{SiP}$ with the number of silicon atoms in the ring of phosphacyclopoly-silanes.

going from a five-membered ring (^2J = 1.0 Hz) to a six-membered ring (^2J = 14.5 Hz) and the fact that, in six- and seven-membered rings $^2J_{SiP} >$ $^3J_{SiP}$; which is not normally the case for carbon–phosphorus coupling (Stothers, 1972). In fact, if the assignments for ^2J and ^3J are interchanged, the correlation between $^2J_{SiP}$ and ring size becomes much better (Fig. 1).

The effect of a wider range of substituents has been examined in di- and trisilanes. Phenylmethyldi- and phenylmethyltrisilanes have been studied with particular attention to $^1J_{SiSi}$ (Williams et al., 1980). For disilanes $\phi_3Si^ASi^BMe_2R$, $^1J_{SiSi}$ was found to be primarily dependent on the inductive effect of the alkyl group as measured by the Taft σ^* constant (Table IV).

It is possible to extrapolate to compounds where more than one methyl group has been replaced by alkyl substituents, the $^1J_{SiSi}$ then being correlated with with $\Sigma\sigma^*$.

The decrease in $^1J_{SiSi}$ with increasing electron releasing ability of the alkyl group(s) suggests that the effective nuclear charge at the silicon atom is the main factor in determining $^1J_{SiSi}$ for the disilanes studied. A similar conclusion has been drawn for a different series of disilanes (Engelhardt et al., 1981b). For 1,2-disubstituted tetramethyldisilanes there appears to be no need to invoke the involvement of silicon d orbitals to account for the observed substituent effects (Table V). For penta-methyldisilanes ($Me_3Si^ASi^BMe_2X$) there is a clear relationship between δ_{Si} and the electronegativity of the substituent X (δ_{SiB}, r = 0.999; δ_{SiA}, r = 0.941). In mono- and 1,2-dialkoxytetramethyldisilanes there is an obvious shift to higher field as the steric crowding increases.

TABLE IV

Nuclear Magnetic Resonance Data for
$\phi_3Si^ASi^BMe_2R^a$

R	Relative σ^*	δ_{Si^A} (ppm)	δ_{Si^B} (ppm)	J_{SiSi} (Hz)
Me	0	−20.22	−18.37	86.5
Pr	0.12	−20.36	−17.05	85.3
Oct	0.15	−20.51	−16.81	84.5
i-Pr	0.19	−20.60	−11.92	81.5
t-Bu	0.30	−20.37	−7.95	80.0
Vinyl	—	−21.28	−24.70	86.2

[a] Williams et al., 1980.

TABLE V

Nuclear Magnetic Resonance Parameters for Disilanes[a]

XSi^AMe_2Si^BMe_2Y		δ_{Si^A} (ppm)	δ_{Si^B} (ppm)	$^1J_{SiSi}$ (Hz)	$^1J_{Si^AC}$ (Hz)	$^1J_{Si^BC}$ (Hz)
X	Y					
Me	OMe	−22.41	15.68	99.4	43.9	48.0
Me	OEt	−22.49	12.98	99.0	43.6	48.0
Me	O-n-Pr	−22.74	12.80	—	—	—
Me	O-i-Pr	−22.87	10.09	—	—	—
Me	O-n-Bu	−22.44	13.12	99.5	44.6	47.7
Me	O-t-Bu	−22.67	3.24	103.0	44.0	47.6
Me	NMe_2	−21.75	−0.66	94.6	42.9	45.8
OMe	OMe	11.70	—	—	48.4	—
OEt	OEt	8.39	—	—	—	—
O-n-Pr	O-n-Pr	9.10	—	—	48.0	—
O-i-Pr	O-i-Pr	6.56	—	—	48.0	—
O-n-Bu	O-n-Bu	9.05	—	—	48.4	—
O-t-Bu	O-t-Bu	−0.19	—	—	47.7	—
Oϕ	Oϕ	12.15	—	—	50.5	—
OAc	OAc	12.08	—	—	53.2	—
Cl	Cl	16.95	—	—	49.1	—
NHEt	NHEt	−9.19	—	—	—	—
Cl	OEt	9.18	17.46	109.8	49.1	52.1
Cl	O-n-Pr	9.30	17.31	107.0	49.1	51.1
Cl	O-n-Bu	9.23	17.27	108.5	47.7	51.3
OSiMe_2SiMe_2O		2.67	—	—	—	—
SSiMe_2SiMe_2S		−5.44	—	—	—	—
HNSiMe_2SiMe_2NH		−9.75	—	—	—	—

[a] Engelhardt et al., 1981b.

B. Strained Rings

The ring silicon in siliranes resonates in the range -49 to -60 ppm (Marsmann, 1981a). However, for (**I**) the resonance is found to be -31.6 ppm. (Fritz *et al.*, 1981).

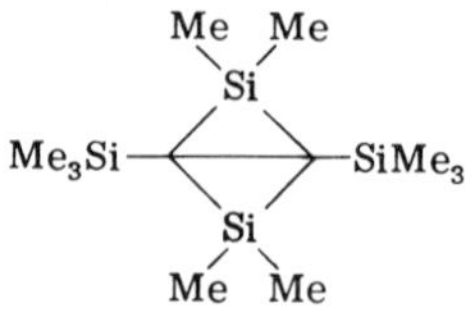

I

The cause of the shift to lower field was not discussed.

The ^{29}Si-NMR spectra of 1,2-disilacyclobutanes have been reported, the data for which are summarized in Table VI.

For the first three entries the chemical shifts can be compared with the resonance for the SiMe group in $[(Me_3Si)_2MeSi]_2$ at -81.4 ppm (Ishikawa

TABLE VI

Nuclear Magnetic Resonance Parameters for 1,2-Disilacyclobutanes

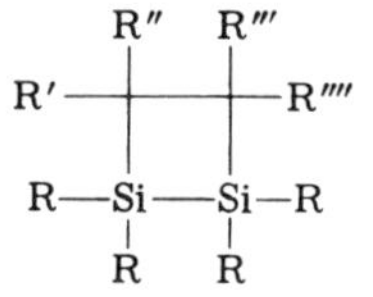

R	R'	R'''	R''	R''''	$\delta_{Si(ring)}$ (ppm)	$^1J_{SiF}$ (Hz)	$^2J_{SiF}$ (Hz)	Ref.
Me$_3$Si	OSiMe$_3$	OSiMe$_3$	ϕ	ϕ	-40.39	—	—	Brook *et al.* (1979b)
Me$_3$Si	OSiMe$_3$	OSiMe$_3$	3,5-diMeϕ	3,5diMeϕ	-40.9 -41.76	— —	— —	Brook *et al.* (1979b)
Me$_3$Si	OSiMe$_3$	OSiMe$_3$	CMe$_3$	CMe$_3$	-43.78	—	—	Brook *et al.* (1979b)
Me	Me	Me	Me	Me	-11.6	—	—	Seyferth *et al.* (1980)
F	H	H	H	H	13.8	±408	±33.0	Thompson *et al.* (1981)
F	Me	H	H	H	7.4 12.0	±405 ±413	±32.0 ±30.1	Thompson *et al.* (1981)
F	Me	H	H	Me	4.9	±413	±31.7	Thompson *et al.* (1981)
F	Me	H	Me	H	6.0	±409 ±416	±32.1 ±29.1	Thompson *et al.* (1981)

TABLE VII
Nuclear Magnetic Resonance Data for $\left[\begin{array}{c} R^2\ \ R^3 \\ | \quad | \\ C-C-SiF_2SiF_2 \\ | \quad | \\ R\ \ H \end{array}\right]_n$ Polymers[a]

R	R^2	R^3	δ_{Si} (ppm)	$^1J_{SiF}$ (Hz)	$^2J_{SiF}$ (Hz)
Me	H	H	−8.3	Not determinable	
Me	H	Me	−7.6	±354.9	±48.5
H	Me	Me	−7.5	±353.2	±48.2

[a] Reynolds et al., 1980a.

et al., 1981). The permethyl compound can be compared with hexamethyldisilane which resonates at −19.7 ppm. The downfield shift could be the result of ring strain. The chemical shifts and coupling constants for the fluoro compounds can be compared with those of the corresponding polymer (Reynolds et al., 1980a) and 1,2-disilacyclohexane (Reynolds et al., 1980b), which are given in Tables VII and VIII, respectively.

TABLE VIII
Nuclear Magnetic Resonance Data
for 1,2-Disilacyclohexanes[a]

Structure	δ_{Si} (ppm)	$^1J_{SiF}$ (Hz)	$^2J_{SiF}$ (Hz)
	−4.9	±348	±39.4
	−4.4	±347	±37.4
	−2.8	±348	±36.6
	−6.8	±351	±35.0

[a] Reynolds et al., 1980b.

TABLE IX

Nuclear Magnetic Resonance Parameters for Silacyclobutanes[a]

R[1]	R[2]	δ_{Si} (ppm)	$^1J_{SiC(Me)}$ (Hz)	$^1J_{SiC(ring)}$ (Hz)	$^2J_{SiC(ring)}$ (Hz)	$^1J_{SiC\phi}$ (Hz)	$^1J_{SiH}$ (Hz)
Me	Me	18.4	45	42	16	—	—
Me	H	6.0	45	42	—	—	187.0
H	H	−5.0	—	42	—	—	194.2
Me	Cl	32.5	49	45	17	—	—
Cl	Cl	18.1	—	52	19	—	—
Me	OEt	14.1	—	—	—	—	—
Me	O-n-C$_7$H$_{15}$	14.5	—	—	—	—	—
Me	Oϕ	17.0	54	47	—	—	—
Me	OAc	22.7	—	—	—	—	—
Me	O[b]	7.0	53	48	—	—	—
EtO	EtO	−17.1	—	55	19	—	—
Me	Et$_2$N	7.8	52	45	—	—	—
Me	HN[c]	7.6	50	45	—	—	—
Me	ϕ	11.9	46	43	—	60	—
ϕ	ϕ	7.0	—	44	—	62	—
ϕ	Cl	21.4	—	—	—	—	—

[a] Krapivin et al., 1980.
[b] Siloxane.
[c] Silazane.

Again there is a downfield shift of ~15 to 20 ppm for the 1,2-disilacyclo-butane resonances compared to the models. There is also a big difference in $^1J_{SiF}$ between the four-membered ring compounds and the models: For the former $^1J_{SiF}$ is ~410 Hz, whereas for either type of model $^1J_{SiF}$ is ~350 Hz.

Contrary to what has been observed for disilanes (Engelhardt et al., 1981b), it has been found that there is a correlation ($r = 0.93$) between δ_{Si} for silacyclobutanes and the net atomic charge on silicon (q_{Si}) obtained from CNDO/2 calculations using an spd basis set (Krapivin et al., 1980). There is no correlation ($r = 0.51$) when only an sp basis set is used. The reason for the dependence on q_{Si} is not known. In order to highlight the effect of ring strain the NMR parameters for silacyclobutanes (Table IX) and disilacyclobutanes (Table X) have been compared with those for the corresponding silacyclopentanes and disilacyclopentanes (Table XI).

TABLE X

Nuclear Magnetic Resonance Parameters for 1,3-Disilacyclobutanes[a]

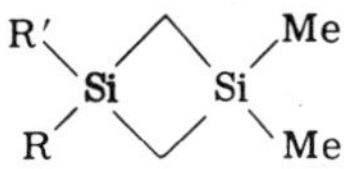

R	R′	$\delta_{Si(1)}$ (ppm)	$\delta_{Si(3)}$ (ppm)	$^1J_{Si(1)C(Me)}$ (Hz)	$^1J_{Si(3)C(Me)}$ (Hz)	$^1J_{Si(1)C(ring)}$ (Hz)	$^1J_{Si(3)C(ring)}$ (Hz)
Me	Me	2.1	2.1	49	49	35	35
Cl	Me	25.5	0.0	54	51	40	36
Cl	Cl	17.6	−4.9	—	52	48	32

[a] Krapivin *et al.*, 1980.

The effect of substituents is almost the same in the silacyclopentanes as in acyclic dimethylsilanes but substantially different from that in the silacyclobutanes. The difference between the chemical shifts for silicon in a four-membered and a five-membered ring Δ ($\delta_{Si(4\text{-ring})} - \delta_{Si(5\text{-ring})}$) should be an indication of the degree of strain (Table XII).

TABLE XI

Nuclear Magnetic Resonance Data
for Silacyclopentanes[a]

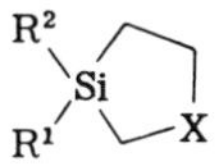

R¹	R²	X	δ_{Si} (ppm)	$^1J_{SiMe}$ (Hz)	$^1J_{Si(ring)}$ (Hz)
Me	Me	CH_2	16.4	50	51
Me	Cl	CH_2	45.4	55	57
Me	EtO	CH_2	29.9	—	—
Me	O[b]	CH_2	24.2	—	—
Me	NEt_2	CH_2	20.0	55	56
Me	Me	$SiMe_2$	14.0	50	52
					42 (C-2)
Me	Cl	$SiMe_2$	41.8	58	50 (C-2), 58 (C-5)
			12.8 (Si_x)	51	43 (C-2), 53 (C-4)

[a] Krapivin *et al.*, 1980.
[b] Siloxane.

TABLE XII

The Effect of Substituents on $\Delta\delta =$
$\delta_{Si(4\text{-ring})} - \delta_{Si(5\text{-ring})}$[a]

R[1]	R[2]	$\Delta\delta_{Si}$ (ppm)
Me	Me	2.0
Me	Cl	−12.9
Me	OEt	−15.8
Me	O[b]	−17.2
Me	NEt$_2$	−12.2

[a] Krapivin et al., 1980.
[b] Siloxane.

Because it has been argued that upfield shifts are associated with an increased positive charge on silicon it seems that the silicon in a silacyclobutane is made more positive than in a five-membered ring when substituted by an electronegative group. This then seems to argue against the involvement of d orbitals as Krapivin and associates (1980) claim.

In general the one-bond coupling constants between silicon and the ring carbons are smaller in four-membered rings (30–50 Hz) than in the five-membered rings (>50 Hz).

C. π-Bonding in Organosilicon Compounds

Stable compounds containing a carbon–silicon double bond (silene) have been made by Brook et al. (1979). Silenes are made by photolysis of acylsilanes, and NMR spectra of the photolysis mixtures have been reported. Silicon-29 resonances have been assigned to the silenes generated (Table XIII).

The shifts at ~41 ppm and 54 ppm are at unusually low fields, which is consistent with an sp^2-hybridized silicon atom.

The interaction of saturated silicon with unsaturated carbon systems has been much studied. Both involvement of silicon d orbitals and $\sigma^*–\pi$ hyperconjugation would lead to an increase in electron density at silicon and might be detectable by ^{29}Si NMR. The initial large substituent effect (−10.68 ppm) when a vinyl group replaces one of the methyl groups in tetramethylsilane (Table XIV), and thereafter an almost constant effect (−6.7 ppm) for additional vinyl groups, was considered consistent with "{d, $\sigma^*–\pi$} hyperconjugation" (Delmulle and Van der Kelen, 1980a). The increase in $^1J_{SiC}$ on going from mono- to tetravinyl substitution is ascribed

TABLE XIII

Silicon-29 Chemical Shifts for sp^2 Silicon in Silenes[a]

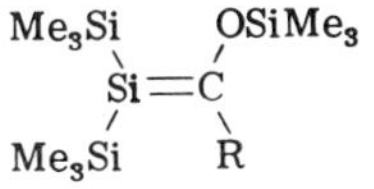

Me$_3$Si⧵ /OSiMe$_3$ Si=C /Me$_3$Si ⧵R	
R	$\delta_{Si(sp^2)}$ (ppm)
Me$_3$C	41.4
Et$_3$C	54.3
Adamantyl	41.8

[a] Brook et al., 1979b.

to a decrease in the relative importance of σ–π conjugation and an increase in $\{d, \sigma^*-\pi\}$ hyperconjugation.

When a methylene group is interposed between the silicon atom and the π system (as in allylmethylsilanes) hyperconjugative interaction is effectively attenuated, leading to an almost constant chemical shift (~0 to −2 ppm) for all allylmethylsilanes [Me$_x$Si(CH$_2$CH=CH$_2$)$_{4-x}$] (Delmulle and Van der Kelen, 1980b). In allyl halogenosilanes, the halogens dominate the ^{29}Si chemical shift (Delmulle and Van der Kelen, 1981).

A study of trimethylsilyl-substituted carbocations has been made by Olah et al. (1982), and the data are summarized in Table XV.

The small chemical shift change ($\Delta\delta$) for diphenyl(trimethylsilyl) methanol on going to the carbocation is the result of extensive charge delocali-

TABLE XIV

Nuclear Magnetic Resonance Data for Methylvinylsilanes[a]

Me$_x$Si(CH=CH$_2$)$_{4-x}$				
x	δ_{Si} (ppm)	$^1J_{SiC(Me)}$ (Hz)	$^1J_{SiC(1)}$ (Hz)	$^2J_{SiH(Me)}$ (Hz)
4	0	50.7	—	6.6
3	−10.68	52.0	64.0	6.6
2	−17.35	53.7	66.0	6.5
1	−24.06	55.9	68.0	6.7
0	−30.31	—	72.8	—

[a] Delmulle and Van der Kelen, 1980a.

TABLE XV

Nuclear Magnetic Resonance Data for Silyl-Substituted Carbocations and
Their Precursors[a]

Precursor, P	Carbocation, C	$\delta_{Si(P)}$ (ppm)	$\delta_{Si(C)}$ (ppm)	$\Delta\delta = \delta_{Si(C)} - \delta_{Si(P)}$
$\phi_2\overset{\text{OH}}{\underset{\,}{C}}-SiMe_3$	$\phi_2\overset{+}{C}-SiMe_3$	4.7	5.4	0.7
$Me-\overset{O}{\overset{\|}{C}}-SiMe_3$	$Me-\overset{OH}{\overset{+}{C}}-SiMe_3$	−10.1	11.0	21.1
$\phi-\overset{O}{\overset{\|}{C}}-SiMe_3$	$\phi-\overset{OH}{\overset{+}{C}}-SiMe_3$	−15.1	2.9	18.0
$Me_3Si-C\equiv C-\overset{\phi}{\underset{\phi}{C}}-OH$	$Me_3Si-C\equiv C-\overset{+}{C}\overset{\phi}{\underset{\phi}{\diagdown}}$	−19.8	−10.5	9.3
$Me_3Si-C\equiv C-\overset{\triangledown}{\underset{\triangle}{C}}-OH$	$Me_3Si-C\equiv C-\overset{+}{C}$	−18.6	−11.1	7.5
$Me_3Si-C\equiv C-\overset{Me}{\underset{Me}{C}}-OH$	$Me_3Si-C\equiv C-\overset{+}{C}\overset{Me}{\underset{Me}{\diagdown}}$	−18.4	−6.2	12.2
$Me_3Si-C\equiv C-\overset{\phi}{\underset{\triangle}{C}}-OH$	$Me_3Si-C\equiv C-\overset{+}{C}\overset{\phi}{\diagdown}$	−18.0	−10.5	7.5

[a] Olah *et al.*, 1982.

zation into the aromatic rings. It is thus surprising that protonation of
phenyl trimethylsilyl ketone causes so large a downfield shift (18 ppm)
compared with methyl trimethylsilyl ketone ($\Delta\delta$ 21.1), where there is no π
system.

In ethynyl carbocations delocalization puts charge on the carbon α to
the silicon atom:

$$Me_3Si-C\equiv C-\overset{\oplus}{C}R_2 \longleftrightarrow Me_3\,Si-\overset{\oplus}{C}=C=CR_2$$

TABLE XVI

Nuclear Magnetic Resonance Data for Silyl Anions and Their Precursors[a]

Precursor, P	Salt, S	$\delta_{Si(P)}$ (ppm)	$\delta_{Si(S)}$ (ppm)	$\Delta\delta = \delta_{Si(S)} - \delta_{Si(P)}$ (ppm)
ϕ_3SiCl	ϕ_3SiLi	−15.3	−30.0	−14.7
ϕ_2MeSiCl	ϕ_2MeSiLi	10.0	−20.5	−30.5
ϕMe$_2$SiCl	ϕMe$_2$SiLi	19.8	−22.1	−41.9
Me$_3$SiCl	Me$_3$SiK	29.8	−34.4	−64.2

[a] Olah and Hunadi, 1980.

and this is reflected in intermediate values of $\Delta\delta$ between 7.5 and 12.2 ppm. The largest change is again observed with the methyl analog where the charge is least delocalized.

Silyl anions have been studied by ^{29}Si NMR (Olah and Hunadi, 1980). Because $\delta_{Si(S)}$ changes very little with increasing methyl substitution (Table XVI), the charge must be principally located on the silicon atom for all the anions. The value of $\Delta\delta$ seems to follow a linear relationship: There is increased shielding of the ^{29}Si nucleus with increasing methyl substitution.

D. Higher Coordination

The normal coordination number of silicon is 4, but anionic, cationic, and neutral complexes are known in which silicon is penta- and hexacoordinate. Silicon-29 NMR is promising as an indicator for higher coordination silicon (Cella *et al.*, 1980). Hexacoordinate complexes have a ^{29}Si resonance between −140 and −200 ppm with the majority between −187 and −197 ppm (Table XVII).

In general the silicon resonance is at lower field when the chelate ring is five-membered than when it is six-membered. Pentacoordinate complexes show a lower symmetry than hexacoordinate ones and are thus more sensitive to charge: There is an almost 50-ppm difference between the ^{29}Si resonance for anionic catechol and cationic tropolone pentacoordinate complexes. For the corresponding hexacoordinate complexes the difference is 0.1 ppm. Liepins *et al.* (1980) have interpreted the NMR data (Table XVIII) for N-substituted 2,2-diaryl-1,3-dioxa-6-aza-2-silacyclooctanes in solution as an indication that there is transannular interaction

TABLE XVII

Nuclear Magnetic Resonance Data for Penta- and
Hexacoordinate Silicon Complexes[a]

Complex				δ_{Si} (ppm)

Cationic hexacoordinate

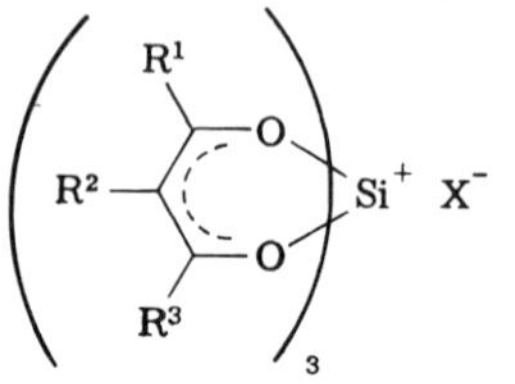

R^1	R^2	R^3	X	
Me	H	Me	HCl_2^-	-194.4
Me	H	Me	$ZnCl_3^-$	-193.7
ϕ	H	ϕ	SbF_6^-	-191.4
ϕ	H	Me	HCl_2^-	-192.4
Me	Me	Me	HCl_2^-	-195.7
Me	ϕ	Me	HCl_2^-	-195.3
Me	Cl	Me	HCl_2^-	-196.4
$-CH_2(CH_2)_2CH_2-$		Me	HCl_2^-	-195.0
$-CH_2CH_2CH_2-$		Me	HCl_2^-	-191.2
t-Bu	H	t-Bu	HCl_2^-	-193.8
Me	2,4-Dinitrophenyl	Me	SbF_6^-	-195.9
Me	$COCH_3$	Me	HCl_2^-	-197.2
EtO	H	ϕ	HCl_2^-	-187.9

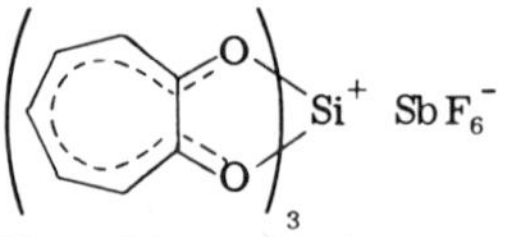

-139.4

Neutral hexacoordinate

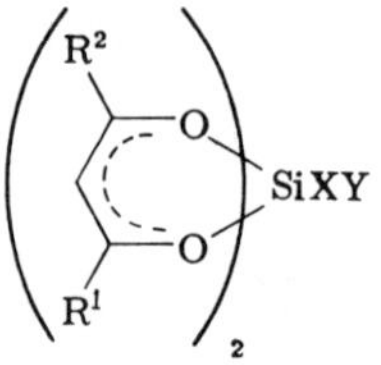

R^1	R^2	X	Y	
Me	Me	OAc	OAc	-196.8
ϕ	ϕ	OAc	OAc	-195.4
Me	Me	Me	Cl	-149.5

TABLE XVII (*Continued*)

Complex	δ_{Si} (ppm)
Anionic hexacoordinate	

| | -139.3 |
| Cationic pentacoordinate | |

| | -175.8 |

R = CH=CH$_2$	-141.3
R = ϕ	-141.8
R = (CH$_2$)$_4$CH$_3$	-127.1
R = Me	-128.6
Anionic pentacoordinate	

| | -87.1 |

[a] Cella *et al.*, 1980.

between silicon and nitrogen. The shift to higher field compared to acyclic model compounds has been attributed to reversible coordination of the nitrogen to the silicon. The coordination would be expected to be more important at lower temperatures where there is less molecular motion. For 2-furyl and 4-chlorophenyl compounds, cooling a solution to −60°C causes an upfield shift in the [29]Si resonance of 6–7 ppm compared to the same solution at 30°C.

TABLE XVIII

Nuclear Magnetic Resonance Data for
1,3-Dioxa-6-aza-2-silacyclooctanes and
Model Compounds[a]

$$Ar_2Si \underset{OCH_2CH_2}{\overset{OCH_2CH_2}{\diagup\diagdown}} NR$$

Ar	R	δ_{Si} (ppm)	$\delta_{Si(ref)}$ (ppm)	Reference compound
ϕ	H	−44.7	−32.4	$\phi_2Si(OEt)_2$
ϕ	Me	−43.9	−31.6	$\phi_2Si(OC_2H_4NMe_2)_2$
4-$CH_3C_6H_4$	Me	−41.5	—	—
4-ClC_6H_4	Me	−47.0	−33.4	$(4\text{-}ClC_6H_4)_2Si(OEt)_2$
(furanyl)	Me	−69.4	−50.6	$(2\text{-}C_4H_3O)_2Si(OEt)_2$
ϕ	Et	−40.0	—	—
ϕ	n-Pr	−39.6	—	—
ϕ	n-Bu	−39.6	—	—
ϕ	t-Bu	−35.2	—	—

[a] Liepins *et al.*, 1980.

E. Aqueous Silicates

Simple inorganic silicates are the salts of silicic acid $[Si(OH)_4]$ and its oligomers. The simple empirical formulas for solid salts often disguise complex chemistry. For instance, sodium metasilicate dissolves in water to give a complex mixture of ions based on the orthosilicate structure of silicic acid. The orthosilicate ion polymerizes to give a wide variety of species in dynamic equilibrium (Marsmann, 1981a).

These phenomena have been partially clarified with the aid of ^{29}Si-enriched compounds and high-field spectrometers which allow detection of signals from the minor components (Harris *et al.*, 1980a, 1981). The use of ^{29}Si-enriched compounds has another advantage: ^{29}Si—^{29}Si coupling can be observed, and so the connectivity of the molecule can be seen and checked by decoupling experiments (Harris *et al.*, 1981).

The structures of species identified in aqueous potassium silicate solution by Harris *et al.* (1981), together with their chemical shifts, are given in Fig. 2.

This list of structures has some surprising omissions. For instance, a linear trimer ($Q^1Q^2Q^1$) and a linear tetramer ($Q^1Q^2Q^2Q^1$) were identified in

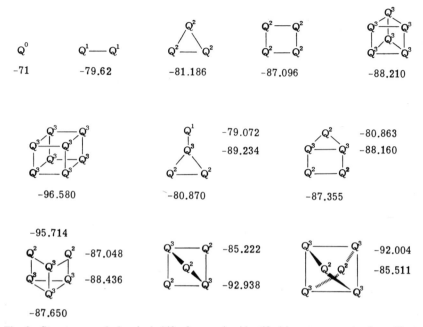

Fig. 2 Structures and chemical shifts for species identified in aqueous potassium silicate solution. Q = $Si(O_{1/2})_4$, and the superscript refers to the number of neighboring silicon atoms (Harris *et al.* 1981.)

earlier work (Harris and Newman, 1977) but not in the study using enriched material.

However, it is dangerous to compare spectra unless the concentrations, pH, and temperature are exactly the same (Harris *et al.*, 1980a), because these factors influence the equilibria and hence chemical shifts. Presumably, linear polysilicates cyclize very rapidly, and because of the low concentrations were not identified in the enriched samples. So far no rings larger than four-membered ones have been observed in alkali metal silicate solutions. However, there were still unidentified peaks in the spectrum of the enriched sample (Harris *et al.*, 1981), and it is quite possible that a whole range of interesting structures awaits discovery. The general chemical shift ranges for alkali metal silicates are given in Table XIX.

When tetraethylammonium (TEA) is the counterion, the predominating structures are different (Hoebbel *et al.*, 1980). For concentrated solutions with TEA/Si molar ratios ranging from 3.7 to 1 the major structure is a double three-membered ring Q_6^3 (δ_{Si} − 88.7).

For concentrated solutions with TEA/Si ratios of 0.8 to 0.6 the main structures are double three-, double four- (−98.2 ppm), double five-, and

TABLE XIX
Chemical Shift Ranges for Q^n in
Alkalimetal Silicates[a]

Q^n	(ppm)
Q^0	-71
Q^1	-79
Q^2 in a three-membered ring	-81
Q^2 in a bicyclo[1.1.1] system	-85
Q^2	-87
Q^3 in a three-membered ring	-88 to -89
Q^3 in a four-membered ring	-92 to -97

[a] Harris *et al.*, 1981.

probably double six-membered ring anions. For a concentrated ($[SiO_2] =$ 3.07 M) solution with TEA/Si $= 0.81$, there are also weak signals for monosilicate (δ -70), Q^1 end groups ($\delta \sim -79$), Q^2 in three-membered rings ($\delta \sim -81$), and Q^2 in chains and rings of more than three silicon atoms ($\delta \sim -87$ to -90), as well as Q^3 branching points (δ -96 to -100). For concentrated solutions with an even lower TEA/Si ratio there are very broad signals for Q^4 ($\delta \sim -105$ to -125). This predominance of double ring anions is probably due to the tendency of tetraalkyl-ammonium salts to form clathrate structures on crystallizing from aqueous solutions.

In addition to the structural complexity of the aqueous silicates, great care must be taken to ensure that the solution being studied is at equilibrium. The condensation of silicate species is slow, on the order of days (Harris *et al.*, 1980b). Thus 52.5 h after tetramethoxysilane had been hydrolyzed at pH 3.5 there was still a significant amount of monomer present (Harris *et al.*, 1980b). The only other species identified were dimer and linear trimer, and possibly cyclic tetramer.

F. Transition-Metal Complexes

Silyl complexes are being increasingly investigated to determine the cause of their greater stability compared to their carbon analogs. The chemical shift range for silicon directly bonded to a transition metal is from -89.6 ppm [for *trans*-(IPt[P(Et)$_3$]$_2$SiH$_2$TeSiH$_3$) (Marsmann, 1981b)†] to $+173$ ppm for (**II**) (Bikovetz *et al.*, 1980).

† Marsmann cites this chemical shift and refers to Ebsworth, E. A. V., Edward, J. M., and Rankin, D. W. H. (1976). *J. Chem. Soc. Dalton Trans.* 1667–1672. We have been unable to find any ^{29}Si chemical shifts in this article.

$$\underset{\textbf{II}}{\overset{\displaystyle Me_2Si\begin{matrix}\diagup\ Fe(CO)_4 \\ \big| \\ \diagdown\ Fe(CO)_4\end{matrix}}{}}$$

Only in the last 2 yr have reports appeared dealing systematically with the change in chemical shift as the nature of the ligands in σ-silyl transition-metal complexes changes.

Absolute values of the chemical shift for $(\eta^5\text{-}C_5H_5)FeL_2R$ complexes offer little useful information directly (Table XX). However, the difference between the chemical shifts for the complex and for the silyl ligand where the $(\eta^5\text{-}C_5H_5)FeL_2$ group is replaced by a methyl group gives general chemical shift changes. For dicarbonyl complexes σ-Si Δδ is between 36.5 and 42.5 ppm. For the phosphine monocarbonyl complex Δδ is slightly less, whereas for a bisphosphine–chelate complex Δδ is only 25.2 ppm. The corresponding difference for the silicon β to iron is 8.3–12.9 ppm regardless of the atom between. The γ silicon is essentially unaffected by the presence of the iron group.

TABLE XX

Chemical Shift Data for $(\eta^5\text{-}C_5H_5)FeL_2R$ Complexes and MeR Compounds[a]

L	L	R	Chemical shift (ppm)			Δδ (ppm)		
			α	β	γ	α	β	γ
CO	CO	SiMe₃	41.3	—	—	41.3	—	—
		MeSiMe₃	0.0	—	—	—	—	—
CO	CO	SiMe₂SiMe₃	16.95	−11.3	—	36.53	8.28	—
		MeSiMe₂SiMe₃	−19.58	−19.58	—	—	—	—
CO	CO	SiMe₂SiMe₂SiMe₃	21.22	−36.54	−15.07	37.2	11.92	0.81
		MeSiMe₂SiMe₂SiMe₃	−15.9	−48.45	−15.9	—	—	—
CO	CO	CH₂SiMe₃	—	9.28	—	—	7.28	—
		MeCH₂SiMe₃	—	2.0	—	—	—	—
CO	CO	CH₂SiMe₂SiMe₃	—	−6.7	−19.92	—	12.88	−0.34
		MeSiMe₂SiMe₃	—	−19.58	−19.58	—	—	—
CO	CO	SiMe₂CH₂SiMe₃	42.86	—	0.36	42.54	—	0.04
		MeSiMe₂CH₂SiMe₃	0.32	—	0.32	—	—	—
CO	Pφ₃	SiMe₃	34.75	—	—	34.75	—	—
CO	Pφ₃	SiMe₂CH₂SiMe₃	36.99	—	0.30	36.67	—	0.02
CO	Pφ₃	CH₂SiMe₃	—	9.17	—	—	7.17	—
φ₂PCH₂CH₂Pφ₂		SiMe₃	25.16	—	—	25.16	—	—

[a] Pannell and Bassindale, 1982.

The NOEs for all silicons in the chain increase markedly when an (η^5-C_5H_5)Fe(CO)$_2$ group is substituted for a methyl group. Dipole–dipole relaxation is most significant for the α-Si atom, with spin–rotation relaxation becoming increasingly more important the further a silicon atom is from the iron. When a carbonyl group is replaced by triphenylphosphine, the molecule becomes so ponderous that dipole–dipole relaxation predominates (>90%).

The T_1 values for R = SiMe$_3$ and CH$_2$SiMe$_3$ are 79 and 75 s, respectively, which within experimental error are the same, probably because the increased mobility resulting from being farther away from the heavy metal is countered by the increased mass of the complex.

The effect on the chemical shift of the σ-bonded silicon of increasing the length of the σ-silyl group is -24.3 ppm for the addition of an extra silicon to a trimethylsilyl complex (cf. $\Delta\delta = -19.58$ ppm for Si$_2$Me$_6$ versus SiMe$_4$).

For one extra silicon in the chain the change in chemical shift for the σ-bonded silicon is 4.27 ppm (cf. 3.68 ppm for Si$_3$Me$_8$ versus Si$_2$Me$_6$).

TABLE XXI

Nuclear Magnetic Resonance Data for Silicon Hydrides and Related Complexes[a]

R	R¹	R²	R³	δ_{Si} (ppm)	J_{SiH} (Hz)	J_{SiF} (Hz)	Δδ for ligand— hydride	Δδ for anion— hydride
H	φ	φ	φ	18.5	64.7	—	—	—
Anion	φ	φ	φ	57.0	—	—	39.6	38.5
HSiφ₃				−21.1	198	—	—	—
H	φ	1-Naphthyl	H	7.5	208, 69	—	—	—
Anion	φ	1-Naphthyl	H	42.6	—	—	42.9	35.1
φ(1-Naphthyl)SiH₂				−35.4	198, 6.2	—	—	—
H	φ	1-Naphthyl	F	57.3	—	348	—	—
Anion	φ	1-Naphthyl	F	98.9	—	287.5	58.8	41.6
φ(1-Naphthyl)SiHF				−1.5	229.4, 6.2	287.5	—	—
H	φ	1-Naphthyl	Cl	58.3	—	—	—	—
Anion	φ	1-Naphthyl	Cl	80.6	—	—	63.4	22.3
φ(1-Naphthyl)SiHCl				−5.1	238.5, 6.0	—	—	—
Me	φ	φ	φ	38.7	—	—	—	—

[a] Colomer et al., 1982.

TABLE XXII

Silicon-29 NMR Data for π-Organosilyl Metal Complexes[a]

	Chemical shift (ppm)		$\Delta\delta$	J_{SiM}	$-\eta$	
Complex	Complex	Ligand	(ppm)	(Hz)	Complex	Ligand
AcacRh(CH$_2$=CHSiMe$_3$)$_2$	−0.62	−7.6	6.98	2.3	2.0	0.2
AcacRh(CH$_2$=CHSiMe$_2$OEt)$_2$	11.1	4.12	6.98	2.4	2.4	0.3
AcacRh(CH$_2$=CHSi(OEt)$_3$)$_2$	−53.2	−59.0	5.84	2.0	—	—
AcacRh(CH$_2$=CHCH$_2$SiMe$_3$)$_2$	−1.39	0.39	−1.78	1.9	—	—
(η^5-C$_5$H$_5$)Rh(CH$_2$=CHSiMe$_3$)	−0.18	−7.6	7.42	2.2	2.2	0.2
(CO)$_4$Fe(Me$_3$SiC≡CSiMe$_3$)	−8.54	−19.25	10.71	—	—	—
(CO)$_{12}$Co$_2$(Me$_3$SiC≡CSiMe$_3$)	1.06	−19.25	20.31	—	—	—
(CO)$_6$Co$_2$(Me$_3$SiC$_4$SiMe$_3$)						
α-Si	2.45	−15.99	18.44	—	1.5	0.5
β-Si	−16.91	−15.99	−0.92	—	1.0	0.5
(CO)$_{12}$Co$_4$(Me$_3$SiC$_4$SiMe$_3$)	1.06	−15.99	17.05	—	2.4	0.5
(CO)$_6$Co$_2$(Me$_2$ClSiC$_2$SiMe$_2$Cl)	19.17	−0.61	19.77	—	—	—

[a] Pannell et al., 1981.

Large downfield shifts have been reported (Colomer et al., 1982) for a series of manganese hydrides and their derived anions. Insertion of the metal into the silicon–hydrogen bond causes a downfield shift of between 40 and 64 ppm for the coordinated silicon (Table XXI).

The value of $^2J_{SiH}$ across H—Mn—Si is 65 and 69 Hz in the two cases where it was measured, indicating that there is perhaps a large contribution from structures where a silicon hydride is π-bonded to the manganese. These values must be compared with the value of $^2J_{SiH}$ across H—C—Si, which is ~6 Hz.

A collection of π-organosilyl complexes has been studied with ^{29}Si NMR (Pannell et al., 1981); the data are collected in Table XXII. In general there is a shift of 6–11 ppm to low field when a vinylsilane becomes complexed to a single metal atom. The shift is larger for metallic clusters which normally have shifts in the range 18–20 ppm. For silanes with the silicon atom remote from the complexing site the change on complexation is much smaller—in the range 0.5–2.0 ppm upfield.

Silicon-29 NMR has recently been used to correct the conclusion based on ^{1}H- and ^{13}C-NMR data that η^3-1-silapropenyl complexes are formed by the reaction of vinyldisilanes with enneacarbonyl diiron (Sakurai et al., 1980):

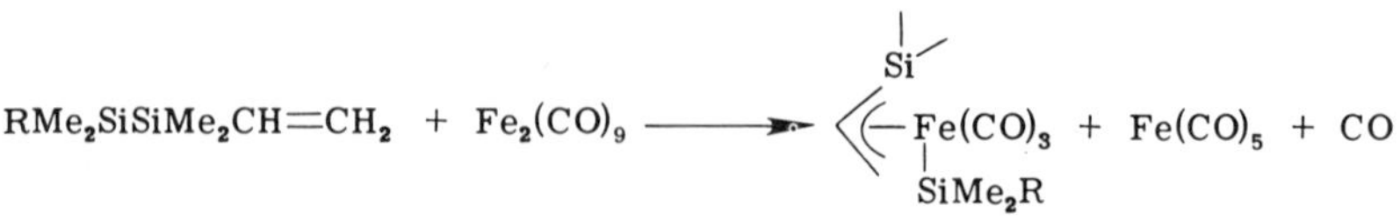

$$RMe_2SiSiMe_2CH{=}CH_2 \; + \; Fe_2(CO)_9 \; \longrightarrow \; \left\langle\!\!\left\langle\!\!\begin{array}{c} \overset{\displaystyle |}{\underset{\displaystyle |}{Si}}\diagup \\ \!\!\!\!-Fe(CO)_3 \\ SiMe_2R \end{array}\right. \; + \; Fe(CO)_5 \; + \; CO \right.$$

R = Me, vinyl

The proposed reaction requires the formation of an Fe—Si bond which would be expected to have a positive ^{29}Si chemical shift.

The chemical shift changes on complexation (Table XXIII) are consistent only with the formation of η^2-vinyldisilane complexes.

III. Solid State Studies

A. Solid Silicones

Although well-known for soluble silicones, ^{29}Si NMR has only recently been applied to analysis of the structure of solid silicones (Engelhardt and Jancke, 1981; Engelhardt et al., 1981a; Maciel et al., 1981a). Comparison of spectra from solid samples and from their solutions has shown that there are no specific solid state effects complicating the spectra and that chemical shifts for a molecule in the solid state are almost identical with those in solution (Engelhardt and Jancke, 1981; Engelhardt et al., 1981a) (Table XXIV).

TABLE XXIII

Nuclear Magnetic Resonance Data for $RMe_2Si^ASi^BMe_2CH{=}CH_2$ and Related $(\eta^2\text{-Vinyl})Fe(CO)_4$ Complexes[a]

	Ligand L		Complex C		$\Delta\delta = \delta_C - \delta_L$	
R	δ_{Si^A} (ppm)	δ_{Si^B} (ppm)	δ_{Si^A} (ppm)	δ_{Si^B} (ppm)	Si^A (ppm)	Si^B (ppm)
Me	−23.96	−19.16	−11.40	−18.58	12.56	0.58
—CH=CH$_2$	−24.14	−24.14	−11.70	−23.77	12.44	−1.77
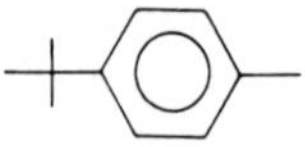 +⟨O⟩—	−24.08	−22.05	−11.55	−21.55	12.53	0.50

[a] Sakurai et al., 1980.

TABLE XXIV

Comparison of Chemical Shifts for Silicones in the Solid State and in Solution[a]

		δ_{Si} (ppm)					
		Me | —OSiO— | Me	ϕ | HOSiO— | ϕ	OH | —OSiO— | Me	O | —OSiO— | Me	OH | —OSiO— | ϕ	O | —OSiO— | ϕ
Silicone	State						
Methyl	Powder	−19.4	—	—	−65.6 and −55.8	—	—
	Solution	−19.3	—	—	−65.3 and −55.9	—	—
Methyl/ phenyl	Powder	−17.9	−37.5	−55.2	−64.3	−70.0	−79.3
	Solution	−17.8 and −21.9	−37.0	−54.6	−64.2	−69.7	−78.7

[a] Engelhardt *et al.*, 1981a.

When there is a large amount of molecular motion such as in poly(dimethylsiloxane) side chains in methylsilicone, the signals are not observable with cross-polarization magic angle spinning (CP-MAS) but can be seen using direct fourier transform (FT) techniques (Engelhardt *et al.*, 1981a).

Of more interest are insoluble silicones, because it is difficult to obtain structural information about them by other means. It has been shown that ^{29}Si and ^{13}C CP-MAS NMR are complementary for the study of methylsilicones (Maciel *et al.*, 1981a). For methylsilicone prepared from MeSiCl$_3$ the ^{29}Si NMR spectrum consisted of a strong signal at −65 ppm corresponding to CH$_3$Si(OSi≡)$_3$ groups, and an additional peak at −55 ppm assigned to CH$_3$Si(OH)(Osi≡)$_2$ groups. The ^{13}C CP-MAS NMR spectrum consisted of only one line. When the polymer was made from CH$_3$Si(OEt)$_3$, the ^{29}Si spectrum was one line with an unresolved shoulder, whereas the ^{13}C CP-MAS NMR spectrum showed signals that have been assigned to CH$_3$Si(OEt)(OSi≡)$_2$ groups.

B. Amorphous Silicon–Hydrogen Films

Nuclear magnetic resonance studies of amorphous Si—H films have concentrated on the use of proton NMR and have been reviewed by Reimer (1981). Silicon-29 CP-MAS NMR at low field shows only one line (Reimer *et al.*, 1981); high fields must be used if chemical shift resolution

TABLE XXV
Nuclear Magnetic Resonance Data for
Silica Gel[a]

Group	δ_{Si} (ppm)
$(HO)_2Si(OSi\lessdot)_2$	−90.6
$HOSi(OSi\lessdot)_3$	−99.8
$Si(OSi\lessdot)_4$	−109.3

[a] Maciel and Sindorf, 1980.

is to be obtained (Lamotte *et al.*, 1981b). There appears to be some confusion about assignment of the resolved lines: More recent work (Lamotte *et al.*, 1981b) seems to contradict the results of previous studies (Lamotte *et al.*, 1981a) on the chemical shifts of the mono- and dihydride species. The shifts for these species are −69 and −83 ppm, respectively (Lamotte *et al.*, 1981b). Peaks at −99 ppm and −111 ppm previously assigned (Lamotte *et al.*, 1981a) to mono- and dihydrides are now considered to be from hydrolysis products.

C. Silica Gel and Derivatives

The surface of silica gel has been characterized by CP-MAS NMR (Maciel and Sindorf, 1980; Lippmaa *et al.*, 1981a). It is possible to distinguish $Si(OSi\lessdot)_4$, $HOSi(OSi\lessdot)_3$, and $(HO)_2Si(OSi\lessdot)_2$ groups (Table XXV).

The spectra are far from quantitative: The bulk of the ^{29}Si is far removed from the surface and from the only available protons, so the efficiency of cross-polarization for these nuclei is low at short contact times (Lippmaa *et al.*, 1981a). On the basis of NMR spectra of trimethylsilylated polysilicic acid, it has been concluded that polysilicic acid possesses very few geminal $Si(OH)_2$ groups (Engelhardt *et al.*, 1981c). Absorption of water onto the surface of silica gel does not lead to hexacoordinate silicon, because no signals could be seen at shifts higher than −120 ppm (Lippmaa *et al.*, 1981a). (See also Sections II,D and III,D).

Organic derivatives of silica gel are used as stationary phases in liquid chromatography, and a series of such phases has been studied using solid state NMR (Maciel *et al.*, 1981b). As with the study of silicones, here too ^{13}C and ^{29}Si CP-MAS NMR complement each other.

When $(CH_3O)_3SiCH_2CH_2CH_2SH$ was used to derivatize silica gel, ^{13}C CP-MAS NMR showed that there were still methoxy groups present.

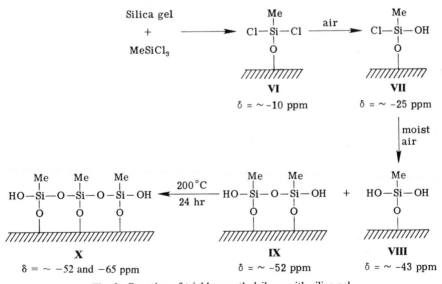

Structure (**III**) was favored from signal intensities. Silicon-29 CP-MAS NMR, however, besides signals for the silica gel surface and for structure (**III**), also showed signals for structures (**IV**) and (**V**).

Sindorf and Maciel (1981) studied the reaction of trichloromethylsilane with silica gel followed by exposure to air and claimed to have detected the reactions shown in Fig. 3.

Surprisingly, none of the trichloromethylsilane became doubly σ-bonded directly to the silica gel surface, even though such a doubly bonded species was claimed to be one of the major products from dichlorodimethylsilane (Sindorf & Maciel, 1981).

Heating the silica gel with attached (**VIII**) and (**IX**) to 200°C for 24 h caused substantial rearrangement: The ratio of surface groups to $Si(OSi\equiv)_4$ groups actually increased and a "trimeric" species (**X**) was formed.

Fig. 3 Reaction of trichloromethylsilane with silica gel.

TABLE XXVI

Chemical Shift Ranges for Q^n in
Solid Silicates[a]

Group	Chemical shift range (ppm)
Q^0	-66.4 to -73.5
Q^1	-77.9 to -82.6
Q^2	-86.3 to -88.0
Q^3	-90.4 to -99.3
Q^4	-107.4 to -109.9

[a] Lippmaa et al., 1980.

D. Silicates

Thaumasite $[Ca_3(CO_3)(SiOH_6)SO_4 \cdot 12H_2O]$ is a natural silicate which contains hexacoordinate silicon. Silicon-29 MAS NMR shows only one sharp line at -180 ppm (Lippmaa et al., 1981a). This is the only example of a ^{29}Si chemical shift for hexacoordinate silicon in the solid state so far reported. Other natural silicates have been studied (Lippmaa et al., 1980), and the chemical shift ranges for the different types of silicate group are given in Table XXVI.

Although most silicates contain only one type of group, xonotlite $[Ca_6(OH)_2 (Si_6O_{17})]$ has two, and indeed in the ^{29}Si MAS NMR spectrum there are signals at -86.8 ppm (Q^2) and -97.8 ppm (Q^3).

There is only one report of ^{29}Si MAS NMR analysis of synthetic silicates (Hoebbel et al., 1980). Two different structures have been reported for TEA silicates. When the TEA/Si ratio is 0.67, the solid crystallizes in a double four-membered ring structure which resonates at -99.5 ppm. When the TEA/Si ratio is 1.26, the structure was assumed to be a double three-membered ring from the broad resonance at ~ -90.5 and -90.6 ppm, and in solution such a structure resonates at -88.7 ppm (Section II,E). There are also ill-defined shoulders that have not yet been assigned to specific structures.

E. Aluminosilicates and Zeolites

Natural aluminosilicates have also been studied by ^{29}Si MAS NMR (Lippmaa et al., 1980). Known X-ray structures were used to calibrate the chemical shift ranges for the natural minerals given in Table XXVII. These chemical shifts are peculiar to the solid state, because the structures do not exist in solution. Similar chemical shift ranges have been defined for synthetic zeolites (Table XXVIII), where almost all the silicon is present in Q^4 groups bonded to various numbers of AlO_4 tetrahedra. These ranges are almost identical to those for the more complicated natural aluminosilicates.

The major signal for Na A zeolite (Si/Al ratio about unity) is at -89.6 ppm which is in the range for Si(3Al). The assumption (Engelhardt *et al.*, 1980) that this is the main structural unit is a contradiction of Loewenstein's rule (Loewenstein, 1954). This rule basically states that in aluminosilicates linked tetrahedra are not arranged adjacent to alumina tetrahedra. This rule has been extensively used to interpret often confusing X-ray data, and because of its importance it was reassuring that support for the conclusions derived from the ^{29}Si-NMR data rapidly appeared (Lodge *et al.*, 1980). The chemical shift was confirmed by a second NMR group who also assumed there was 3 : 1 coordination (Thomas *et al.*, 1981a). After long accumulations the spectrum of Na A zeolite shows small peaks attributed to small amounts of Si(4Al) and Si(2Al) in less-ordered regions of the Na A zeolite lattice. However, the validity of the conclusions based on ^{29}Si NMR have been questioned (Adams and Haselden, 1982; Cheetham *et al.*, 1982). Adams and Haselden (1982) used neutron diffraction to study dehydrated Na A zeolite and found no rhombohedral distortions, in contradiction to the earlier results. Cheetham and associates (1982) reassessed the ^{29}Si chemical shift data for the same zeolite and concluded that there is no violation of Loewenstein's rule. This is based on the fact that for ZK-4 zeolite (a high-silica variant of Na A zeolite) the signal at -88.4 ppm must be assigned to Si(4Al) to allow correct calculation of the Si/Al ratio. The chemical shifts in Na A zeolite are now believed to be, because of ring strain, exceptions to the general

TABLE XXVII

Chemical Shift Ranges for the Different Groups in Aluminosilicates[a]

	Q^3		Q^4				
	1 Al	0 Al	4 Al	3 Al	2 Al	1 Al	0 Al
Code in text	—	—	Si(4Al)	Si(3Al)	Si(2Al)	Si(1Al)	Si(0Al)
Chemical shift	-84.6	-91.5	-83.1	-87.7	-92.5	-96.7	-107.4
range (ppm)	to	to	to	to	to	to	
	-86.7	-95.0	-84.8	-88.4	-95.7	-104.2	

[a] Lippmaa *et al.*, 1980.

TABLE XXVIII

Chemical Shift Ranges for Synthetic Zeolites

Zeolite structure	Si(4Al)	Si(3Al)	Si(2Al)	Si(1Al)	Si(0Al)	Reference
Na A	−83.5 to −85.0	−87.7 to −94.0	−94.2 to −99.4	−98.6 to −105.3	−103.1 to −112.8	Lippmaa et al. (1981b)
Faujasite	−83.8 to −85.3	−89.0 to −89.6	−94.0 to −95.9	−98.8 to −101.5	−103.4 to −107.8	Engelhardt et al. (1981b)

ranges in Tables XXVII and XXVIII. This does not invalidate the use of these data for other zeolites, because zeolite A is the only structure to have sodalite cages joined at highly strained double four-membered ring structures (Cheetham *et al.*, 1982).

The chemical shifts for Na A zeolite are independent of the method of synthesis, and thus presumably the structure is as well (Thomas *et al.*, 1981b). Sodalite and cancrinite, however, can have either Si(4Al) or Si(3Al) as the major structural arrangement depending on the source of the material (Thomas *et al.*, 1981b).

The partial overlap of the five chemical shift ranges for zeolites means that ^{29}Si MAS NMR by itself cannot rigorously determine the structure of aluminosilicates. Two groups have studied faujasites and agree on the most likely structures (Engelhardt *et al.*, 1981c; Ramdas *et al.*, 1981), but in one instance it was not possible definitely to rule out a structure with Al—O—Al bonds on the basis of the ^{29}Si NMR data (Ramdas *et al.*, 1981).

Since the first reports of ^{29}Si NMR being used to show violation of Loewenstein's rule, the technique has been used to demonstrate that other minerals also violate the rule (Klinowski *et al.*, 1981a). It is to be hoped that these chemical shifts do not also prove to be exceptions to the general range for Si(4Al).

Silicon-29 has been used to study reactions and transformations of zeolites. Treatment of synthetic Na Y zeolite with $SiCl_4$ produces an essentially aluminum-free faujasite structure (Klinowski *et al.*, 1981b). The ^{29}Si-NMR spectrum consists of a single resonance at -107.4 ppm, which is characteristic of Si(OAl). The peak is very narrow ($\sim$0.5 ppm at half-height), indicating a highly symmetric environment.

The range in which Si(OAl) resonates has been extended from a study on zeolites with a very high silicon/aluminum ratio (>12) (Nagy *et al.*, 1981). In such species the Si(OAl) groups are present in five-membered rings, and this leads to a ^{29}Si resonance in the -112 to -114 ppm region.

Two groups of workers have used ^{29}Si solid state NMR to study the dealumination of zeolites (Engelhardt *et al.*, 1982; Maxwell *et al.*, 1982). Exchange of NH_4^+ for the Na^+ in Na Y zeolite caused no change in the zeolite framework (Engelhardt *et al.*, 1982). Heating either 50% (Engelhardt *et al.*, 1982) or 97.5% (Maxwell *et al.*, 1982) ammonium-exchanged zeolites in the presence of steam caused a loss of aluminum from the framework, as evidenced by the increase in intensity of Si(OAl) signals at ~ -107 ppm. The ammonium exchange was shown to be necessary by heating Na Y under similar conditions; the broad featureless signal showed that only amorphous material was produced (Maxwell *et al.*, 1982).

When a zeolite has been made with a high silicon/aluminum ratio much of the silicon is present in hydroxylated form on the surface or at defect sites (resulting from incomplete aluminum replacement). Cross-polarization can be used to enhance selectively the ^{29}Si-NMR signals of the SiOH and Si(OH)$_2$ groups but does not give details of the surrounding environment (Engelhardt et al., 1981b and 1982).

The validity and use of ^{29}Si MAS NMR in the analysis of aluminosilicates has been discussed (Bursill et al., 1981), and it is claimed that three techniques are required for a full structure analysis: electron microscopy, neutron diffraction, and ^{29}Si MAS NMR (Bursill et al., 1981a; Thomas et al., 1981).

References

Adams, J. M., and Haselden, D. A. (1982). J. Chem. Soc. Chem. Commun., 822–823.

Bikovetz, A. L., Kuzmin, O. V., Vdovin, V. M., and Krapivin, A. M. (1980). J. Organ. Chem. **194,** C33–34.

Boberski, W. G., and Allred, A. L. (1974). J. Organomet. Chem. **74,** 205–208.

Brook, A. G., Harris, J. W., Lennon, J., and El Sheikh, M. (1979a). J. Am. Chem. Soc. **101,** 83–95.

Brook, A. G. et al. (1979b). J. Am. Chem. Soc. **101,** 6750–6752.

Bursill, L. A., Lodge, E. A., Thomas, J. M., and Cheetham, A. K. (1981). J. Phys. Chem. **85,** 2409–2421.

Cella, J. A., Cargioli, J. D., and Williams, E. A. (1980). J. Organomet. Chem. **186,** 13–17.

Cheetham, A. K., Fyfe, C. A., Smith, J. V., and Thomas, J. M. (1982). J. Chem. Soc. Chem. Commun., 823–825.

Colomer, E., Corriu, R. J. P., Marzin, C., and Vioux, A. (1982). Inorg. Chem. **21,** 368–373.

Delmulle, L. and Van der Kelen, G. P. (1980a). J. Mol. Struct. **66,** 309–314.

Delmulle, L. and Van der Kelen, G. P. (1980b). J. Mol. Struct. **66,** 315–318.

Delmulle, L., and Van der Kelen, G. P. (1981). J. Mol. Struct. **70,** 207–211.

Doddrell, D. M., Pegg, D. T., Brooks, W., and Bendall, M. R. (1981). J. Am. Chem. Soc. **103,** 727–728.

Engelhardt, G., and Jancke, H. (1981). Polym. Bull. **5,** 577–584.

Engelhardt, G., Zeigan, D., Lippmaa, E., and Mägi, M. (1980). Z. Anorg. Allg. Chem. **468,** 35–38.

Engelhardt, G., Jancke, H., Lippmaa, E., and Samoson, A. (1981a). J. Organomet. Chem. **210,** 295–301.

Engelhardt, G., Radeglia, R., Kelling, H., and Stendel, R. (1981b). J. Organomet. Chem. **212,** 51–58.

Engelhardt, G., Lohse, U., Lippmaa, E., Tarmak, M., and Mägi, M. (1981c). Z. Anorg. Allg. Chem. **482,** 49–64.

Engelhardt, G. et al., (1982). Zeolites **2,** 59–62.

Fritz, G., Wartanessian, S., Matern, E., Hönle, W., and Von Schnering, H. G. (1981). Z. Anorg. Allg. Chem., **475,** 87–108.

Harris, R. K., and Newman, R. H. (1977). J. Chem. Soc. Faraday Trans. 2 **73,** 1204–1215.

Harris, R. K., Kennedy, J. D., and McFarlane, W. (1978). "NMR and the Periodic Table" (R. K. Harris and B. E. Mann, eds.), p. 309–377. Academic Press, London.

Harris, R. K., Jones, J., Knight, C. T. G., and Pawson, D. (1980a). *J. Mol. Struct.* **69**, 95–103.

Harris, R. K., Knight, C. T. G., and Smith, D. N. (1980b). *J. Chem. Soc. Chem. Commun.*, 726–727.

Harris, R. K., Knight, C. T. G., and Hull, W. E. (1981). *J. Am. Chem. Soc.* **103**, 1577–1578.

Hoebbel, D. *et al.* (1980). *Z. Anorg. Allg. Chem.* **465**, 15–33.

Ishikawa, M. *et al.* (1981). *J. Am. Chem. Soc.* **103**, 4845–4850.

Klinowski, J., Thomas, J. M., Fyfe, C. A., and Hartman, J. S. (1981a). *J. Phys. Chem.* **85**, 2590–2594.

Klinowski, J. *et al.* (1981b). *J. Chem. Soc. Chem. Commun.*, 570–571.

Kowalewski, J., and Morris, G. A. (1982). *J. Magn. Reson.* **47**, 331–338.

Krapivin, A. M. *et al.* (1980). *J. Organomet. Chem.* **190**, 9–33.

Lamotte, B., Rousseau, A., and Chenevas-Paule, A. (1981a). *Recent Dev. Condens. Matter Phys.* **2**, 247–251.

Lamotte, B., Rousseau, A., and Chenevas-Paule, A. (1981b). *J. Phys. Colloq.* **C4**, 839–841.

Liepins, E. *et al.* (1980). *J. Organomet. Chem.* **201**, 113–121.

Lippmaa, E., Mägi, M., Samoson, A., Engelhardt, G., and Grimmer, A. R. (1980). *J. Am. Chem. Soc.* **102**, 4889–4893.

Lippmaa, E., Samoson, A. V., Brei, V. V., and Gorlov, Y. I. (1981a). *Dokl. Akad. Nauk. SSSR* **259**, 403–408 (Eng. trans., 639–643).

Lippmaa, E., Mägi, M., Samoson, A., Tarmak, M. and Engelhardt, G. (1981b). *J. Am. Chem. Soc.* **103**, 4992–4996.

Lodge, E. A., Bursill, L. A., and Thomas, J. M. (1980). *J. Chem. Soc. Chem. Commun.*, 875–876.

Loewenstein, W. (1954). *Am. Mineral.* **39**, 92.

Maciel, G. E., and Sindorf, D. W. (1980). *J. Am. Chem. Soc.* **102**, 7606–7607.

Maciel, G. E., Sullivan, M. J., and Sindorf, D. W. (1981a). *Macromolecules* **14**, 1608–1611.

Maciel, G. E., Sindorf, D. W., and Bartuska, V. J. (1981b). *J. Chromatogr.* **205**, 438–443.

Marsmann, H. (1981a). *NMR: Basic Princ. Prog.* **17**, 65–235.

Marsmann, H. (1981b). *NMR: Basic Princ. Prog.* **17**, 226.

Maxwell, I. E. *et al.* (1982). *J. Chem. Soc. Chem. Commun.*, 523–524.

Morris, G. A., and Freeman, R. (1979). *J. Am. Chem. Soc.* **101**, 760–762.

Nagy, J. B., Gilson, J. P., and Derouane, E. G. (1981). *J. Chem. Soc. Chem. Commun.* 1129–1131.

Newman, T. H., West, R., and Oakley, R. T. (1980). *J. Organomet. Chem.* **197**, 159–168.

Olah, G. A., and Hunadi, R. J. (1980). *J. Am. Chem. Soc.* **102**, 6989–6992.

Olah, G. A., Berrier, A. L., Field, L. D., and Prakash, G. K. S. (1982). *J. Am. Chem. Soc.* **104**, 1349–1355.

Pannell, K. H., and Bassindale, A. R. (1982). *J. Organomet. Chem.* **229**, 1–9.

Pannell, K. H., Bassindale, A. R., and Fitch, J. W. (1981). *J. Organomet. Chem.* **209**, C65–68.

Ramdas, S., Thomas, J. M., Klinowski, J., Fyfe, C. A., and Hartman, J. S. (1981). *Nature* **292**, 228–230.

Reimer, J. A. (1981). *J. Phys. Colloq.* **C4**, 715–724.

Reimer, J. A., Murphy, P. D., Gerstein, B. C., and Knights, J. C. (1981). *J. Chem. Phys.* **74**, 1501–1503.

Reynolds, W. F., Thompson, J. C., and Wright, A. P. G. (1980a). *Can. J. Chem.* **58**, 425–435.

Reynolds, W. F., Thompson, J. C., and Wright, A. P. G. (1980b). *Can. J. Chem.* **58,** 436–446.

Sakurai, H. *et al.* (1980). *J. Organ. Chem.* **201,** C14–18.

Seyferth, D., Annarelli, D. C., Vick, S. C., and Duncan, D. P. (1980). *J. Organ. Chem.* **201,** 179–195.

Sindorf, D. W., and Maciel, G. E. (1981). *J. Am. Chem. Soc.* **103,** 4263–4265.

Stanislawski, D. A., and West, R. (1981a). *J. Organ. Chem.* **204,** 295–305.

Stanislawski, D. A., and West, R. (1981b). *J. Organ. Chem.* **204,** 307–314.

Stothers, J. B. (1972). "Carbon-13 NMR Spectroscopy", p. 376. Academic Press, New York.

Thomas, J. M., Bursill, L. A., Lodge, E. A., Cheetham, A. K., and Fyfe, C. A. (1981a). *J. Chem. Soc. Chem. Commun.* 276–277.

Thomas, J. M., Klinowski, J., Fyfe, C. A., Hartman, J. S., and Bursill, L. A. (1981b). *J. Chem. Soc. Chem. Commun.* 678–679.

Thompson, J. C., Wright, A. P. G., and Reynolds, W. F. (1981). *J. Fluorine Chem.* **17,** 509–529.

Williams, E. A., and Cargioli, J. D. (1979). *Annu. Rep. NMR Spectrosc.* **9,** 221–318.

Williams, E. A., Cargioli, J. D., and Donahue, P. E. (1980). *J. Organomet. Chem.* **192,** 319–327.

8 Magnesium-25 and Calcium-43

Klaus J. Neurohr,* Torbjörn Drakenberg, and Sture Forsén

Department of Physical Chemistry 2
University of Lund
Chemical Center
Lund, Sweden

I. Introduction

In the past few years, ^{25}Mg- and ^{43}Ca-NMR studies have provided valuable information on ion binding phenomena, particularly in biochemistry (Andersson *et al.*, 1982, and ref. cited therein; Forsén and Lindman, 1981a). The combined use of isotopically enriched ^{25}Mg or ^{43}Ca, Fourier transform (FT) techniques, high magnetic fields, and a solenoid-type

---

* Present address: Department of Molecular Biophysics and Biochemistry, Yale University, P.O. Box 6666, 260 Whitney Avenue, New Haven, Connecticut 06511.

229

probe design makes NMR studies on these cations now feasible at milli-molar concentrations.

Studies employing ^{25}Mg- and ^{43}Ca-NMR have become useful in the investigation of ion binding to proteins (Forsén and Lindman, 1981[a]). In principle, the method is capable of providing information on (a) associa-tion constants in the range $1-10^4\ M^{-1}$, (b) the competition of other cations for Mg^{2+}- and/or Ca^{2+}-binding site(s), (c) the effects of other ligands (drugs, etc.) on the protein, (d) the apparent pK values for the groups involved in Mg^{2+} or Ca^{2+} binding, and (e) dynamic parameters, namely, exchange rates and correlation times. Moreover, it is now possible to observe directly ^{43}Ca resonances of Ca^{2+} ions bound to the small Ca^{2+}-binding proteins parvalbumin, troponin C, and calmodulin, yielding infor-mation on the correlation time and the quadrupole coupling constant in the bound environment (Andersson $et\ al.$, 1982).

Although ^{25}Mg- and ^{43}Ca-NMR studies on ion binding to macro-molecules are still limited, the potential of such work has been clearly demonstrated. Both elements are widely distributed in nature. In the form of divalent ions, calcium and magnesium are involved in a large variety of physiological processes and are thus of fundamental importance in biol-ogy (Williams, 1970). Nuclear magnetic resonance studies on these cat-ions should prove particularly valuable because Mg^{2+} and Ca^{2+} lack other useful spectroscopic properties, such as EPR, UV, and visible spectra, and luminescence properties.

In biological and other macromolecular systems, slow motions gener-ally influence the magnetic relaxation, and the extreme narrowing condi-tion does not apply at the site with the slow motion. The interpretation of relaxation measurements is then complicated by the multiexponential re-laxation behavior of nuclei with spin $I \geqslant \frac{3}{2}$. However, for the regime of "nearly exponential" relaxation, i.e., for $\omega_0\tau_c \approx 1$, simple analytical ex-pressions for the apparent relaxation rates have been derived (Halle and Wennerström, 1981).

The principles of ^{25}Mg and ^{43}Ca NMR, as well as some chemical and biochemical applications, have been reviewed by Forsén and Lindman, 1981a). In this chapter, we shall focus our attention mainly on recent advances in this field, with emphasis on the relaxation behavior and sec-ond-order dynamic frequency shifts in nonextreme narrowing situations. Furthermore, we shall present new chemical shift and relaxation data, as well as quadrupolar coupling constants, for ^{25}Mg and ^{43}Ca in some small complexes. It is hoped that this brief look at the state of the art of ^{25}Mg and ^{43}Ca NMR will serve to illustrate the potential as well as the limita-tions of this technique in the study of ion binding phenomena in chemistry and biology.

TABLE I

Atomic and Nuclear Properties of ^{25}Mg and ^{43}Ca

Property	^{25}Mg	^{43}Ca
Spin, $I\hbar$	$\frac{5}{2}$	$\frac{7}{2}$
Natural abundance (%)	10.1	0.145
NMR resonance frequency at 8.45 T (MHz)	22.03	24.23
Electric quadrupole moment (10^{-28} m^2)	0.20[a]	−0.05[b]
Relative sensitivity of equal number of nuclei at constant field (^{1}H = 1.00)	2.68×10^{-3}	6.39×10^{-3}
Relative receptivity at natural isotopic abundance (^{13}C = 1.00)	1.54	0.053
Spin–lattice relaxation rate $1/T_1$ in aqueous solution at infinite dilution (s^{-1} at 25°C)	4.5[c]	0.75[d]
Sternheimer antishielding factor of M^{2+} ion[e]	4.32	13.12
Effective ionic radii of M^{2+} ions for different coordination numbers (CN) (Å)[f]		
6	0.72	1.00
8	0.89	1.12
12	—	1.34
Ionization potentials (kJ mol^{-1})[g]		
First	734	587
Second	1443	1140
Third	9700	4930
Absolute standard enthalpy of hydration of cation at 298 K (kJ mol^{-1})[h]	−1922	−1592
Absolute standard entropy of hydration of cation at 298 K (J mol^{-1} K^{-1})[h]	−268	−209

[a] Bauche et al., 1974.
[b] Olsson and Salomonson, 1982.
[c] Holz et al., 1979.
[d] For 0.2 M CaCl$_2$ solution, Forsén et al., unpublished.
[e] Lucken, 1969; Sen and Narasimhan, 1974.
[f] Shannon, 1976.
[g] Franklin et al., 1969.
[h] Data discussed in Chapter 7 of Burgess, 1978. On the scale chosen here the standard enthalpy of hydration of H$^+$ is −1090.8 kJ mol^{-1} and the standard molal entropy of H$^+$ (aq) is zero.

II. Physical Properties and NMR Parameters

Some of the important physical properties and NMR parameters of ^{25}Mg and ^{43}Ca are summarized in Table I. The NMR sensitivity of ^{43}Ca, for an equal number of nuclei, is ~2.5 times higher than that of ^{25}Mg. However, since the natural abundance of ^{43}Ca is only 0.15%, compared to 10.1% for ^{25}Mg, the receptivity of ^{43}Ca at natural isotopic abundance is one of the lowest of all elements with nonzero nuclear spin in the periodic table; only [57]Fe, ^{15}N, and a few other isotopes have lower values. The relative receptivity of [25]Mg at natural isotopic abundance is higher than that of [13]C (Table I). With the use of isotope-enriched material, the NMR sensitivity can be improved considerably (Section III). The Larmor frequencies and spin–lattice relaxation rates given in Table I are for ions in dilute aqueous solutions. The values of the nuclear electric quadrupole moments deserve a brief comment. Accurate values of quadrupole moments are, in general, difficult to obtain. They are usually determined by the analysis of a hyperfine splitting in the atomic spectrum of a particular isotope, caused by nuclear electric quadrupole–electron interactions. This analysis yields the electric quadrupole interaction constant with dimensions of frequency. In order to obtain the nuclear quadrupole moment from this parameter one must know the electric field gradient at the nucleus in the particular electronic state studied. In the case of [43]Ca, the interaction constant is known to four significant figures. However, the electric field gradient must be calculated theoretically and is associated with considerable uncertainty. The most refined theoretical calculations performed so far, using either multiconfigurational Hartree–Fock (Olsson and Salomonson, 1982) or many-body perturbation theory (Grundvik et al., 1979), indicate that the correct value is approximately -0.050×10^{-28} m^2. A similar value, -0.06×10^{-28} m^2, was obtained by comparing the experimental relaxation with that predicted by the theory of Hertz (1973) and treating the nuclear electric quadrupole moment as an unknown parameter (Forsén and Lindman, 1978; Lindman et al., 1977).

If one considers the typical NMR parameters of importance in solution studies, namely, chemical shifts, spin–spin coupling, relaxation times, and the diffusion coefficient, one can for these nuclides rule out spin–spin coupling, and the diffusion coefficient because of sensitivity and relaxation reasons. Although self-diffusion studies can be of great value, spin–echo NMR is ruled out for [25]Mg and [43]Ca because of the low magnetogyric ratios of these nuclei. In addition, the short quadrupolar relaxation times are also unfavorable for such studies.

The chemical shift range of [25]Mg is very small. Chemical shift effects on binding in aqueous solutions are therefore usually unobservably small compared with quadrupole relaxation effects, particularly for macromolecular systems. For [43]Ca, the chemical shift range is larger, and chemical shift changes can be observed on [43]Ca binding to small ligands (Section V), although the interaction with macromolecules is much more difficult to monitor in this way.

There are a considerable number of studies concerned with the solvent and concentration dependence of [25]Mg and [43]Ca chemical shifts and/or relaxation times and with the effects of ion pairing. In addition, Hertz (1973) and Deverell (1969) have developed theoretical approaches to ion quadrupole relaxation. These subjects are beyond the scope of this chapter and have been discussed in considerable detail in earlier reviews (Forsén and Lindman, 1981a; Lindman and Forsén, 1978; Lindman et al., 1977).

III. Experimental Aspects

As indicated in Table I, the resonance frequencies of ^{25}Mg and ^{43}Ca are quite low, particularly at the relatively low magnetic fields that can be obtained with iron core magnets. In fact, few spectrometers with iron core magnets have been adequately equipped for the detection of magnetic resonances at 6 to 7 MHz or lower frequencies. Because of this problem, and because of the low NMR sensitivity of ^{25}Mg and ^{43}Ca, few NMR studies on these nuclei were reported before 1977. A discussion of these early investigations of ^{25}Mg and ^{43}Ca NMR can be found in Forsén and Lindman (1981a).

The detection sensitivity of ^{25}Mg and ^{43}Ca is, of course, improved considerably by using superconducting magnets. At the same time, the resonance frequencies are significantly increased. For example, in a magnetic field of 8.45 T (corresponding to an ^{1}H resonance frequency of 360 MHz), the ^{25}Mg and ^{43}Ca resonate at 22.0 and 24.2 MHz, respectively. By using isotope-enriched material, the NMR sensitivity can be further improved, by a factor of ~10 for ^{25}Mg and of ~500 to 600 for ^{43}Ca. This is particularly valuable for physicochemical and biological studies, which usually involve much more than the mere detection of a signal. In fact, nearly all recent NMR studies concerned with ^{25}Mg or ^{43}Ca binding to various proteins (Andersson et al., 1982; Forsén and Lindman, 1981a,b; Marsh et al., 1979) and nucleic acids (Rose et al., 1980) have employed enriched samples. In the case of ^{25}Mg, the cost involved is not significant. For ^{43}Ca, however, the cost can be considerable, and the ^{43}Ca isotope generally must be recovered.

An additional improvement in the NMR sensitivity of ^{25}Mg and ^{43}Ca is possible through the use of sideways solenoid transmitter–receiver rf coils. Commercial NMR spectrometers with superconducting magnets are equipped in a standard way with the saddle-shaped Helmholtz coils around the cylindrical sample tube, positioned on the symmetry axis of the B_0 field. Although Helmholtz coils are superior to solenoids in terms of correcting B_0 field inhomogeneity, they are inferior to solenoids in terms of the achievable signal/noise ratio by a factor of about 2–2.5

(Hoult, 1978). To gain the full advantage of the solenoid coil it should have its axis perpendicular to the B_0 field, which implies that the sample must be inserted sideways into the probe. In order to change samples, the probe may have to be taken out of the magnet. Sideways solenoid coils have for many years been successfully employed in our laboratory in Lund as well as in other laboratories (Oldfield and Skarjune, 1978). As an example, Fig. 1 shows a probe designed and constructed at Lund University. This probe has an insert with a horizontally arranged 17-mm-ID transmitter–receiver solenoid coil mounted in a Teflon tube with an OD of 25 mm. The tube screws onto a 500-mm-long aluminum rod with four holes, one for the transmitter–receiver cable, one for a stick with a fixed capacitor, and the other two for adjustment of the tune and match capacitors. The aluminum rod with the transmitter–receiver coil can be easily removed to change the sample. An outer part of the probe (Fig. 1B), fixed to the normal Nicolet probe stock and variable-temperature unit, contains saddle-shaped (Helmholtz) decoupling and lock coils. The samples are provided in sample tubes with a volume of ~3 ml. Two such inserts have been made, one for the frequency range 10–38 MHz with 11 turns, and another for the range 35–100 MHz with 4 turns.

IV. Quadrupolar Effects

Nuclear magnetic relaxation of the quadrupolar nuclei ^{25}Mg and ^{43}Ca is due to the interaction of the nuclear electric quadrupole moment eQ with fluctuating electric field gradients at the nucleus. The field gradients may be intra- or intermolecular in origin, and the efficiency of the quadrupole relaxation depends on both the magnitude of the field gradients and the time scale of their fluctuations. The appearance of the NMR spectrum depends on whether the nucleus moves in an isotropic or in an anisotropic system. In the following, we shall first consider relaxation of ^{25}Mg and ^{43}Ca in isotropic systems. We shall briefly discuss relaxation under extreme narrowing conditions and then focus our attention mainly on the nonextreme narrowing case. Finally, we shall present some equations valid for anisotropic systems.

A. Relaxation in Isotropic Systems under Extreme Narrowing Conditions

When the correlation time describing the reorientation of the electric field gradient at the nucleus τ_c is much smaller than the inverse Larmor

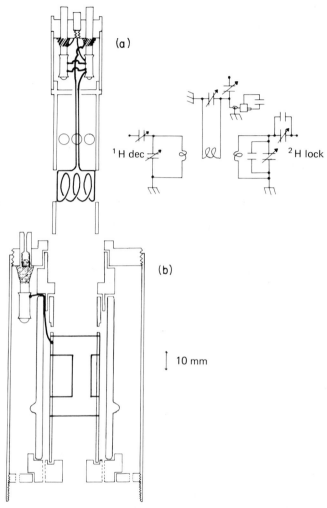

Fig. 1 (a) An insert with a horizontally arranged transmitter–receiver coil accommodating 17-mm-OD sample tubes. Tuning to the actual frequency is obtained by changing a fixed capacitor, and the fine tuning is made with a variable capacitor. (b) A probe head which is fully compatible with the Nicolet probes. It contains the ^{2}H lock and ^{1}H decoupling coils which are both Helmholz coils. All variable capacitors are 1–10 pF (JMC7585).

frequency ($\omega_0\tau_c \ll 1$), i.e., in the so-called extreme narrowing case, the decays of the longitudinal and transverse magnetizations are equal and exponential. As a result of fast molecular motion, the field gradients take up all directions with equal probability and only a single Lorentzian absorption line is observed. The relaxation rates $R_1 = 1/T_1$ and $R_2 = 1/T_2$ are equal and are given by (Abragam, 1961)

$$R_1 = R_2 = \frac{3\pi^2}{10} \frac{2I + 3}{I^2(2I - 1)} \chi^2 \left(\frac{1 + \eta^2}{3}\right) \tau_c \qquad (1)$$

where χ is the nuclear quadrupole coupling constant (in hertz) defined by

$$\chi = e^2 q_{zz} Q/h$$

where q is the electric field gradient at the nucleus with q_{zz} as the largest component, e the electron charge, Q the nuclear quadrupole moment, and η an asymmetry parameter for the field gradient. The asymmetry parameter lies in the range $0 < \eta < 1$ and in molecules rarely exceeds 0.5, such that $(1 + \eta^2/3 \approx 1.08$. For Mg^{2+} and Ca^{2+} complexes, the situation is not likely to be much different, so that η in Eq. (1) may be neglected for all practical purposes.

In studies on ion binding to biological macromolecules, extreme narrowing conditions are usually valid for the free site (normally the bulk solution). Moreover, relaxation of ^{25}Mg and ^{43}Ca in low-molecular-weight complexes occurs under extreme narrowing conditions. As seen from Eq. (1), a determination of the relaxation rates from the line width or from pulse experiments under extreme narrowing conditions yields only the product of the correlation time and the square of the quadrupole coupling constant. An independent measurement of τ_c is therefore required to calculate χ. This can be done, for example, by measuring spin–lattice relaxation times of ^{13}C nuclei of the ion complex (Section V), assuming that it reorients as a rigid body.

B. Relaxation in Isotropic Systems under Nonextreme Narrowing Conditions

In biological studies, for example, ion binding to macromolecules, extreme narrowing conditions are usually not valid for the bound site, with the result that the relaxation becomes multiexponential. In this case, the concept of relaxation times is no longer valid and the decays of the longitudinal and transverse magnetizations must be considered explicitly. Hubbard (1970) has shown that for half-integer spins the longitudinal (M_1) and the transverse (M_2) magnetizations decay as weighted sums of $I + \frac{1}{2}$ exponentials:

$$M_1(t) = M_1(\infty) \left\{1 - k \left[\sum_{i=1}^{I+1/2} C_{1,i} \exp(-R_{1,i}t)\right]\right\} \qquad (2)$$

$$M_2(t) = M_2(0) \left[\sum_{i=1}^{I+1/2} C_{2,i} \exp(-R_{2,i}t) \right] \tag{3}$$

where $\sum_{i=1}^{I+1/2} C_{\alpha,i} = 1$ and $k = 2$ for an inversion recovery experiment.

For the case where spin $I = \frac{3}{2}$, the preexponential factors in Eqs. (2) and (3) do not depend on the parameters that determine the relaxation (ω, τ_c), and analytical expressions have been derived for the magnetization decays (Bull, 1972). In contrast, for spin-$\frac{5}{2}$ and spin-$\frac{7}{2}$ nuclei like ^{25}Mg and ^{43}Ca, the relaxation behavior is much more complicated, because in this case the preexponential factors in Eqs. (2) and (3) (namely, the intensities of the different components) also depend on the relaxation parameters. Therefore, exact analytical solutions of the relaxation equations cannot be obtained, but they can be solved numerically for each set of values for ω_0 and τ_c (Bull et al., 1979; Reuben and Luz, 1976).

The numerical solutions show that in many regions the relaxation is dominated by a single exponential [i.e., by one of the relaxation components (Bull et al., 1979)]. In practice, the relaxation usually appears exponential, even when the contribution from the bound state is considerable and the longitudinal and transverse relaxation rates are unequal and frequency-dependent. Pronounced deviations from simple exponential decays and Lorentzian absorption line shapes have been observed for ^{23}Na ($I = \frac{3}{2}$) by several workers (cf. Forsén and Lindman, 1981b). In the case of ^{25}Mg ($I = \frac{5}{2}$) Rose et al. (1980) have observed non-Lorentzian line shapes. In the ^{43}Ca-NMR studies on Ca^{2+} binding to proteins reported so far, the observed line shapes were virtually Lorentzian (Andersson et al., 1982; Forsén and Lindman, 1981a). Nonextreme narrowing effects were, however, clearly indicated by the fact that T_1, as obtained from pulse experiments, was not equal to the apparent T_2 obtained from the line width. Unlike the situation in extreme narrowing, where only the product of the quadrupole coupling constant and the correlation time is obtained [Eq. (1)] (requiring independent measurement of τ_c to calculate χ), nonexponential relaxation behavior in principle allows separate determination of the two parameters. For spin-$\frac{3}{2}$ nuclei, useful expressions for the apparent T_1 and T_2 have been derived by a linearization of the magnetization decays, which are valid for $\omega_0\tau_c \leq 1.5$ (Bull, 1972).

Halle and Wennerström (1981) employed a perturbation treatment to derive approximate analytical expressions for the longitudinal and transverse relaxation rates in the regime of nearly exponential relaxation for spin-$\frac{5}{2}$ and spin-$\frac{7}{2}$ nuclei.

In this treatment, the extreme narrowing situation is regarded as being perturbed by allowing for a frequency dependence in the spectral density. The main results of this treatment will be summarized in the following discussion; for details the reader is referred to Halle and Wennerström (1981).

To first order, the longitudinal and transverse relaxation rates of the major relaxation component are given by

$$\langle R_1 \rangle = \frac{3\pi^2}{10} \chi^2 \frac{2I + 3}{I^2(2I - 1)} (0.2J_1 + 0.8J_2) \tag{4}$$

$$\langle R_2 \rangle = \frac{3\pi^2}{10} \chi^2 \frac{2I + 3}{I^2(2I - 1)} (0.3J_0 + 0.5J_1 + 0.2J_2) \tag{5}$$

where χ is the quadrupole coupling constant. It should be mentioned that these equations are analogous to those derived by McLachlan (1964) for the case of electron spin relaxation. In order to assess the accuracy of the description of the relaxation behavior as a simple exponential decay with a rate given by Eqs. (4) and (5), Halle and Wennerström (1981) compared the relaxation rates obtained via these equations with experimentally accessible apparent relaxation rates for a fast-exchange two-state model. For rapid exchange of a quadrupolar nucleus between a bound state (B) and a free state (F), the effective spectral density is a weighted average of the individual spectral densities:

$$J_q = P_F J_q^F + P_B J_q^B \tag{6}$$

where P_F and P_B are the relative populations of the two states. For the free state, extreme narrowing conditions are assumed to be valid, whereas the spectral density in the bound state is frequency-dependent. Assuming isotropic motion in the bound state with a correlation time τ_{cB}, one obtains

$$J_q^F = J_0^F$$

$$J_q^B = \frac{1}{10} (V_{zz}^M)^2 \left(1 + \frac{\eta^2}{3}\right) \frac{\tau_{cB}}{1 + (q\omega_0\tau_{cB})^2} \tag{7}$$

where V_{zz}^M stands for the maximum component of the field gradient tensor.

Because of the frequency-independent first term in Eq. (6), the relative variation of the spectral density with frequency is smaller and the first-order Eqs. (4) and (5) are expected to be valid over a fairly wide range of conditions. On inserting Eqs. (6) and (7) into Eqs. (4) and (5), the following "reduced" relaxation rates are obtained:

$$\langle R_1 \rangle / P_F R_F = 1 + Q(0.2\tilde{J}_1^B + 0.8\tilde{J}_2^B) \tag{8}$$

$$\langle R_2 \rangle / P_F R_F = 1 + Q(0.3 + 0.5\tilde{J}_1^B + 0.2\tilde{J}_2^B) \tag{9}$$

with

$$\tilde{J}_q{}^B \equiv J_q{}^B/J_0{}^B = [1 + (q\omega_0\tau_{cB})^2]^{-1} \tag{10}$$

$$Q \equiv (P_B/P_F)(\chi_{eB}{}^2/\chi_{eF}{}^2)(\tau_{cB}/\tau_{cF}) \tag{11}$$

where the effective quadrupole coupling constant is defined by

$$\chi_e = \frac{|eQV_{zz}{}^M|}{h}\left(1 + \frac{\eta^2}{3}\right)^{1/2} \tag{12}$$

The symbol $\langle R_a \rangle$ denotes the rates given by Eqs. (4) and (5). Reduced apparent relaxation rates $R_a^*/P_F R_F$ (with $a = 1, 2$) were obtained by numerical diagonalization of the full-relaxation matrices for given values of Q and $\omega_0\tau_{cB}$. Apparent relaxation rates were then calculated from the line width of the $I + \frac{1}{2}$ superimposed Lorentzians (R_2^{*a}) or from a least squares fit of a simple exponential decay (R_1^{*e}, R_2^{*e}) (Halle and Wennerström, 1981). The error made by equating the apparent relaxation rate R_1^{*e}, R_2^{*e}, or R_2^{*a} with the first-order rate $\langle R_a \rangle$, as given by Eqs. (4) and (5), is illustrated in Fig. 2. As seen from the figures for $I = \frac{5}{2}$ and $I = \frac{7}{2}$, the error in the longitudinal rate (dotted curves) is a maximum for $\omega_0\tau_{cB} \approx 1$. Although the longitudinal decay is more nearly exponential (because the dominating amplitude is closer to unity) than the transverse decay, the first-order expressions may be less accurate for the longitudinal rates (dotted curves) than for the transverse rates (dashed curves) because the minor components are associated with larger relaxation rates in the longitudinal case.

It may be mentioned that the approximation improves considerably on going from $I = \frac{3}{2}$ to $I = \frac{5}{2}$, with a further slight improvement for $I = \frac{7}{2}$. As seen from Fig. 2 for spin-$\frac{5}{2}$ and spin-$\frac{7}{2}$ nuclei, the first-order rates are accurate to within a few percent for Q values on the order of unity or less. It seems therefore that the analytical expressions, Eqs. (4) and (5), should be useful in all cases, except possibly for large populations of nuclei rigidly bound to large macromolecules. Furthermore, if the correlation time τ_{cB} in the bound state is calculated from the ratio of the excess ($R_{aB} - R_{aF}$) longitudinal and transverse rates at a single frequency, a cancellation effect in the error occurs, as illustrated in Fig. 3.

The analytical expressions for the apparent relaxation rates, Eqs. (4) and (5), have recently proved useful in the study of ^{43}Ca resonances of Ca^{2+} ions bound to the three proteins parvalbumin, troponin C, and calmodulin (Andersson et al., 1982). These Ca^{2+}-binding proteins are rather small, with molecular weights in the range 12,000–18,000, and the relaxation of the bound ^{43}Ca nuclei occurs near the extreme narrowing limit. Equations (4) and (5) yielded correlation times in the range 4–11 ns

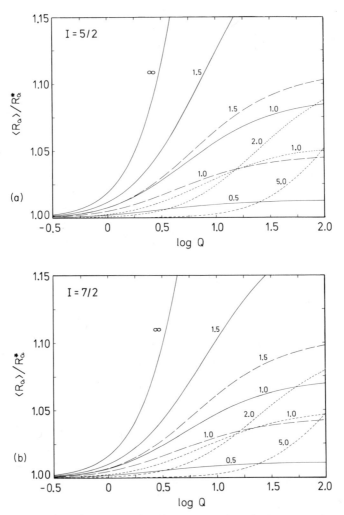

Fig. 2 The ratio of the first-order relaxation rate R_a, given by Eqs. (4) and (5) and the apparent rates R_1^{*e} (dotted curves), R_2^{*e} (dashed curves), or R_2^{*a} (solid curves), for indicated spin I. The dimensionless parameters are Q, which is defined by Eq. (11) and $\omega_0 \tau_{cB}$, the value of which appears beside each curve. (From Halle and Wennerström, 1981.)

and quadrupolar coupling constants in the range 1.05–1.3 MHz. The agreement between the apparent relaxation rates obtained from Eqs. (4) and (5) and those obtained numerically is good up to $\omega_0 \tau_c \approx 1$.

In practical applications of ^{25}Mg and ^{43}Ca NMR, it is often very difficult to achieve a sufficient signal/noise ratio to record the line shape reliably as a whole, and one can only measure the chemical shift and the line width

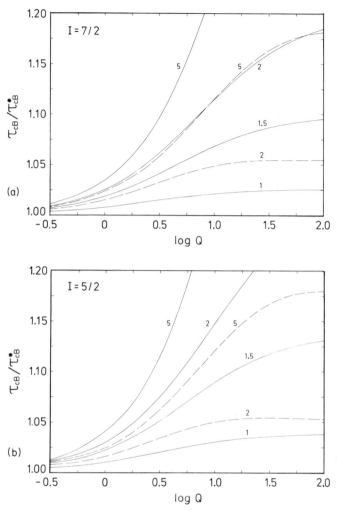

Fig. 3 The ratio of the actual correlation time τ_{cB} and the apparent correlation time τ_{cB}^* obtained by equating the first-order relaxation rates with the corresponding apparent rates: R_1^{*e} and R_2^{*e} (dashed curves) or R_1^{*e} and R_2^{*a} (solid curves), for indicated spin I. The dimensionless parameters are Q, which is defined by Eq. (11), and $\omega_0\tau_{cB}$, the value of which appears beside each curve. (From Halle and Wennerström, 1981.)

at half-height. Moreover, receiver dead times are generally so long, particularly at low frequencies, that the most rapidly decaying components are partially lost before the free induction decay (FID) is accumulated. As an example, Andersson *et al.* (1982) used a dead time of 300 μs. The effect of the long dead time is to increase the relative amplitude of the dominant relaxation component, thus making it difficult to observe experimentally deviations from Lorentzian line shapes. In addition, a long dead time results in a considerable loss of signal intensity due to relaxation during the dead time.

C. Second-Order Dynamic Frequency Shifts

An additional complication in the description of the relaxation behavior for nuclei with spin $I \geq \frac{3}{2}$ arises from the fact that the quadrupole interaction not only gives rise to relaxation but also generates a second-order dynamic shift. An illustration of the origin of this shift is found by considering the rigid lattice limit. For the $m = \frac{1}{2} \rightarrow m = -\frac{1}{2}$ transition, the first-order quadrupole interaction is zero and a second-order quadrupole shift, with a nonzero isotropic average, is obtained (Wennerström et al., 1974).

The dynamic shift has usually been neglected because it is smaller than the broadening (Abragam, 1961). However, although the dynamic shifts are clearly smaller than the width of the broadest component, they are often larger than the width of the narrowest component in multiexponential decay situations and influence the spectra to an extent that is important in a quantitative description (Werbelow and Pouzard, 1981).

It is interesting to note that the modulation of the quadrupole interaction in quadrupole NMR relaxation is analogous to the modulation of the zero-field splitting in ESR for systems with $S \geq 1$. Nonextreme narrowing is much more common in ESR, because of the difference in Larmor frequencies, and multiexponential relaxation has been discussed quite extensively, including the consequences of dynamic shifts (see, for example, Poupko et al., 1974, and references cited therein; Rubenstein et al., 1971).

For quadrupolar relaxation, the effects of second-order dynamic frequency shifts have been discussed in detail for the spin-$\frac{3}{2}$ case (Fouques and Werbelow, 1979; Werbelow, 1979; Werbelow and Marshall, 1981). For a spin-$\frac{3}{2}$ nucleus, under nonextreme narrowing conditions, the two transverse relaxation components not only have different relaxation rates but also differ in their absorption frequencies. Consequently, the NMR line shape becomes asymmetric, which makes adjustment of the phase a difficult problem. Clearly, if one were unaware of the asymmetry of the absorption line shape under these conditions, an incorrect phase adjustment could obscure the effect and result in erroneous determination of the relaxation rates.

Westlund and Wennerström (1982) have derived expressions for the line shapes of nuclei with spin $\frac{5}{2}$ and $\frac{7}{2}$, both in the presence and in the

Def. of spectral densities

$$\begin{cases} J(\omega) \equiv \int_0^\infty \langle v_o(0)v_o(\tau)\rangle \cos\omega\tau \, d\tau \quad\Longrightarrow\quad 0.3\,V_{zz}^2 \cdot \dfrac{\tau_c}{(1+\omega^2\tau_c^2)} \\[2ex] Q(\omega) \equiv \int_0^\infty \langle v_o(0)v_o(\tau)\rangle \sin\omega\tau \, d\tau \quad\Longrightarrow\quad 0.3\,V_{zz}^2 \cdot \dfrac{\omega\tau_c^2}{(1+\omega^2\tau_c^2)} \end{cases}$$

(A) <u>Slow motion limit</u> $(J(o) \gg J(\omega_o), J(2\omega_o))$:

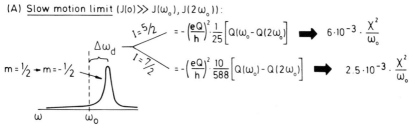

$$I = \tfrac{5}{2} \qquad = -\left(\frac{eQ}{h}\right)^2 \frac{1}{25}\left[Q(\omega_o) - Q(2\omega_o)\right] \quad\Longrightarrow\quad 6\cdot10^{-3}\cdot\frac{\chi^2}{\omega_o}$$

$$I = \tfrac{7}{2} \qquad = -\left(\frac{eQ}{h}\right)^2 \frac{10}{588}\left[Q(\omega_o) - Q(2\omega_o)\right] \quad\Longrightarrow\quad 2.5\cdot10^{-3}\cdot\frac{\chi^2}{\omega_o}$$

$m = \tfrac{1}{2} \rightarrow m = -\tfrac{1}{2}$

(B) <u>Near extreme narrowing</u> $(J(o) \gtrsim J(\omega) \gtrsim J(2\omega))$:

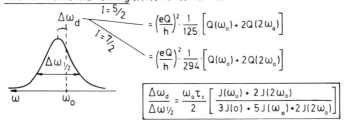

$$I = \tfrac{5}{2} \qquad = \left(\frac{eQ}{h}\right)^2 \cdot \frac{1}{125}\left[Q(\omega_o) + 2Q(2\omega_o)\right]$$

$$I = \tfrac{7}{2} \qquad = \left(\frac{eQ}{h}\right)^2 \cdot \frac{1}{294}\left[Q(\omega_o) + 2Q(2\omega_o)\right]$$

$$\boxed{\frac{\Delta\omega_d}{\Delta\omega_{1/2}} = \frac{\omega_o\tau_c}{2}\left[\frac{J(\omega_o) + 2J(2\omega_o)}{3J(o) + 5J(\omega_o) + 2J(2\omega_o)}\right]}$$

(C) Extreme narrowing

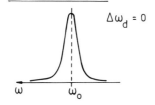

$\Delta\omega_d = 0$

Fig. 4 Dynamic frequency shifts: some general characteristics.

absence of chemical exchange. Since these expressions are rather lengthy, they will not be repeated here. Rather, we shall discuss the effects of the second-order dynamic shifts on the NMR line shapes and point out complications that may be encountered in practical applications. The main results are summarized in Fig. 4. In extreme narrowing, there is only a single Lorentzian line, and the dynamic shift is zero. In the slow-motion limit, where $J(0) >> J(\omega_0)$, $J(2\omega_0)$, the narrow $m = \frac{1}{2} \rightarrow m = -\frac{1}{2}$ transition dominates the spectrum and the signal asymmetry is small. Furthermore, the signal is shifted to lower frequency or higher field, and the dynamic shift approaches a constant value proportional to χ^2/ω_0, where χ is the quadrupole coupling constant (Fig. 4). In the near extreme narrowing limit ($\omega_0\tau_c \approx 1$), where $J(0) \gtrsim J(\omega_0) \gtrsim J(2\omega_0)$, there is essentially only one component in the signal, and the line shape is virtually Lorentzian with only slight asymmetry. In this case, there is a dynamic shift to higher frequency (or lower field) (Fig. 4), and it is also possible to derive a relationship between the dynamic shift $\Delta\omega_d$ and the line width $\Delta\omega_{1/2}$ valid for both $I = \frac{5}{2}$ and $I = \frac{7}{2}$:

$$\frac{\Delta\omega_d}{\Delta\omega_{1/2}} = \frac{\omega_0\tau_c}{2} \left[\frac{J(\omega_0) + 2J(2\omega_0)}{3J(0) + 5J(\omega_0) + 2J(2\omega_0)} \right] \tag{13}$$

This expression depends only on ω_0 and τ_c. In the transition region, however, markedly asymmetric line shapes appear, even when there is only one site.

In practical applications of ^{25}Mg and ^{43}Ca NMR, one often deduces the apparent T_2 from the line width of the signal at half-height. As pointed out by Westlund and Wennerström (1982), this procedure can be misleading when the actual line shape is not known. Furthermore, if one follows changes in line width as a function of temperature, the effects observed can be caused by changes in the line shape and one can arrive at qualitatively incorrect conclusions. Westlund and Wennerström (1982) have described some of the effects that can be encountered for a system in the absence of chemical exchange and for a two-site exchange situation.

For a system with one site, in the slow-motion limit at low temperature, the relatively narrow $m = \frac{1}{2} \rightarrow m = -\frac{1}{2}$ transition is observed. When the temperature is increased, the broader components start to contribute to the line width. Furthermore, the different components are shifted relative to one another, and there is also a shift contribution to the line width. This behavior can lead to an increase in line width with increasing temperature in a certain temperature range. At still higher temperatures, the system approaches the extreme narrowing limit (where there is only a single Lorentzian), and the line width decreases again. If one followed the variation in line width as a function of temperature, such behavior could easily be mistaken for a two-site exchange situation, where the exchange goes from slow to fast. In addition, in a two-site exchange situation, where one site represents the ion in the bulk solution, the temperature dependence of the line width does not follow what is expected for a spin-$\frac{1}{2}$ nucleus. This is schematically shown in Fig. 5. The detailed shape of the theoretical

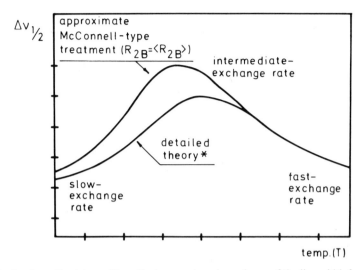

Fig. 5   A schematic picture of how the temperature dependence of the line width for a two-site exchange of a spin-$\frac{5}{2}$ or -$\frac{7}{2}$ nucleus may deviate from what is observed for a spin-$\frac{1}{2}$ nucleus. The effect is always such that the temperature dependence becomes less pronounced. The magnitude of the effect depends on the correlation time and the quadrupole coupling constant. (*Westlund and Wennerström, 1982.)

temperature dependence depends strongly on the correlation time and the quadrupole coupling constant, however, this temperature dependence is always less pronounced than that given by the approximate treatment.

D. Quadrupole Splittings for Spin-$\frac{5}{2}$ and Spin-$\frac{7}{2}$ Nuclei

In an anisotropic system (e.g., a solid or a liquid crystalline phase) with a nonzero field gradient, there is a combined effect of magnetic Zeeman and electric quadrupole interactions, the latter being different for the different magnetic quantum numbers m_I. The NMR spectrum (usually obtained under the condition $\Delta m_I = 1$) then consists of $2I$ lines (Abragam, 1961). To first order, the lines are equidistant and the frequency separation of two adjacent lines is called the quadrupole splitting. Quadrupole splittings can give information on orientation effects both on the microscopic and the macroscopic levels and, for a quadrupole ion, on the mode of ion binding. The general principles for the spin-$\frac{3}{2}$ case have been summarized in a review (Forsén and Lindman, 1981b), and theoretical treatments of ion quadrupole splittings in liquid crystals are available (Wennerström *et*

al., 1974; Lindblom *et al.*, 1976). For spin-$\frac{5}{2}$ and spin-$\frac{7}{2}$ nuclei the principles valid for the $I = \frac{3}{2}$ case apply directly, but in the case of relaxation
under nonextreme narrowing conditions, important complications arise.
We shall therefore in this chapter consider only quadrupole splittings for
the case where ion exchange between different sites is rapid compared to
the splittings. Furthermore, it is assumed that extreme narrowing conditions are valid for all sites. For a macroscopically oriented (lamellar)
liquid crystal, the quadrupole splitting is given by the equation

$$\Delta(\theta_{LD}) = \left| (3 \cos^2 \theta_{LD} - 1) \sum \frac{3P_i \chi_i S_i}{4hI(2I - 1)} \right| \tag{14}$$

whereas for a powder sample, the splitting is given by

$$\Delta_p = \left| \sum \frac{3P_i \chi_i S_i}{4hI(2I - 1)} \right| \tag{15}$$

where θ_{LD} is the angle between the normal to the lamellae (the director)
and the magnetic field, P_i the fraction of ions at site i, and χ_i the quadrupole coupling constant for the ion at site i. The S_i are order parameters
describing the partial orientation of the field gradient with respect to the
director, given by

$$S_i = \tfrac{1}{2}(3 \cos^2 \theta_{DM,i} - 1) \tag{16}$$

where $\theta_{DM,i}$ is the angle between the director and the electric field gradient
at site i. The relaxation rates are given by

$$\frac{1}{T_1} = \frac{1}{T_2} = \frac{3\pi^2}{10} \frac{2I + 3}{2I - 1} \sum P_i \chi_i^2 \tau_{ci} \tag{17}$$

The definitions of the angles θ_{LD} and θ_{DM} are given in Fig. 6.

V. ^{43}Ca- and ^{25}Mg-NMR Studies on Small Ca and Mg Complexes

A few early ^{25}Mg- and ^{43}Ca-NMR experiments demonstrated the possibility of using NMR in studying, for example, complex formation (Magnusson and Bothner-By, 1969; Bryant, 1969). These studies and other early
investigations using ^{25}Mg and ^{43}Ca are discussed in a review (Forsén and
Lindman, 1981a) and will not be dealt with further in this chapter.
    Evidently very little ^{25}Mg NMR work has been done on magnesium

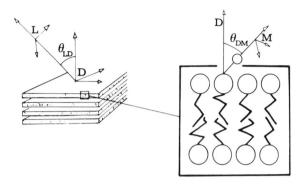

Fig. 6 Schematic drawing of the mesomorphous structure in a lamellar phase. The different coordinate systems used are outlined in the figure, laboratory frame (L), directory frame (D), and molecular frame (M). θ_{LD} and θ_{DM} are angles between the z axes in laboratory–director systems and directory–molecular systems, respectively. (From Wennerström *et al.*, 1974).

binding to small ligands. The only direct observation is, to our knowledge, one by Bouhoutsos-Brown *et al.* (1981). They reported the pH dependence of the ^{25}Mg-NMR spectrum from a solution containing Mg^{2+} and EDTA in a 1:1 ratio. At high pH values one signal from magnesium bound to EDTA is observed, whereas for pH values from 7 to 9 two overlapping lines are observed: a narrow line from free Mg^{2+} and a broad line from complexed ions. They did not observe chemical shift difference, which is in line with the observations by Robertson *et al.* (1978) for the Mg-Z-Gla-Gla-OMe complex and by Kraft *et al.* (1981) for the interaction between Mg^{2+} and acetylacetone.

It has been shown that Ca^{2+} bound to small ligands can give rise to well-resolved ^{43}Ca-NMR signals (Bouhoutsos-Brown *et al.*, 1981; Drakenberg, 1982; Farmer and Popov, 1981). In all three publications the ^{43}Ca chemical shift from the Ca–EDTA complex was reported (~20 ppm), and Drakenberg also reported the quadrupole coupling constant (0.5 MHz). Very few complexes have been studied by ^{43}Ca NMR, presumably partly because of the low receptivity of the ^{43}Ca nucleus and partly because there are not too many Ca complexes where the metal exchange is slow on the NMR time scale. Figure 7 summarizes the ^{43}Ca-NMR data on calcium complexes known as of this writing. Two things, which can be seen from Fig. 7, are worth mentioning. First, all ligands with only oxygens coordinated to the calcium ion cause a low-frequency shift of the ^{43}Ca-NMR signal, whereas other ligands with nitrogen coordination as well give rise to high-frequency shifts. Second, the quadrupole coupling constant for ^{43}Ca

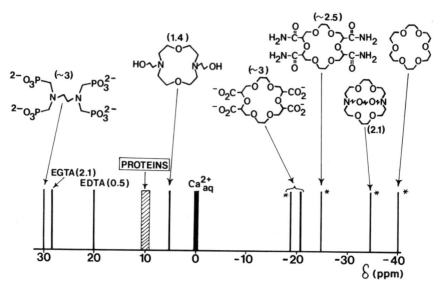

Fig. 7 A collection of ^{43}Ca chemical shifts of various small complexes. Shifts are given in parts per million relative to the water-solvated ion and are positive to higher frequency. An asterisk indicates that the value is for a methanol solution, otherwise the solution was water. The numbers in parentheses are quadrupole coupling constants in megahertz.

bound to EDTA is much smaller than for any of the other ligands. The reason for this is not well understood, but it must in some way be due to a higher effective symmetry or a lower field gradient. Even the calcium-binding proteins, which presumably have close to an octahedral symmetry with six oxygen ligands, have a quadrupole coupling constant twice as large as the one for the EDTA complex (Andersson *et al.*, 1982). This may indicate that the symmetry around the Ca^{2+} ions in the calcium-binding proteins is not as high as normally assumed.

In addition to the data in Fig. 10 there are a few reports of chemical shifts of the ^{43}Ca-NMR signal caused by the addition of weakly complexing ligands in fast chemical exchange. Lutz *et al.* (1975) studied the concentration dependence of the ^{43}Ca shift for several calcium salts. They found that the oxyanions formate and lactate caused shifts to lower frequencies, whereas the halide ions gave rise to shifts to higher frequencies. Robertson *et al.* (1978) reported a high-frequency shift (~2 ppm) for the ^{43}Ca-NMR signal for calcium bound to the peptide Z-Gla-Gla-OMe, and Kraft *et al.* (1981) found a high-frequency shift for calcium bound to acetylacetone.

VI. Ca^{2+} and Mg^{2+} Chemical Exchange: Strategy in Evaluating Experimental Data

Very often in studies on Mg^{2+} and Ca^{2+} binding to small ligands or macro-molecules the metal ion exchange is not slow on the NMR time scale. Bouhoutsos-Brown *et al.* (1981) have shown that the calcium exchange rate for the Ca^{2+}–EDTA system is in the intermediate range at pH 4.7 and room temperature. They have also shown that the magnesium ion exchange varies from fast at pH 5 to slow at pH 7 for the Mg^{2+}–EDTA system. It has also been shown on several occasions that the calcium exchange for Ca^{2+} ions bound to various proteins is in the intermediate range at room temperature (Andersson *et al.*, 1981a, 1981b, 1982). To treat data obtained from, for example, variable-temperature ^{43}Ca-NMR studies on metal binding, a calculation scheme based on the total band shape equation has been developed (Drakenberg *et al.*, unpublished). The strategy used in this calculation scheme is as follows.

(a)   Bloch equations modified for exchange in the way outlined by McConnell are used to derive the band shape $G(v)$ for a system of N exchanging sites (Sutherland, 1971).

(b)   If it is assumed that the temperature dependence of the exchange is in accordance with transition state theory (Glasstone *et al.*, 1941), the exchange rate is obtained as a function of temperature T and the enthalpy $\Delta H\ddagger$ and entropy $\Delta S\ddagger$ of activation.

(c)   The perturbation expression given in Section IV,B [Eq. (4)] (valid for $\omega\tau_c \lesssim 1.5$) is used to calculate the longitudinal relaxation time as a function of the correlation time τ_c and the quadrupole coupling constant χ.

(d)   The correlation time can be obtained from, for example, ^{13}C spin–lattice relaxation times or, for larger systems with sufficiently long correlation times (so that the condition $\omega\tau_c \ll 1$ is not fulfilled), from measurements of both T_1 and T_2 for the metal nucleus. Furthermore, τ_c is temperature-dependent and can be assumed to follow

$$1/\tau_c = (kT/h) \exp(-\Delta G\ddagger/RT)$$

(e)   Temperature dependence of binding constants can also be included as a temperature dependence in the populations, but usually this is not necessary because the conditions are mostly such that the binding sites are saturated with metal ions.

(f)   Nuclear magnetic resonance spectra for ~10 temperatures are used for an iterative least squares fit between experimental and calculated band shapes. This results in values for the quadrupole coupling constant χ for the bound ion and for the enthalpy and entropy of activation for the metal ion exchange.

The same scheme can also be used to obtain binding constants in the range $1–10^4 \, M^{-1}$, when the metal ion exchange is fast to intermediate. In this case the binding constant, the quadrupole coupling constant for the bound ions, and the metal exchange rate are varied to obtain the best fit between experimental and calculated band shapes. This calculation scheme has been applied to several studies on calcium binding to proteins. Some of the results are discussed in Vol. I, Chapter 7.

VII. Future Prospects

Although this chapter hopefully has outlined some of the basic characteristics of ^{25}Mg and ^{43}Ca NMR, it is certainly by no means exhaustive. Indeed, FT NMR spectroscopic investigation of these two nuclei has just begun, and much work remains to be done.

On the instrumentation side undoubtedly some additional increases in sensitivity may in the future be gained by using very high magnetic fields ($B_0 > 11.4 \, T$), but the sensitivity of ^{25}Mg and ^{43}Ca enriched to high isotopic content is already such that millimolar and even submillimolar concentrations can be studied with no great effort.

More detailed studies are needed on quadrupole coupling constants as well as on other NMR parameters for small-molecule complexes and on how these parameters relate to the structural and chemical features of the complexes. Such work will be of value not the least as a basis for the interpretation of ^{25}Mg- and ^{43}Ca-NMR data on biological systems.

As will be further discussed in Chapter 8 ^{25}Mg and ^{43}Ca NMR has already provided valuable information such as chemical exchange rates, activation parameters, and mobility characterizing the ion-binding site(s) in the study of Mg^{2+}- and Ca^{2+}-binding macromolecules. This area holds considerable promise for the future, and new applications for both proteins and nucleic acids as well as other biological polyelectrolytes may be foreseen.

A word of caution may, however, be mentioned at this point. As we have seen, quadrupolar effects in NMR spectra of spin $I > 1$ nuclei can become quite complex, in particular when nonextreme narrowing conditions prevail. The combined effects of nonexponential relaxation and second-order dynamic frequency shifts will, for example, make the extraction of exchange rates through variable temperature studies a difficult matter. This is not to say that these effects will be a hindrance—on the contrary they increase the information content in the spectrum—but rather that considerable care must be exercised in analysis of the experimental data. It is to be expected that ^{25}Mg-NMR studies in macromolecular systems will be somewhat more difficult than ^{43}Ca-NMR studies because of the larger quadrupole moment in the former case.

An area that we have not touched on in this chapter is the solid state NMR of ^{25}Mg and ^{43}Ca complexes. At the present time it is difficult to predict how useful these techniques may become in the future. So far it has been demonstrated that $m = \frac{1}{2} \rightarrow m = -\frac{1}{2}$ transitions in the solid state can be made reasonably narrow for spin-$\frac{3}{2}$ nuclei, like ^{23}Na, but the general utility of this technique for spin-$\frac{5}{2}$ and spin-$\frac{7}{2}$ nuclei still has not been substantiated.

References

Abragam, A. (1961). "The Principles of Nuclear Magnetism." Oxford Univ. Press, London.

Andersson, T., Drakenberg, T., Forsén, S., Thulin, E., and Swärd, M. (1982). *J. Am. Chem. Soc.* **104**, 576–580.

Andersson, T., Drakenberg, T., Wieloch, T., and Lindström, M., (1981a). *FEBS Lett.* **125**, 115–117.

Andersson, T., Drakenberg, T., Forsén, S., and Thulin, E. (1981b). *FEBS Lett.* **125**, 39–43.

Bauche, J., Couarrage, G., and Labarthe, J.-J. (1974). *Z. Phys.* **270**, 311–318.

Bouhoutsos-Brown, E., Rose, D. M., and Bryant, R. G. (1981). *J. Inorg. Nucl. Chem.* **43**, 2247–2248.

Bryant, R. G. (1969). *J. Amer. Chem. Soc.* **91**, 1870–1871.

Bull, T. E. (1972). *J. Magn. Res.* **8**, 344–353.

Bull, T. E., Forsén, S., and Turner, D. L. (1979). *J. Chem. Phys.* **70**, 3106–3111.

Burgess, J. (1978). "Metal Ions in Solution." Horwood, Chichester.

Deverell, G. (1969). *Prog. NMR Spectrosc.* **4**, 235–334.

Drakenberg, T. (1982). *Acta Chem. Scand.* **A36**, 79–82.

Farmer, R. M., and Popov, A. I. (1981). *Inorg. Nucl. Chem. Lett.* **17**, 51–56.

Forsén, S., and Lindman, B. (1978). *Chem. Br.* **14**, 29–35.

Forsén, S., and Lindman, B. (1981a). *Ann. Rep. NMR Spectrosc.* **11A**, 183–226.

Forsén, S., and Lindman, B. (1981b). "Methods of Biochemical Analysis" (D. Glick, ed.), Vol. 27, pp. 290–486. Wiley, New York.

Fouques, C. E. M., and Werbelow, L. G. (1979). *Can. J. Chem.* **57**, 2329–2332.

Franklin *et al.* (1969). "Ionization Potentials, Appearance Potentials and Heats of Formation of Gaseous Positive Ions." National Bureau of Standard, Reference Data Series 26, U.S. Department of Commerce.

Glasstone, S., Laidler, K. J. and Egring (1941). "The Theory of Rate Processes." McGraw–Hill, New York.

Grundvik, P. *et al.* (1979). *Phys. Rev. Lett.* **42**, 1528–1531.

Halle, B., and Wennerström, H. (1981). *J. Magn. Reson.* **44**, 89–100.

Hertz, H. G. (1973). *Ber. Bunsenges. Phys. Chem.* **77**, 531–540.

Holtz, M., Gunther, S., Lutz, O., Nolle, A., and Schrade, P.-G. (1979). *Z. Naturforsch.* **34a**, 944–949.

Hoult, D. (1978). "Progress in NMR Spectroscopy" (J. Emsley, J. Feeney, and L. Sutcliffe, eds.), Vol. 12, p. 45. Pergamon, Oxford, U.K.

Hubbard, P. (1970). *J. Chem. Phys.* **53**, 985–987.

Kraft, H. G., Peringer, P., and Rode, B. M. (1981). *Inorg. Chim. Acta* **48**, 135–137.

Lindblom, G., Wennerström, H., and Lindman, B. (1976). *ACS Symp. Ser.* **34**, 372–396.

Lindman, B., and Forsén, S. (1978). "NMR and the Periodic Table" (R. Harris, and B. Mann, eds.), pp. 183–194. Academic Press, New York.

Lindman, B., Forsén, S., and Lilja, H. (1977). *Chem. Scripta* **11**, 91–92.

Lucken, E. A. C. (1969). "Nuclear Quadrupole Coupling Constants." Academic Press, London.

Lutz, O., Schwenk, A., and Uhl, A. (1975). *Z. Naturforsch.* **30a**, 1122–1127.

McLachlan, A. D. (1964). *Proc. R. Soc. London* **280A**, 271–288.

Magnusson, J. A., and Bothner-By, A. A. (1971). "Magnetic Resonance in Biological Research" (C. Franconi, ed.), pp. 365–376. Gordon and Breach, New York.

Marsh, H. C., Robertson, P., Jr., Scott, M. E., Koehler, K. A., and Hiskey, R. G. (1979). *J. Biol. Chem.* **254**, 10268–10275.

Oldfield, E., and Skarjune, R. P. (1978). *J. Magn. Reson.* **31**, 527–531.

Olsson, G., and Salomonson, S. (1982). *Z. Physik A,* **307**, 99–107.

Poupko, R., Baram, A., and Luz, Z. (1974). *Mol. Phys.* **27**, 1345–1352.

Reuben, J., and Luz, Z. (1976). *J. Phys. Chem.* **80**, 1357–1361.

Robertson, P., Jr., Hiskey, R., and Koehler, K. A. (1978). *J. Biol. Chem.* **253**, 5880–5883.

Rose, D. M., Bleam, M. L., Record, M. T., Jr., and Bryant, A. G. (1980). *Proc. Natl. Acad. Sci. USA* **77**, 6289–6292.

Rubinstein, M., Baram, A., and Luz, Z. (1971). *Mol. Phys.* **20**, 67–80.

Sen, K. D., and Narasimhan, P. T. (1974). "Advances in Nuclear Quadrupole Resonance" (J. A. S. Smith, ed.), Vol. 1, p. 277. Heyden, London.

Shannon, R. D. (1976). *Acta Cryst.* **A32**, 751–767.

Sutherland, I. O. (1971). "Ann. Reports NMR Spectroscopy" (E. F. Mooney, ed.), Vol. 4, pp. 71–235. Academic Press, London.

Wennerström, H., Lindblom, G., and Lindman, B. (1974). *Chem. Scr.* **6**, 97–103.

Werbelow, L. G. (1979). *J. Chem. Phys.* **70**, 5381–5383.

Werbelow, L. G., and Marshall, A. G. (1981). *J. Magn. Res.* **43**, 443–448.

Werbelow, L., and Pouzard, G. (1981). *J. Phys. Chem.* **85**, 3887–3891.

Westlund, P.-O., and Wennerström, H. (1982). *J. Magn. Reson.* **50**, 451–466.

Williams, R. J. P. (1970). *Q. Rev. Chem. Soc. London* **24**, 331–365.

9 Cobalt-59

Pierre Laszlo

Institut de Chimie Organique et de Biochimie
Université de Liège, Liège, Belgium

Among all the nuclides in the periodic table, ^{59}Co is unique in its range of chemical shifts—more than 18,000 ppm—so that one is tempted to express ^{59}Co chemical shifts in percentages rather than in parts per million. But, as with other sorts of natural beauty, an asset can easily turn into a liability—such an enormous spread in chemical shift comes from low-lying excited states contributing profusely to the paramagnetic part of the shielding constant in diamagnetic cobalt complexes. However, many cobalt complexes are themselves paramagnetic, and therefore not amenable to NMR studies. Complementing such extensive chemical shifts, their temperature dependence is also impressive—typically on the order of parts per million per degree—which can cause difficulties in obtaining good spectra. To compound this already somewhat less than rosy picture, the cobalt nucleus has a very large electric quadrupole moment (more than 0.4 barn) which leads to such large quadrupolar interactions that in practice only highly symmetric tetrahedral and octahedral complexes are amenable to high-resolution NMR work. Yet, ^{59}Co NMR has been exploited for a large number of applications to chemical and biochemical problems, and its potential, in my opinion, is still largely untapped.

I. Magnetic Properties of the ^{59}Co Nucleus

Like a large number of isotopes having an odd atomic number and an even number of neutrons (^{19}F, ^{23}Na, and ^{31}P are other examples familiar to NMR spectroscopists), ^{59}Co has a natural abundance of 100%: It is the sole stable cobalt isotope (^{60}Co is a radioactive isotope, as is well known both from its medical applications and from its presence in the atmospheric fallout from nuclear weapons).

Its gyromagnetic ratio, $\gamma = 6.3472 \times 10^7$ rad T^{-1} s^{-1}, places its resonance frequency in the neighborhood of that for ^{13}C: At 2.3488 T, ^{59}Co resonates at 23.727 MHz, whereas ^{13}C resonates at 25.144 MHz (at 11.7440 T, the corresponding resonance frequencies are 118.635 and 125.721 MHz, respectively) (see Brevard and Granger, 1981).

The receptivity of ^{59}Co, attributable both to greater intrinsic sensitivity (0.277 versus 1.59×10^{-2} for an equal number of nuclei at constant field)

TABLE I

Values of the IQ Term for Selected Quadrupolar Nuclei

Nucleus	IQ^a
D (^{2}H)	0.0375
^{14}N	25.2
^{23}Na	19.2
^{25}Mg	15.5
^{27}Al	71.0
^{35}Cl	85.3
^{39}K	40.3
^{43}Ca	33.3
^{45}Sc	65.9
^{47}Ti	26.9
^{51}V	0.37
^{53}Cr	1.20
^{55}Mn	96.8
^{59}Co	21.8

^{a} $IQ = 10^3 \times Q^2(2I + 3)/I^2(2I - 1)$.

and to its greater natural abundance (100 versus 1.1%), is better than that of ^{13}C by a factor of 1570. If this factor were the only one to be considered, it would make ^{59}Co an extremely "easy" nucleus.

However, the quadrupole moment Q has the very large value of 0.404 barn (Ehrenstein *et al.*, 1960). Because, in the relaxation equations, a relaxation rate typically depends on the product of Q^2 and the $(2I + 3)/I^2(2I - 1)$ term, the impact of such a high Q value is fortunately somewhat dampened by ^{59}Co having a spin quantum number $I = \frac{7}{2}$ (Table I). Therefore, with a moderate IQ comparable to those of ^{14}N and ^{23}Na, one can expect ^{59}Co resonances to be very sensitive to the presence of electrostatic field gradients.

II. Chemical and Magnetic Properties of Cobalt Complexes

Cobalt has been a key element in the history of inorganic chemistry and in the development of new concepts of structural theory. As long ago as 1798, B. M. Tassaert observed color changes on air oxidation of a cobalt(II) salt in aqueous ammonia (Purcell and Kotz, 1977). Cobalt(III) compounds were crucial in Alfred Werner's formulation of the principles of coordination chemistry: He postulated octahedral structures in 1893, synthesized with his co-workers approximately 700 cobalt compounds, and published 127 articles dealing with their properties (Morral, 1967; Werner, 1893).

Besides Co(0) and Co(I) oxidation states, found in organometallic compounds, many Co(II) and Co(III) complexes exist. The hexaaquo-cobalt(III) species is a powerful oxidant capable of oxidizing water according to

$$4Co^{3+}(aq) + 2H_2O \rightarrow 4Co^{2+}(aq) + 4H^+ + O_2 \qquad (1)$$

In the presence of coordinating ligands other than H_2O molecules, it is possible, however, to diminish the reduction potential in order to tone down the oxidizing ability of Co(III): Although the reduction potential E^0 is $+1.84$ V for $Co(H_2O)_6^{3+}$, it is decreased to $E^0 = 0.108$ V in $Co(NH_3)_6^{2+}$ and to -0.8 V in $Co(CN)_6^{3-}$.

This stabilization by ligands of the Co(III) oxidation state is of importance in NMR applications, because all cobalt(II) (d^7) complexes are paramagnetic. Depending on the stabilities of the ligands as σ or π donors or acceptors, Co(III) complexes can be either low spin or high spin. High-resolution NMR studies being constrained to diamagnetic species, only

O_h low-spin Co(III) (d^6) complexes can be thus studied. The ground state configuration of the octahedral low-spin d^6 structure is $^1A_{1g}$, and the first excited state configuration is $t_2^5e^1$, corresponding to the two states $^1T_{1g}$ and $^1T_{2g}$ (Purcell and Kotz, 1977).

A temperature-dependent spin transition has been reported for the hexacoordinate O_h complex [Co(III)L$_2$]PF$_6$, where L is the tridentate oxygen tripod ligand [(C$_5$H$_5$)Co(P(O)(OC$_2$H$_5$)$_2$)$_3$].$^-$ At low temperatures the cation is in the $^1A_1$ low-spin state, and as the temperature is raised, the spin equilibrium shifts toward the $^5T_2$ (O_h) high-spin state, as evidenced by magnetic susceptibility measurements (Kläui, 1979) or by ^{31}P NMR (Gütlich et al., 1980). This occurs because the difference between the ligand field strength Δ and the mean spin pairing energy $|\Delta - P|$ is on the order of KT.

III. Electrostatic Field Gradients in Diamagnetic Cobalt Complexes

With a point-charge model, the electric field gradient at the central cobalt nucleus in octahedral complexes CoX$_n$Y$_{6-n}$ can be calculated as a function of the metal–ligand bond lengths r_1 and r_2 and of the ligand charges e_1 and e_2 (Valiev and Zaripov, 1966). The values of the electric field gradient tensor invariant g^2 are listed in Table II. In spite of its approximate

TABLE II

Electric Field Gradient at the Central Cobalt Nucleus in a Series of Octahedral CoX$_n$Y$_{6-n}$ Complexes[a]

Complex Symmetry Point Group	$\dfrac{g^2}{e^2/r_2^3 - e_1/r_1^3}$
CoX$_6$ O_h	0
CoX$_5$Y C_{4v}	9
trans-CoX$_4$Y$_2$ D_{4h}	36
cis-CoX$_4$Y$_2$ C_{2v}	16
mer-CoX$_3$Y$_3$ C_{2v}	27
fac-CoX$_3$Y$_3$ C_{3v}	0
trans-CoX$_2$Y$_4$ D_{4h}	36
cis-CoX$_2$Y$_4$ C_{2v}	16
CoXY$_5$ C_{4v}	9
CoY$_6$ O_h	0

[a] After Valiev and Zaripov, 1966.

character, such an approach accounts semiquantitatively for the NMR line widths observed with O_h complexes of other elements, such as ^{93}Nb (Tarasov et al., 1978) and ^{27}Al (Wehrli and Wehrli, 1981). In the case of ^{59}Co, the predictions of Table II are also borne out by the experimental data: For instance, fac-[Co(NH$_3$)$_3$(H$_2$O)$_3$]$^{3+}$ displays a much narrower line width than the mer isomer, and the trans-X$_4$Y$_2$ and trans-X$_2$Y$_4$ complexes show broader ^{59}Co spectra than the cis isomers in most cases (Yamasaki et al., 1968). These observations suggest that a quadrupolar mechanism is the predominant mode of relaxation in cobalt(III) complexes devoid of full octahedral symmetry.

IV. Mechanisms for ^{59}Co Relaxation in O_h Complexes

The dominant relaxation mechanism for ^{59}Co in fully symmetric octahedral complexes Co(III)L$_6$ has been a matter of controversy, with quadrupolar relaxation, spin–rotation relaxation, and scalar coupling of the second kind as leading contenders. Cobalt-59 relaxation in aqueous cobalticyanide Co(CN)$_6^{3-}$ was studied under the assumption of dominant quadrupolar relaxation by Ader and Loewenstein (1971). These authors used a correlation time they inferred from the ^{14}N relaxation rate to derive the ^{59}Co quadrupolar coupling constant. The inconsistencies in such a treatment have been pointed out, inter alia, by Doddrell et al., 1979 and by Dwek et al., 1970.

Indeed, with a strict O_h instantaneous geometry the electrostatic field gradient vanishes at the nucleus, and ^{59}Co line widths ought to be zero. Hence, even in cases in which quadrupolar relaxation can be proven to predominate, the origin of the fluctuating electrostatic field gradient for Co(III)L$_6$ complexes must be found either in instantaneous distortions with respect to the ideal O_h geometry or in deviations from this perfect symmetry due to the presence of, for example, counterions in the outer sphere. Bryant and his co-workers studied ion pairs in which a highly charged ion was placed in the second coordination sphere of the Co(en)$_3^{3+}$ and Co(pn)$_3^{3+}$ complexes (en, ethylenediamine; pn, propylenediamine), and the contribution to the cobalt quadrupole relaxation from the anion (Br$^-$, Cl$^-$, ClO$_4^-$, SO$_4^{2-}$, CO$_3^{2-}$, PO$_4^{3-}$) in the second coordination sphere was insignificant (Craighed et al., 1975; Rose and Bryant, 1979). There remains the former possibility, that excitation of the vibrational modes of the octahedron provides a relaxation mechanism for ^{59}Co spins in the symmetric Co(III)L$_6$ complexes (Hertz, 1973; Hertz et al., 1974; Valiev,

1960; Valiev and Zaripov, 1962). This was investigated in detail with tris(tropolonato)cobalt(III) as an example (Doddrell *et al.*, 1979), and it was concluded that experimental evidence fits such a vibrational mechanism. A comparable study has been made for $K_3Co(CN)_6$ in the crystalline state accounting for the temperature dependence of the quadrupolar coupling constant, based on the stretching and rotational internal modes of vibration of the $Co(CN)_6$ octahedron (Lourens and Smit, 1980).

Another mechanism besides quadrupolar relaxation is present in cobalt-(III) complexes having nitrogens bonded to the metal: scalar relaxation of the second kind (Abragam, 1961). This arises from the scalar coupling J between the ^{59}Co nucleus and the ^{14}N nuclei, which relax efficiently by a nucleus–electric quadrupole mechanism. Such an interaction leads to a *transverse* ^{59}Co relaxation rate proportional to the product of the square of the interaction energy J and the correlation time τ_N descriptive of reorientation of the nitrogen spin (i.e., $\tau_N = T_1$ for the ^{14}N if nitrogen relaxation is rapid compared to chemical exchange). The differences between T_1 and T_2 in some CoN_6 complexes have been shown to arise from unresolved ^{59}Co—^{14}N scalar couplings: Examples include hexanitrocobaltate, in which J_{CoN} is estimated to be 46 ± 4 Hz (Rose and Bryant, 1979), and cobalt hexammine $Co(NH_3)_6^{3+}$ (Doddrell *et al.*, 1979). In the latter case, the ^{59}Co spectrum on 95% ^{15}N-enriched $Co(NH_3)_6Br_3$ shows a ^{59}Co—^{15}N scalar coupling of 64 Hz, which (neglecting isotope effects) corresponds to a ^{59}Co—^{14}N scalar coupling of approximately 50 Hz.

For the same cobalt hexammine species $Co(NH_3)_6^{3+}$, a spin–rotation mechanism has also been suggested for ^{59}Co relaxation (Jordan, 1980). The evidence is the near temperature invariance of the longitudinal relaxation rate extrapolated to infinite dilution: 26.3 Hz (25°C), 42.9 Hz (35.5°C), 34.7 Hz (45°C), 44.9 Hz (60°C). This may occur because the temperature dependence of the spin–rotation mechanism is opposite that of the quadrupolar mechanism. From his results, Jordan estimates a quadrupolar coupling constant $e^2qQ/h \simeq 2.7$ MHz for the $Co(NH_3)_6^{3+}$ cation, and ≈ 3.5 MHz for the $Co(NH_3)_6^{3+}$–ClO_4^- ion pair, in a dimethyl sulfoxide (DMSO) solution. He also estimates a spin–rotation coupling constant C_{SR} of 72 kHz and finds that the product IC_{SR}, where I is the moment of inertia, increases 30-fold on going from the cation to the ion pair.

With hexacoordinate Co(III) complexes lacking full O_h symmetry, quadrupolar relaxation is likely to be predominant. Based on such a premise, after determining from NMR measurements on single crystals the magnitudes of the quadrupolar coupling constants, the correlation times for three complexes in aqueous solution are obtained (Hartmann and Sillescu, 1964), as shown in the accompanying table.

	$[Co(NH_3)_5H_2O]^{3+}$	$[Co(NO_2)_4(NH_3)_2]^-$	$trans$-$Co(en)_2Cl_2$
e^2qQ/h (MHz)	57.6	32.7	75.8
τ_c (ps)	9.95	14.25	18.35

An interesting comparison, illustrative of the contrast between inner-sphere and outer-sphere perturbations, results: The quadrupolar coupling constant has a value of 2.7 MHz in $Co(NH_3)_6^{3+}$, increases slightly to 3.5 MHz in the $Co(NH_3)_6^{3+}$–ClO_4^- ion pair (Jordan, 1980), and is multiplied about 20-fold on replacement of one water molecule by one ammonia ligand (Hartmann and Sillescu, 1964).

V. The ^{59}Co Chemical Shift: Theory

As for many other nuclei, most formalisms were devised within the framework of the Ramsey partition of the shielding constant into a diamagnetic and a paramagnetic term: $\sigma = \sigma_d + \sigma_p$ (Ramsey, 1950; see also Chapter 4, Vol. I). Griffith and Orgel devised, based on the example of the low-spin d^6 complexes of cobalt(III), a general application of crystal field theory to the Ramsey shielding theory, in the local term approximation of Saika and Slichter. This made it a general treatment for the chemical shift of transition metals (Griffith and Orgel, 1957). The chemical shift of cobalt(III) complexes was thus predicted to be dominated by a paramagnetic term expressed as

$$\sigma_p = \frac{\mu_0}{4\pi} 4\mu_B^2 \langle r^{-3}\rangle_{3d} \frac{\langle 0|L^2|0\rangle}{3\Delta_0} \tag{2}$$

where r is the radius of the d orbitals and Δ_0 the splitting in the octahedral crystal field. The prediction of a linear dependence between the ^{59}Co chemical shift and the excitation energy for the $d \rightarrow d$ transition was soon borne out: A good correlation was found between the cobalt resonance frequencies and the spectrochemical series for 14 complexes of cobalt(III) (Freeman et al., 1957). Another prediction of the Griffith–Orgel theory is that the shielding will decrease slightly with increasing temperature, because populating higher vibrational modes reduces Δ_0 slightly. Such an effect was also found experimentally (Freeman et al., 1957). Griffith and Orgel also pointed out that neglect of configuration interaction tends to compensate for the neglect of covalent bonding in their treatment, which overestimates the orbital angular momentum L because electron delocalization is underestimated. These effects have indeed been found to amount at most to a few percent (Betteridge and Golding, 1969).

In the simple crystal field model of Griffith and Orgel, the e_g and t_{2g} orbitals are assumed to be pure d orbitals. An improvement is to allow covalency by introducing an orbital reduction factor $k_{\sigma\pi}$ between the t_{2g} and the e_g orbitals, so that the paramagnetic shielding term, according to Freeman *et al.* (1957), becomes

$$\sigma_p = -32\mu_B^2\, k_{\sigma\pi}^2\, \langle r^{-3}\rangle_{3d}/\Delta_0 \tag{3}$$

where μ_B is the Bohr magneton.

The problem then becomes determining the two unknowns $k_{\sigma\pi}$ and $\langle r^{-3}\rangle_{3d}$ from a single observable (the chemical shift δ). One can assume that the $3d$-orbital size remains constant, because its relative importance peaks at the nucleus, just where the ligands have the smallest effect on it. The orbital reduction factor $k_{\sigma\pi}$ was calculated, from δ and Δ_0 values, as 0.85 for first-row ligands (C, N, O), 0.72 for second-row ligands (Si, P, S, Cl), and 0.67 for third-row ligands (As, Se) (Betteridge and Golding, 1969).

For Co(III) complexes, one can safely assume that only the lowest $d \rightarrow d$ transition contributes significantly to the paramagnetic term (Buckingham

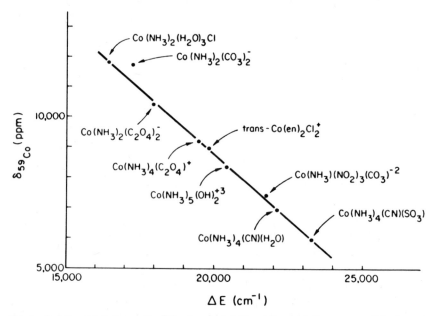

Fig. 1   Relationship between the ^{59}Co chemical shift and the excitation energy of the lowest $d \rightarrow d$ electronic transition. (After Rinaldi *et al.*, 1980.)

and Stephens, 1964; Freeman *et al.*, 1957; Griffith and Orgel, 1957). As we have already indicated, the octahedral low-spin d^6 complexes admit the two spin-allowed excited states $^1T_{1g}$ ($t_{2g}^5 e_g^1$) and $^1T_{2g}$ ($t_{2g}^5 e_g^1$). Only the former interacts with the ground state through the angular momentum operator (Buckingham and Stephens, 1964; Griffith and Orgel, 1957). Hence the significance of the Δ_0 splitting is that it measures the promotion energy ΔE from the ground state $^1A_{1g}$ to the $^1T_{1g}$ excited state. In other terms, *the ^{59}Co chemical shifts arise predominantly from second-order paramagnetism on application of the magnetic field, which mixes the $^1A_{1g}$ ground state with the $^1T_{1g}$ excited state.*

Quite another effect has been predicted by Ramsey: He has suggested that molecules having abnormally large shielding constants, such as some ^{59}Co compounds, may display a dependence of the shielding on the magnetic field (Ramsey, 1970). Doddrell has reported small effects, at the limit of detection, for $Co(^{15}NH_3)_6Cl_3$ and $Co(acac)_3$ which, however, are in the wrong direction for the field dependence predicted on the basis of paramagnetic currents and fourth-order perturbation theory (Bendall and Doddrell, 1979).

VI. The Cobalt Chemical Shift: Correlation with $d \rightarrow d$ Electronic Transitions

According to the prediction from Eq. (3), the ^{59}Co chemical shift displays excellent linear correlations with the lowest $d \rightarrow d$ transition ($^1A_{1g} \rightarrow {}^1T_{1g}$). Such a correlation with oxygen and nitrogen ligands, and cyanide (Dharmatti and Kanekar, 1959; Freeman *et al.*, 1957; Fujiwara *et al.*, 1969), is shown in Fig. 1. Note that the ^{59}Co chemical shifts span ~6000 ppm in this correlation. If the correlation is extended to include second-row ligand atoms (P, S), then two distinct regression lines will be obtained (Fig. 2). Both the slopes and the intercepts of these linear plots are of importance: The slope $B =$ constant $\times k_{\sigma\pi}^2 \langle r^{-3} \rangle_{3d}$ is most conveniently reported in reciprocal centimeters. The intercept at $\Delta E^{-1} = 0$ corresponds to the diamagnetic part of the shielding constant, and for the complexes listed in Table III this intercept varies greatly between -3625 and $-15,395$ ppm.

From the correlations shown in Fig. 2 and summarized in Table III, it appears that first-row and second-row ligators differ in the slopes and intercepts of these δ-versus-ΔE^{-1} correlations. However, Juranic has shown that it is possible to obtain a single correlation to which all the available data conform by taking into account the dependence of B on the nephelauxetic ratio β_{35} (Jørgensen, 1962, 1966, 1968), semiempirically

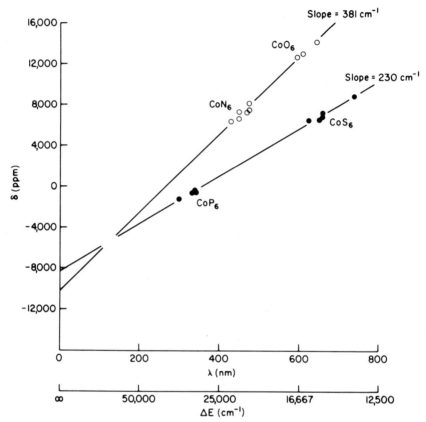

Fig. 2 Cobalt-59 chemical shift dependence on ΔE^{-1} for octahedral cobalt(III) complexes. (After Kidd, 1980.)

reducing B in proportion to the magnitude of β_{35} (Juranic, 1980). The complexes studied are listed, together with their properties, in Table IV, and the resulting single linear regression corresponds to a slope of 615 cm^{-1}, very close to the theoretical value of 620 cm^{-1} for localized free-ion d orbitals, and to an intercept $\gamma_0 = 10.060$ MHz T^{-1}, from which the most reliable value of the gyromagnetic ratio presently available for ^{59}Co can be obtained: For ^{59}Co $\gamma/2\pi = 10.060 \times 10^7$ T^{-1} s^{-1} or $\gamma = 6.3209 \times 10^7$ rad T^{-1} s^{-1} [as compared to the value of 6.3472 given by Brevard and Granger (1981)].

In another clarification (Juranic et al., 1979), Juranic has shown convincingly that, for tris chelate $CoN_{6-x}O_x$ complexes, deviations from the linear relationship between δ_{59Co} and the $d \rightarrow d$ wavelength are due to the

TABLE III

Paramagnetic Term Dependence of ^{59}Co on ΔE and B^a

Co(III) Complexes	Slope B (cm^{-1})	Intercept δ (ppm)
C, N, O ligands	495	$-15,395$
	450	$-13,995$
Si, N, O ligands		$-5,000$
	400	$-11,000$
S, Se, As ligands	310	$-13,842$
Si, P, S, Cl ligands	315	$-14,000$
As, Se, Br ligands	275	$-14,000$
Tris(β-diketonates)	270	$-3,625$
S, P ligands	230	$-8,260$

[a] From Kidd, 1980.

anisotropy of the chemical shielding of Co(III). The complexes showing deviations from the linear dependence [Eq. (3)] have very asymmetric first absorption bands. This asymmetry is due to the decrease in ligand field symmetry and the attendant splitting of the threefold degenerate $^1T_{1g}$ energy level. As a result, the ^{59}Co chemical shift is no longer isotropic

TABLE IV

Cobalt(III) Complexes, Their Longest Wavelength $d \rightarrow d$ Transition, and the Nephelauxetic Ratio $\beta_{35}{}^a$

Cobalt Complex	ΔE for $^1A_{1g} \rightarrow {}^1T_{1g}$ (nm)	β_{35}
Co(C$_2$O$_4$)$_3{}^{3-}$	605	0.48
Co(CO$_3$)$_3{}^{3-}$	645	0.49
Co(NH$_3$)$_6{}^{3+}$	475	0.56
Co(en)$_3{}^{3+}$	465	0.54
Co(CN)$_6{}^{3-}$	311	0.41
Co(S$_2$P(OC$_2$H$_5$)$_2$)$_3$	735	0.37
Co(S$_2$COC$_2$H$_5$)$_3$	617	0.29
Co(S$_2$CNH$_2$)$_3$	637	0.32
Co(P(OCH$_3$)$_3$)$_6{}^{3+}$	345	0.25
Co(OCH$_3$)CCH$_3$)$_6{}^{3+}$	300	0.31
Co(Se$_2$CN(CH$_3$)$_2$)$_3$	667	0.30

[a] After Juranic, 1980.

(Fujiwara *et al.*, 1969; Martin and White, 1969). Hence, in these complexes of lowered symmetry, it is necessary to find values of the paramagnetic term along all three principal axes of the complex according to

$$\sigma_{\alpha\alpha}{}^P = 4\beta^2 \sum_n \langle r^{-3} \rangle_{3d} \left\langle 0 \left| \sum_k 1_{\alpha,k} \right| n \right\rangle^2 (E_n - E_0)^{-1} \qquad \alpha \in x, y, z \quad (4)$$

and to use the trace of the shielding tensor $\sigma^P = \frac{1}{3}(\sigma_{xx}{}^P + \sigma_{yy}{}^P + \sigma_{zz}{}^P)$ as the chemical shift characteristic of these cobalt(III) complexes. In this manner, it is found that, depending on the symmetry group, the paramagnetic terms can be written

$$\begin{aligned}
\sigma_p \, (D_{4h}) &= -32\beta^2 \, \langle r^{-3} \rangle_{3d} \, k_{\sigma\pi}{}^2 \, [\tfrac{1}{3}\Delta E^{-1} \, (^1A_{1g} \to {}^1A_{2g}) \\
&\quad + \tfrac{2}{3}\Delta E^{-1} \, (^1A_{1g} \to {}^1E)] \\
\sigma_p \, (C_{4v}) &= -32\beta^2 \, \langle r^{-3} \rangle_{3d} \, k_{\sigma\pi}{}^2 \, [\tfrac{1}{3}\Delta E^{-1} \, (^1A_1 \to {}^1A_2) \\
&\quad + \tfrac{2}{3}\Delta E^{-1} \, (^1A_1 \to {}^1E)] \\
\sigma_p \, (C_{2v}) &= -32\beta^2 \, \langle r^{-3} \rangle_{3d} \, k_{\sigma\pi}{}^2 \, [\tfrac{1}{3}\Delta E^{-1} \, (^1A_1 \to {}^1A_2) \\
&\quad + \tfrac{1}{3}\Delta E^{-1} \, (^1A_1 \to {}^1B_1) + \tfrac{1}{3}\Delta E^{-1} \, (^1A_1 \to {}^1B_2)]
\end{aligned} \qquad (5)$$

In order to obtain the *individual* excitation energies ΔE involved in Eqs. (5), the first experimental absorption bands in the optical spectra are first deconvoluted into their Gaussian components. For instance, the *trans*-(N)K[Co(gly)$_2$ox] complex (D_{4h}) has component absorptions at 527 nm ($^1A_{1g} \to {}^1E_g$) and 622 nm ($^1A_{1g} \to {}^1A_2$). Taking the wavelengths of these individual components and weighting them with the appropriate factors from Eqs. (5) ($\frac{2}{3}$ and $\frac{1}{3}$, in this case), gives a weighted average of 559 nm. Then the observed chemical shifts correlate very well with the wavelength for the 11 complexes studied (Juranic *et al.*, 1979).

VII. The Chemical Shift: Range and Sensitivity to Minor Changes

With a total chemical shift range of at least ~18,000 ppm, ^{59}Co NMR is a spectroscopist's dream. These shifts are changes in response to major structural alterations in the first coordination sphere. However, minor perturbations in the outer sphere are easily detected by ^{59}Co chemical shifts: For instance, with the cobaltihexacyanide probe Co(CN)$_6{}^{-3}$, the ^{59}Co resonance is shifted by 290 ppm in going from DMSO to water (Eaton *et al.*, 1982). Because ^{59}Co chemical shifts are measured easily to ±0.04 ppm (Delville *et al.*, 1981; see also the next section), it is very tempting to try the method on yet more minute effects. With the same

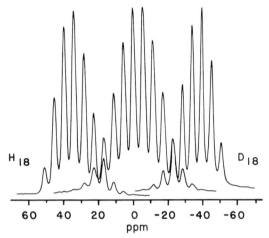

60   40   20   0   -20   -40   -60
ppm

Fig. 3 The 71.77-MHz ^{59}Co-NMR spectra for 0.05 M aqueous solutions of Co(NH$_3$)$_6$Cl$_3$, which are 15, 50, and 85% in D$_2$O, respectively. Rapid exchange of the NH$_3$ protons results in a binomial distribution of isotopes in the complex, where the H$_{18}$ compound is observed at low field in the 15% D$_2$O solution and the D$_{18}$ compound is observed at high field in the 85% D$_2$O solution. The resonance for the H$_9$D$_9$ isomer in the 50% D$_2$O solution has been centered at 0 ppm. The spectra in 15 and 50% D$_2$O solutions have been arranged so that the resonances for the H$_{12}$D$_6$ compound are coincident. Similarly, the spectra from the 50 and 85% D$_2$O solutions have been arranged to overlap the resonances resulting from the H$_5$D$_{13}$ compound.

Although the isotope distribution generally follows a binomial distribution consistent with the number of exchangeable protons and the H$_2$O mole fraction, there is a small deuterium isotope effect in the direction of the more highly deuterated compounds. Finally, there is an average deuterium isotope chemical shift effect on the ^{59}Co resonance of 5.6 ppm per deuterium atom. (Courtesy of R. G. Bryant, S. Philson, J. Russell, University of Minnesota).

Co(CN)$_6$$^{-3}$ anion, a chemical shift difference of 1.05 ± 0.04 ppm is observed between H$_2$O and D$_2$O. It is even possible to measure a solvent isotope effect of ~2 ppm between CH$_3$OD and CD$_3$OD (Laszlo and Stockis, 1980).

According to an ancient Chinese proverb, a picture is worth a thousand words: A good example of the exquisite sensitivity of ^{59}Co chemical shift and its ability to resolve the individual spectra of closely similar chemical species is displayed in Fig. 3.

VIII. The Chemical Shift: Temperature Coefficients

Because of the presence of the low-lying $^1T_{1g}$ level, whose Boltzmann population is nonnegligible at normal temperatures, temperature changes, by varying the relative populations of the $^1A_{1g}$ ground state and of the $^1T_{1g}$ excited state, will show as chemical shifts. Very large temperature coefficients characterize the ^{59}Co chemical shift (Proctor and Yu, 1951). Hence ^{59}Co "thermometers" are easily devised: 0.1 M aqueous solutions of $K_3Co(CN)_6$ and of $Co(acac)_3$ have linear temperature coefficients of 1.5 and 3 ppm deg^{-1} (Levy et al., 1980).

The drawback of such extreme temperature sensitivities is that an uninformed research worker, placing a sample tube in the probe of a NMR spectrometer, quickly discovers that the resulting quite broad resonance is an accurate mapping of the small temperature differences within the sample volume. Because of the great variation in ^{59}Co chemical shifts with temperature, it is mandatory to equilibrate carefully all samples at the probe temperature for at least 30 min prior to recording their spectra. It is only when the temperature remains invariant at ±0.02°C that line widths become reasonably sharp (<5–10 Hz) and chemical shifts can be measured to a precision limit of less than ±0.04 ppm (Delville et al., 1981).

A possible bonus from these large temperature coefficients, not yet realized, would be using such ^{59}Co thermometers as those described previously (Levy et al., 1980) for the accurate measurement of temperature in dynamic NMR studies. This might help to solve the vexing outstanding problem of accurate determination of enthalpies and entropies of activation. The most promising nucleus, in this regard, is ^{13}C, because the proximity of the ^{13}C and ^{59}Co resonance frequencies ($\Delta \simeq 1$ MHz at typical magnetic fields), combined with the high receptivity of ^{59}Co, allows easy detection of the ^{59}Co thermometer signal through the ^{13}C channel with no retuning of the probe.

IX. Substituent Effects on ^{59}Co Chemical Shifts

It is useful to discuss inner-sphere chemical shifts for a series of closely related compounds. I have selected for presentation the cobaloximes listed in Table V: These are hexacoordinate compounds in which the cobalt center makes direct contact with four nitrogens and a methyl carbon. The ^{59}Co shifts are indicated as a function of the sixth ligand L. More generally for $^{59}Co(III)$ complexes, the increasing shielding order is O < N < C, the same as the sequence of these ligators in the spectrochemical series (Jørgensen, 1962; Purcell and Kotz, 1977), by virtue of their increasing ability to cause d-orbital splitting in the cobalt atom. However, if one considers an extensive series of cobalt complexes with C, N, O, P,

TABLE V

Cobalt-59 Chemical Shifts for
Cobaloxime Complexes[a]

L in $CH_3Co(dh)_2L$[b]	$\delta_{^{59}Co}$ (ppm)
CH_3OH	4110
NMe_3	3790
NC_5H_5	3660
β-Picoline	3660
SMe_2	3190
$AsPh_3$	2950
PPh_3	2800
$CNCH_3$	2680
$P(n\text{-}Bu)_3$	2620
$P(OMe)_3$	1580

[a] After Kidd, 1980, and La Rossa and Brown, 1974.
[b] dh, Dimethylglyoximato monoanion.

As, S, and Se ligating atoms, they do not fit a single $\delta_{^{59}Co}$-versus-ΔE^{-1} correlation line (see Section VI; Kanekar *et al.*, 1967; Martin and White, 1969; Weiss and Verkade, 1979). Also, although the halides appear in the spectrochemical series as $I^- < Br^- < Cl^- < F^- < O$, they shield cobalt in the reverse order, $F^- < Cl^- < Br^- < I^-$. The reason for this inversion is that, below the first row of the periodic table, the term $\langle r^{-3}\rangle_{3d}$ cannot be assumed to remain constant. As a function of the nature of the ligating atoms, d-orbital expansion occurs, reducing the $\langle r^{-3}\rangle_{3d}$ term below its free Co^{3+} ion value. A measure of the ability of the various ligators to achieve this volume increase is provided by the nephelauxetic series (Jørgensen, 1962; Purcell and Kotz, 1977), in which the halides indeed appear in the order $F^- \ll Cl^- < Br^- < I^-$ (Kidd, 1980).

Since 1951, when the ^{59}Co spectra of several cobalt(III) complexes were first reported (Proctor and Yu, 1951), there have been approximately 70 reports on ^{59}Co chemical shifts and their explanation or theory (Yamasaki, 1981). A data base, with characteristic chemical shifts for several thousand diamagnetic cobalt(III) complexes, has been constructed (Yamasaki, 1981). The general trend can be described, in going to higher fields (decreasing δ) as $CoO_6 < CoO_5N < CoO_4N_2 < CoO_3N_3 < CoO_2N_4 < CoON_5 < CoN_6 \simeq CoS_6 \simeq CoSe_6 < CoAs_6 < CoC_6 < CoP_6$ (Kidd, 1980). As for cobalt(0) and cobalt($-$I) complexes, whereas the Co(III) compounds span ~14,000 ppm downfield from the usual $Co(CN)_6^{3-}$ reference, they occupy the low-frequency 3400 ppm, with no overlap between the two groups (Kidd, 1980).

TABLE VI

Solvent Dependence of the ^{59}Co
Chemical Shift for 0.1 M solutions of
$(n\text{-Bu}_4\text{N})_3\text{Co(CN)}_6$ at 33.5°C[a]

Solvent	$-\delta_{^{59}\text{Co}}$ (ppm)
Water	29.6
Methanol	92.1
Methanol-d	95.2
Methanol-d_4	97.2
Ethanol	132
1-Propanol	141.1
2-Propanol	169
1-Butanol	145.9
2-Butanol	184
IsoButanol	158.3
1-Pentanol	154
IsoPentanol	154
2-Pentanol	183.5
3-Pentanol	195.1
1-Hexanol	146
1-Heptanol	135.6
1-Octanol	~135

[a] (After Laszlo and Stockis, unpublished).

With such large chemical shifts, it should come as no surprise that they differentiate nicely between cis–trans or mer–fac isomers: in general, cis isomers are more shielded than trans, and fac more shielded than mer (Juranic *et al.*, 1977; Yajima *et al.*, 1971). Yet greater differences are found in ob–lel isomerism than in mer–fac isomerism (Koike *et al.*, 1974).

X. Solvent Effects

We shall now concentrate on studies with the cobaltihexacyanide probe. The Co(CN)_6^{3-} solute, according to ab initio calculations (Sano *et al.*, 1979), has a good part of the negative charge partitioned between the terminal nitrogens. This anion acts as a hydrogen bond acceptor, and the central cobalt atoms respond by electronic donation of the type $\text{Co}-\text{L}:\text{HS}$, where HS denotes a protic solvent molecule. The ^{59}Co chemical shifts for the tetrabutylammonium salt, referred to 0.1 M $\text{K}_3\text{Co(CN)}_6$ in D_2O, in a series of alcoholic and acidic solvents (Laszlo

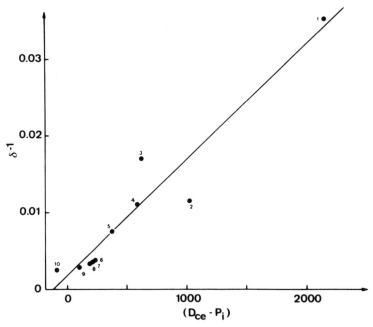

Fig. 4 Plot of the reciprocal chemical shift ($10^6 \, \delta^{-1}$) for $[(n\text{-Bu})_4\text{N}]_3\text{Co(CN)}_6$ against the difference between the density of cohesive energy D_{ce} and the internal pressure P_i (in joules per cubic centimeter). The more strongly associated solvents have the greatest $D_{ce} - P_i$ values ($\rho = 0.96$ for 10 points). The solvents are H_2O (1), formamide (2), ethanediol (3), methanol (4), ethanol (5), propylene carbonate (6), acetonitrile (7), Me_2SO (8), dimethylformamide (9), and hexamethylphosphor-triamide (10). (From Laszlo and Stockis, 1980.)

and Stockis, unpublished), are listed in Table VI. Note the great sensitivity of the cobalt shifts to the detailed molecular structure of these alcohols. For instance, in the case of the pentanols, the two primary alcohols give upfield cobalt shifts of about 154 ppm, whereas the two secondary alcohols give rise to yet larger shifts of 183.5 and 195.1 ppm. Indeed, it is possible to go further and to use ^{59}Co chemical shifts as a quantitative measure of the hydrogen bond strength (Fig. 4). Solvents are ranked as hydrogen bond donors according to the difference between the density of cohesive energy D_{ce} and the internal pressure P_i, and $D_{ce} - P_i$ measures the density of hydrogen bonding (Reichardt, 1979). The rather good linear correlation found between the reciprocal of the ^{59}Co chemical shift and this $D_{ce} - P_i$ parameter (Fig. 4) shows promise for the use of ^{59}Co NMR as an accurate and sensitive probe of hydrogen bond donation (Laszlo and Stockis, 1980).

Acids also show large effects. Under the conditions of Table VI, δ_{59Co} values are $+70$ (formic acid), -13 (acetic acid), -37 (propionic acid), -42 (butyric acid), and -47 (isobutyric acid), a sequence that agrees with that of acid strength (Laszlo and Stockis, unpublished results). Eaton *et al.* have investigated preferential solvation of the hexacyanocobaltate anion in mixtures of DMSO and acids (formic, propionic, trifluoroacetic) (Eaton *et al.*, 1982). Dimethyl sulfoxide solvates the $Co(CN)_6^{3-}$ anion preferentially over water, but stronger acids are preferred to DMSO as second-sphere ligands. The extent of the preference increases in the order of increasing acidity: propionic ($pK_a = 4.87$) < formic ($pK_a = 3.75$) < trifluoroacetic ($pK_a \sim -1.0$) (Eaton *et al.*, 1982).

With the stronger trifluoroacetic acid, the hydrogen-bonded complex is sufficiently long-lived that the ^{59}Co line width reflects *both* the increased reorientational correlation time and the substantial electrostatic field gradient. By analogy with the $Fe(CN)_6^{4-}$ anion, these authors estimate a ^{59}Co quadrupolar coupling constant of 6.4 MHz which, coupled to a correlation time of 49 ps, would lead to a 12.6-kHz line width. They have observed a line width on the order of 3 kHz (Eaton *et al.*, 1982). It seems that qualitatively the main effect here is full protonation of the anion probe by CF_3COOH, leading to the formation of $HCo(CN)_6^{2-}-CF_3COO^-$ with a strong electric field gradient at the cobalt nucleus.

In DMSO–H_2O binary mixtures, the line width goes up quasi linearly with the mole fraction of water to a maximum of ~ 95 Hz (for a mole fraction of water ~ 0.6) and decreases back, also quasi linearly, to a very small value of about 4 Hz. The authors explain these observations by a very short-lived hydrogen bond, existing less than the rotational correlation time (which is on the order of 10^{-11} s). They provide infrared evidence in support of this interpretation (Eaton and Sandercock, 1982).

In a DMSO–HCOOH binary mixture, the line width shoots up with the addition of a little acid, goes through a maximum for 0.2 mole fraction of HCOOH, decreases to a minimum (mole fraction HCOOH ~ 0.8), and then increases again (Eaton *et al.*, 1982). My conjecture, here, if indeed quadrupolar relaxation predominates, is that, at low formic acid concentrations, a monoprotonated species starts to form and is characterized by a very large intrinsic line width, as we saw previously for the CF_3COOH–DMSO results. As more formic acid is added, diprotonation and triprotonation occur. Perhaps the minimum in the line width (mole fraction ~ 0.8) corresponds to the presence of the fac fully neutralized $Co(CN)_6H_3$ species with its vanishing efg at this point. Adding still more acid leads to protonation of a fourth nitrogen lone pair, with a consequence qualitatively similar that of the initial monoprotonation.

XI. Ion Pairing Phenomena

Cobalt-59 NMR can be used for studying ion pairing. Since the efg values of O_h CoL_6 complexes are little affected by the presence of the counterion in the outer sphere (Rose and Bryant, 1979), the line width increases mostly because of the lengthened correlation time. For instance, with the $Co(en)_3^{3+}$ cation, ion pairing with the phosphate anion increases the rotational correlation time by a factor of 4.3. Correspondingly, the longitudinal relaxation rate increases by more than a factor of 6 (Rose and Bryant, 1979). The ^{59}Co chemical shift is another observable with which to monitor ion pairing.

With the cobaltihexacyanide anion in aqueous solution two types of behavior can occur. With protons and with alkali metal counterions only small downfield shifts of a few parts per million are observed (Laszlo and Stockis, 1980). In contrast, quaternary ammonium cations Q^+ cause large *upfield* shifts, and the effect grows in proportion to the size of the alkyl chains in the Q^+ ion. Qualitatively, these effects are similar to those undergone by the same $Co(CN)_6^{3-}$ anion in organic solvents as compared to water (Delville *et al.*, 1981). A detailed quantitative study on the ^{59}Co shift was made for the aqueous systems $K_3Co(CN)_6$, $K_2QCo(CN)_6$, $KQ_2Co(CN)_6$, and $Q_3Co(CN)_6$, where $Q^+ = {}^+N(n\text{-Bu})_4$. In order to account for the chemical shift variations, it is necessary to postulate the coexistence of ion pairing equilibria and of aggregation (or self-association) equilibria. If Co denotes the cobaltihexacyanide trianion, these contributing equilibria are written:

$$Co + Q \underset{}{\overset{K_1}{\rightleftharpoons}} CoQ \qquad \text{(ion pair formation)}$$

$$CoQ + Q \underset{}{\overset{K_2}{\rightleftharpoons}} CoQ_2 \qquad \text{(triple ion formation)}$$

$$CoQ_2 + Q \underset{}{\overset{K_3}{\rightleftharpoons}} CoQ_3 \qquad \text{(neutralized salt formation)}$$

For $CoQ_3 = C_1$,

$$C_1 + C_1 \underset{}{\overset{K}{\rightleftharpoons}} C_2 \qquad \text{(dimer formation)}$$

$$C_2 + C_1 \underset{}{\overset{K}{\rightleftharpoons}} C_3 \qquad \text{(trimer formation)}$$

$$C_3 + C_1 \underset{}{\overset{K}{\rightleftharpoons}} C_4 \qquad \text{(tetramer formation)}$$

etc.

In this manner, for the $^+N(n\text{-Bu})_4$ quaternary ammonium ion, values are found of $K_1 = 50$ to 55, $K_2 = 31$ to 37, and $K_3 = 19$ in water solutions. In methanol, as befits this lower dielectric medium, these values are increased to 516, 152, and 105, respectively. As for aggregate formation in

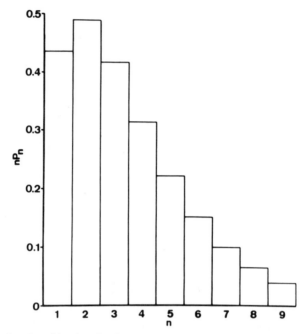

Fig. 5 Calculated partition function for the n-mers formed by self-association of $Q_3Co(CN)_6$ ($Q^+ = n\text{-}Bu_4N^+$) in H_2O solution at $33.05 \pm 0.02°C$ with $[Co]_t = 0.603\ M$. The computed average cluster size is $\langle N \rangle = 2.303$ under these conditions. (From Delville *et al.*, 1981.)

these protic solvents, it is characterized by $K \sim 5$ in water or in methanol. In contrast, $K = 10^{-2}$ in the aprotic propylene carbonate solvent. Aggregation results from an hydrophobic effect in which the n-butyl chains of separate Q^+ ions come together in close proximity (Delville *et al.*, 1981). The calculated distribution of the clusters formed is shown in Fig. 5. Interestingly, this Poisson-like distribution has a relatively slow decay. Appreciable concentrations of nonamers are present, although the *average* cluster size is only slightly greater than two (Delville *et al.*, 1981).

XII. Summary

Like the authors of companion chapters each devoted to a single nucleus or group of nuclei, one should resist the temptation to extol the virtues of the ^{59}Co nuclide. However, the ^{59}Co nucleus is somewhat exceptional and has at least historical importance:

(a) It has the largest range of chemical shifts reported for any nuclide in the periodic table.

(b) It was *the* nucleus on which the theory of the chemical shift for transition metals was based, as early as 1957.

(c) The predicted correlations with wavelength of optical $d \rightarrow d$ transitions hold extremely well.

(d) The chemical shift phenomenon itself was indeed discovered for ^{59}Co (Proctor and Yu, 1951).

Therefore, many useful applications, limited only by the imagination and ingenuity of investigators, are to be foreseen.

References

Abragam, A. (1961). "Principles of Nuclear Magnetism," p. 309. Clarendon, Oxford.

Ader, R., and Loewenstein, A. (1971). *J. Magn. Reson.* **5,** 248–261.

Bendall, M. R., and Doddrell, D. M. (1979). *J. Magn. Reson.* **33,** 659–663.

Betteridge, G. P., and Golding, R. M. (1969). *J. Chem. Phys.* **51,** 2497–2500.

Brevard, C., and Granger, P. (1981). "Handbook of High Resolution Multinuclear NMR," pp. 124–125. Wiley (Interscience), New York.

Buckingham, A. D., and Stephens, P. J. (1964). *J. Chem. Soc.* 2747–2759.

Craighead, K. L., Jones, P., and Bryant, R. G. (1975). *J. Phys. Chem.* **79,** 1868–1874.

Delville, A., Laszlo, P., and Stockis, A. (1981). *J. Am. Chem. Soc.* **103,** 5991–5998.

Dharmatti, S. S., and Kanekar, C. R. (1959). *J. Chem. Phys.* **31,** 1436–1437.

Doddrell, D. M. *et al.* (1979). *Aust. J. Chem.* **32,** 1219–1230.

Dwek, R. A., Luz, Z., and Shporer, M. (1970). *J. Phys. Chem.* **74,** 2232–2233.

Eaton, D. R., and Sandercock, A. C. (1982). *J. Phys. Chem.* **86,** 1371–1375.

Eaton, D. R., Rogerson, C. V., and Sandercock, A. C. (1982). *J. Phys. Chem.* **86,** 1365–1371.

Ehrenstein, D. V., Klopferman, H., and Penselin, S. (1960). *Z. Phys.* **159,** 230–231.

Freeman, R., Murray, G. R., and Richards, R. E. (1957). *Proc. R. Soc.* **A242,** 455–466.

Fujiwara, S., Yajima, F., and Yamasaki, A. (1969). *J. Magn. Reson.* **1,** 203–210.

Griffith, J. S., and Orgel, L. E. (1957). *Trans. Faraday Soc.* **53,** 601–606.

Gütlich, P., McGarvey, B. R., and Kläui, W. (1980). *Inorg. Chem.* **19,** 3704–3706.

Hartmann, H., and Sillescu, H. (1964). *Theoret. Chim. Acta (Berl.)* **2,** 371–385.

Hertz, H. G. (1973). *Ber. Bunsenges. Phys. Chem.* **77,** 688–697.

Hertz, H. G., Holz, M., Keller, G., Versmold, H., and Yoon, C. (1974). *Ber. Bunsenges. Phys. Chem.* **78,** 493–509.

Jordan, R. B. (1980). *J. Magn. Reson.* **38,** 267–275.

Jørgensen, C. K. (1962). "Progress in Inorganic Chemistry" (F. A. Cotton, ed.), Vol. 4, pp. 73–124. Wiley (Interscience), New York.

Jørgensen, C. K. (1966). *Struc. Bonding Berlin* **1,** 3.

Jørgensen, C. K. (1968). *Inorg. Chim. Acta Rev.* **2,** 65–88.

Juranic, N. (1980). *Inorg. Chem.* **19,** 1093–1095.

Juranic, N., Celap, M. B., Vucelic, D., Malinar, M. J., and Radivojsa, P. N. (1977). *Inorg. Chim. Acta* **25,** 229–232.

Juranic, N., Célap, M. B., Vucelic, D., Malinar, M. J., Radivojsa, P. N. (1979). *J. Magn. Reson.* **35**, 319–327.

Kanekar, C. R., Dhingra, M. M., Marathe, V. R., and Nagarajan, R. (1967). *J. Chem. Phys.* **46**, 2009–2010.

Kidd, R. G. (1980). *Ann. Rep. NMR Spectr.* **10A**, 1–79.

Kläui, W. (1979). *J. Chem. Soc. Chem. Commun.*, 700.

Koike, Y., Yajima, F., Yamasaki, A., and Fujiwara, S. (1974). *Chem. Lett.* 177–180.

Laszlo, P., and Stockis, A. (1980). *J. Am. Chem. Soc.* **102**, 7818–7820.

La Rossa, R. A., and Brown, T. L. (1974). *J. Am. Chem. Soc.* **96**, 2072–2081.

Levy, G. C., Bailey, J. T., and Wright, D. A. (1980). *J. Magn. Reson.* **37**, 353–356.

Lourens, J. A. J., and Smit, E. (1980). *Can. J. Phys.* **58**, 68–73.

Martin, R. L., and White, A. M. (1969). *Nature* **223**, 394–396.

Morral, F. R. (1967). *Adv. Chem. Ser.* **62**, 70–77.

Proctor, W. G., and Yu, F. C. (1951). *Phys. Rev.* **81**, 20–30.

Purcell, K. F., and Kotz, J. C. (1977). "Inorganic Chemistry." Saunders, Philadelphia.

Ramsey, N. F. (1950). *Phys. Rev.* **77**, 567; **78**, 699–703.

Ramsey, N. F. (1970). *Phys. Rev. A* **1**, 1320–1322.

Reichardt, C. (1979). "Solvent Effects in Organic Chemistry." Verlag-Chemie, Weinheim.

Rinaldi, P. L., Levy, G. C., and Choppin, G. R. (1980). *Rev. Inorg. Chem.* **2**(1), 63–89.

Rose, K., and Bryant, R. G. (1979). *J. Magn. Reson.* **35**, 223–226.

Sano, M., Yamatera, H., and Hatano, Y. (1979). *Chem. Phys. Lett.* **60**, 257–260.

Tarasov, V. P., Privalov, V. I., and Buslaev, Yu. A. (1978). *Mol. Phys.* **35**, 1047–1055.

Valiev, K. A. (1960). *Sov. Phys.-JETP* **37**, 77–

Valiev, K. A., and Zaripov, M. M. (1962). *Sov. Phys.-JETP* **14**, 545–

Valiev, K. A., and Zaripov, M. M. (1966). *Eh. Strukt. Khim.* **7**, 494–

Wehrli, F. W., and Wehrli, S. (1981). *J. Magn. Reson.* **44**, 197–207.

Weiss, R., and Verkade, J. G. (1979). *Inorg. Chem.* **18**, 529–530.

Werner, A. (1893). *Z. Anorg. Chem.* **3**, 267–330; see the translation and annotation by G. B. Kauffman (1968). "Classics in Coordination Chemistry. Part I. The Selected Papers of Alfred Werner." Dover, New York.

Yajima, F., Koike, Y., Yamasaki, A., Fujiwara, S. (1971). *Bull. Chem. Soc. Jpn.* **47**, 1442–1446.

Yamasaki, A. (1981). *Anal. Chim. Acta* **133**, 741–746.

Yamasaki, A., Yajima, F., and Fjuiwara, S. (1968). *Inorg. Chim. Acta* **2**, 39–42.

10 Selenium-77 and Tellurium-125

H. C. E. McFarlane and W. McFarlane

Chemistry Department
City of London Polytechnic
London, England

I. Introduction

As can be seen from Table I, which summarizes the salient nuclear properties, both ^{77}Se and ^{125}Te have natural abundances and sensitivities which should make them more favorable than ^{13}C for NMR studies. Against this must be set the lower atomic proportions of the two heavier elements in most compounds of interest, and in almost all cases the absence of any significant dipolar relaxation by protons. Relaxation times can therefore be rather long, necessitating slow pulse repetition rates in the absence of added relaxation reagent, and there is seldom any beneficial nuclear Overhauser enhancement (NOE). Overall then, ^{77}Se is in practice rather more difficult to observe than ^{13}C, and ^{125}Te is easier to observe. However, considerably more work has been done on selenium NMR than on tellurium, this disparity being also a reflection of the relative degrees of interest in their chemistry. Developments in multinuclear Fourier transform (FT) instrumentation have now reached the point (Luthra, 1982a) where it is possible to measure ^{77}Se resonances in biological systems containing selenium at sites normally occupied by sulfur atoms.

TABLE I

NMR-Active Selenium and Tellurium Isotopes

Nuclide	Natural abundance (%)	Receptivity[a]	Resonance frequency (MHz)[b]	Reference
^{77}Se	7.58	2.98	19.071523	Me$_2$Se
^{123}Te[c,d]	0.87	0.89	26.169773	Me$_2$Te
^{125}Te[d]	6.99	12.5	31.549802	Me$_2$Te

[a] Relative to ^{13}C.
[b] Of reference compound at a field strength corresponding to a proton resonance in Me$_4$Si of exactly 100 MHz.
[c] Radioactive with a half-life of $\sim 10^{13}$ yr.
[d] γ is negative for this nuclide.

Selenium-77 NMR and ^{125}Te-NMR have been reviewed by McFarlane and McFarlane (1978); there are earlier reviews by Lardon (1973) of selenium NMR and a brief account in Japanese that contains a valuable chart of selenium chemical shifts (Iwamura and Nakanishi, 1981).

II. Selenium-77

Following the earliest measurements of ^{77}Se spectra by direct continuous-wave (CW) methods (Birchall *et al.*, 1965; Lardon, 1970), ^{1}H—^{77}Se double-resonance (INDOR) experiments were used to study chemical shifts and coupling constants in a wide range of organoselenium compounds (McFarlane and Wood, 1972) and to delineate the main factors governing them. The first measurements using proton decoupling and commercially available pulsed FT spectrometers appeared soon after (Gronowitz *et al.*, 1973), and since then most work has been done in this way, although indirect methods including ^{1}H—^{77}Se, ^{1}H—^{31}P,^{77}Se (McFarlane and Rycroft, 1972, 1973), and ^{31}P—^{77}Se,^{1}H$_{noise}$ (Keat *et al.*, 1979) multiple-resonance experiments still find application.

There have been comparatively few measurements of ^{77}Se relaxation times, and these have tended to indicate a variable range of values for T_1. In general it is beneficial to add a relaxation reagent such as Cr(acac)$_3$ for routine FT measurements (Gansow *et al.*, 1978). In addition to the preliminary results of Pan and Fackler (1978) and Gansow *et al.*, (1978) the most comprehensive compilations of ^{77}Se T_1 data are by Odom *et al.* (1979) and Dawson and Odom (1977); some of their results are given in Table II. In no case was any significant NOE observed, even though the theoretical maximum is $\eta = 2.62$, so it is clear that the ^{1}H-^{77}Se dipolar relaxation

TABLE II

Selenium-77 Longitudinal Relaxation Times[a]

Compound	T_1 (s^{-1})	Temperature (°C)	Conditions
Me$_2$Se	5.2	55	CDCl$_3$
	8.6	12	CDCl$_3$
	24.4	−60	CDCl$_3$
Me$_2$Se$_2$	9.0	45	CDCl$_3$
	13.0	0	CDCl$_3$
MeSeH	1.3	40	(CD$_3$)$_2$CO
	3.7	−45	(CD$_3$)$_2$CO
(n-oct)$_2$Se	10.4	41	CDCl$_3$
	14.8	0	CDCl$_3$
	6.0	−42	CDCl$_3$
H$_2$Se	0.7	34	1 M, D$_2$O
	1.0	10	1 M, D$_2$O
DL-MeSe(CH$_2$)$_2$CH(NH$_2$)COOH	13.5	34	0.1 M, D$_2$O
Me$_2$Se(Br)CH$_2$CO$_2$Et	1.8	32	CDCl$_3$
	1.7	−1	CDCl$_3$
	0.7	−30	CDCl$_3$

[a] Odom *et al.* (1979).

mechanism is highly inefficient, which is to be expected in view of the dependence on the inverse sixth power of the internuclear distance and the large selenium covalent radius. However, small ($\eta \leq 0.4$) intermolecular NOEs have been reported (Koch *et al.*, 1978) for the protons of water and ^{77}Se in solutions of selenious acid, and other small effects have been reported by Denney *et al.*, (1981). The temperature dependence of the longitudinal relaxation indicates that for smaller, rapidly tumbling molecules the main mechanism is spin–rotation, whereas for larger molecules the chemical shift anisotropy (CSA) mechanism becomes important and is likely to dominate in molecules of biological interest. A detailed study was undertaken (Jakobsen *et al.*, 1980) on the relaxation behavior of hydrogen selenide, and again spin–rotation was found to dominate. Nothing appears to be known about T_2 for ^{77}Se.

A. Chemical Shifts

Well over 1000 selenium chemical shifts have been reported in the literature, and it is possible to present only a small representative selection in Table III, the references for which provide a guide to larger compilations of selenium chemical shifts. There is considerable diversity of

TABLE III

Selenium-77 Chemical Shifts in Selected Compounds

No.	Compound	Conditions	$\delta_{77_{Se}}{}^a$	Reference
1	H_2Se	Liquid	−226	Birchall *et al.* (1965)
	H_2Se	1 M, D_2O	−288	Birchall *et al.* (1965)
2	$(NH_4)_2Se$	0.5 M, D_2O	−511	Odom *et al.* (1979)
3	Na_2SeO_3	Sat., H_2O	1253	Wachli (1953)
4	K_2SeO_4	Sat., H_2O	1024	Wachli (1953)
5	$SeOCl_2$	Liquid	1479	Birchall *et al.* (1965)
6	$SeCl_4$	Sat., DMF	1154	Birchall *et al.* (1965)
7	SeF_4	Liquid	1092	Birchall *et al.* (1965)
8	SeF_6	Liquid	610	Birchall *et al.* (1965)
9	Se_4^{2+}	30% Oleum	1958	Schrobilgen *et al.* (1978)
10	$Se_2Te_2^{2+}$	68% Oleum	1638	Schrobilgen *et al.* (1978)
11	KSeCN	EtOH	−322	Pan and Fackler (1978)
12	CSe_2	CH_2Cl_2	299	Pan and Fackler (1978)
13	$Cd(Se_2CNBu_2)_2$	$CDCl_3$	717	Pan and Fackler (1978)
14	$Zn(Se_2CN\text{-i-}Bu_2)_2$	$CDCl_3$	668	Pan and Fackler (1978)
15	$K^+(Se_2CN\text{-i-}Bu_2)^-$	D_2O	582	Pan and Fackler (1978)
16	$Pd(Se_2CNEt_2)_2$	$CDCl_3$	365	Pan and Fackler (1978)
17	$(SiH_3)_2Se$	Liquid	−666	Arnold *et al.* (1972)
18	$Li^+(SeSiH_3)^-$	Et_2O	−736	Cradock *et al.* (1976)
19	$(GeH_3)_2Se$	Liquid	−612	Arnold *et al.* (1972)
20	$(Me_3Sn)_2Se$	Liquid	−547	Kennedy and McFarlane (1975)
21	Me_2Se	Liquid	0	McFarlane and Wood (1972)
22	Ph_2Se	Liquid	402	Lardon (1970)
23	MeSeH	Liquid	−116	McFarlane and Wood (1972)
24	*t*-BuSeH	CH_2Cl_2	278	McFarlane and Wood (1972)
25	PhSeH	Liquid	152	McFarlane and Wood (1972)
26	$MeSe^-Na^+$	H_2O	−332	McFarlane and Wood (1972)
27	$Me_3Se^+I^-$	H_2O	253	McFarlane and Wood (1972)
28	*cis*-$(Me_2Se)_2PtCl_2$	CH_2Cl_2	120	McFarlane and Wood (1972)
29	*trans*-$(Me_2Se)_2PtCl_2$	CH_2Cl_2	135	McFarlane and Wood (1972)
30	*trans*-$[Pt(SCN)_2(SeMe_2)_2]$	CH_2Cl_2	134	Anderson *et al.* (1976)
31	Me_2Se_2	Liquid	275	McFarlane and Wood (1972)
32	$Me_2Se_2Cl_2$	CH_2Cl_2	448	McFarlane and Wood (1972)
33	MeSeCN	$CDCl_3$	125	Christiaens *et al.* (1976)
34	$MeSeCl_3$	CH_2CL_2	890	McFarlane and Wood (1972)
35	Me_2SeO	H_2O	812	McFarlane and Wood (1972)
36	$MeSeO_2H$	H_2O	1216	McFarlane and Wood (1972)
37	$MeSeO_3^-K^+$	H_2O	1045	McFarlane *et al.* (1977)
38	Ph_2Se_2	H_2O	460	Lardon (1970)
39	Ph_2SeO	?	738	Fredga (1975)
40	$PhCH_2SeCH_2COOH$	$(CD_3)_2CO$	288	Fredga (1975)
41	*p*-$NO_2C_6H_4SeMe$	Liquid	233	Kalabin *et al.* (1979)
42	*p*-FC_6H_4SeMe	Liquid	200	Kalabin *et al.* (1979)
43	*o*-$MeC(O)C_6H_4SeCl$	$CDCl_3$	1087	Llabrès *et al.* (1981)

TABLE III (Continued)

No.	Compound		$\delta_{77_{Se}}{}^a$	Reference
44	o-HC(O)C$_6$H$_4$SeCl	CDCl$_3$	1097	Llabrès et al. (1981)
45	(MeO)$_2$SeO	Liquid	1339	McFarlane and Wood (1972)
46	(MeO)$_2$SeO$_2$	CH$_2$Cl$_3$	1053	McFarlane and Wood (1972)
47	PhC(Se)OEt	CDCl$_3$	915	Cullen et al. (1981)
48	t-Bu$_2$CSe	CDCl$_3$	2131	Cullen et al. (1981)
49	CF$_3$SeH	?	287	Gombler (1981)
50	Selenophene	CDCl$_3$	605	Christiaens et al. (1976)
51	2-Cl-selenophene	CDCl$_3$	564	Fredga (1975)
52	3-Cl-selenophene	CDCl$_3$	591	Fredga (1975)
53		CD$_3$CN	976	Sándor and Radics (1981)
54	Se-dl-cystine	Acid D$_2$O	293	Pan and Fackler (1978)
55	Se-dl-methionine	Acid D$_2$O	75	Pan and Fackler (1978)
56	Me$_2$PSeMe	CH$_2$Cl$_2$	58	McFarlane and Nash (1969)
57	Me$_2$P(S)SeMe	CH$_2$Cl$_2$	196	McFarlane and Nash (1969)
58	Me$_3$PSe	CH$_2$Cl$_2$	−235	McFarlane and Rycroft (1972, 1973)
59	(MeO)$_3$PSe	CH$_2$Cl$_2$	−396	McFarlane and Rycroft (1972, 1973)
60	Ph$_3$PSe	CH$_2$Cl$_2$	−275	Dean (1979)
61	[Ph$_2$P(Se)]$_2$CH$_2$	CH$_2$Cl$_2$	−256	Dean (1979)
62	cis-MeO(Se)P⟨N(t-Bu)⟩POMe	CDCl$_3$	−173	Keat et al. (1979)
63	trans-MeO(Se)P⟨N(t-Bu)⟩POMe	CDCl$_3$	−51	Keat et al. (1979)

^{a} In parts per million to high frequency of Me$_2$Se. Typical accuracies are plus or minus a few parts per million.

opinion regarding the best reference for selenium chemical shifts. Selenium dioxide solutions (Birchall et al., 1965) have the disadvantage of being strongly pH-dependent (Kolshorn and Meier, 1977). Other references that have been used include selenophene and (p-CH$_3$C$_6$H$_4$)$_2$Se$_2$. Many workers have used Me$_2$Se, which is amenable to both direct and indirect observation and is our choice for the discussion in this chapter. Its selenium resonance frequency $\Xi_{77_{Se}}$ is 19.071523 MHz (McFarlane and Wood, 1972).

The selenium chemical shift range now exceeds 3000 ppm. The normal effects of different solvents and of temperature changes are usually only a few parts per million. In hydrogen selenide the replacement of protium by deuterium leads to an increase in selenium shielding of 7 ppm for each atom substituted (Jakobsen et al., 1980). This is almost certainly due in the main to changes in the extent of hydrogen bonding. However, other smaller isotope effects on selenium shielding have also been noticed. Thus in dimethyl diselenide the selenium chemical shift moves ~0.026 ppm to lower frequency for every increase of two mass units in the other attached selenium atom, and in 2-deuterioselenophene there is a 1.68-ppm two-bond isotope shift (Jakobsen and Hansen, 1978). Detailed solvent studies have been undertaken on the shielding in t-BuSePh, Me_2Se, and several other molecules in a wide range of solvents and show effects up to 14.9 ppm, the largest occurring when the solvent is one capable of interacting coordinatively with the selenium atom (e.g., Me_2SO) (Valeev et al., 1980). In other work (Carr and Colton, 1981) on $[Ph_2P(Se)]_2CH_2$ and Ph_2P-$(Se)CH_2PPh_2$ similar but larger effects were found. The selenium chemical shift of aqueous solutions of selenious acid, H_2SeO_3, varies over a range of almost 50 ppm according to pH (Kolshorn and Meier, 1977), and the titration plot shows steps associated with formation of the ions $HSeO_3^-$ and SeO_3^{2-}. The temperature dependence of selenium chemical shifts is normally very small, but an exception occurs in the case of diorgano-diselenides, where $\delta_{77_{Se}}$ increases by ~0.4 ppm/°C^{-1} (Lardon, 1970). This was attributed to a change in the electronic excitation energy part of the expression for the paramagnetic contribution to the shielding.

Selenium chemical shifts are generally assumed to be dominated by the paramagnetic contribution (Ramsey, 1950), but apart from the temperature dependence of diselenides previously mentioned and results for species with a C=Se linkage, it is hard to find direct evidence for this contention. Empirically, it is found that a decrease in the electron density at selenium is accompanied by decreased shielding, so that derivatives of Se(IV) and Se(VI) tend to have chemical shifts in the high-frequency part of the range. However, this could be understood in terms of changes in either the paramagnetic or the diamagnetic shielding term. Similarly, the Se_4^{2+} ionic cluster discussed later has one of the highest values of $\delta_{77_{Se}}$ of all, +1958 ppm (Schrobilgen et al., 1978). In contrast, the Se^{2-} ion is highly shielded, with $\delta_{77_{Se}} = -511$ ppm (Odom et al., 1979), and when the selenium is bound to an electropositive element such as silicon or tin, there is similarly a substantial increase in shielding (Arnold et al., 1972;

Cradock *et al.*, 1976; Kennedy and McFarlane, 1975). In a series of derivatives CF_3SeX (X = Ag, H, CN, Br, Cl) there was an excellent linear correlation between $\delta_{77_{Se}}$ and the Huggins electronegativity of X (Gombler, 1981). This sensitivity to electronic factors is also exemplified by the organic derivatives of selenium, it being found, for example, that conversion of R_2Se to R_3Se^+ increases $\delta_{77_{Se}}$ by several hundred parts per million, whereas formation of RSeH or RSe^- produces a similar decrease (McFarlane and Wood, 1972). Similarly, $\delta_{77_{Se}}$ for seleninium tetra-fluoroborate is 371 ppm greater than for selenophene (Sándor and Radics, 1981). In this respect the behavior of $\delta_{77_{Se}}$ is quite similar to that of δ_{31_P}, although the sensitivity of the heavier nucleus to a particular type of change is about three times greater.

The selenium chemical shifts of organophosphine selenides (I)

$$R_3P{=}Se \longleftrightarrow R_3P^+{-}Se^-$$

Ia   **Ib**

are well to low frequency of Me_2Se, and from this it has been supposed that the canonical form **(Ib)** makes a substantial contribution to the resonance hybrid (McFarlane and Rycroft, 1972, 1973). Species with a carbon–selenium double bond, where this kind of resonance should be greatly reduced, have chemical shifts well to the high-frequency end of the range (Cullen *et al.*, 1981), being dominated by the paramagnetic contribution (discussed later).

Detailed studies on the dependence of selenium shielding on the finer details of electronic structure have been undertaken for many substituted aromatic and heteroaromatic compounds (Kalabin *et al.*, 1979b). In a range of 4-substituted phenylselenols and other derivatives, correlations of $\delta_{77_{Se}}$ were found with Hammett's σ constants (McFarlane and Wood, 1972). In some 4,4'-disubstituted diphenyl diselenides the selenium chemical shifts ranged from 567 ppm (NH_2 as substituent) to 629 ppm (NO_2 as substituent) and correlated well with Hammett and with Swain–Lupton substituent parameters (Gronowitz *et al.*, 1977, Kalabin *et al.*, 1978). Correlations were also observed with the ^{19}F chemical shifts in 4-substituted fluorobenzenes and with the ^{13}C chemical shifts of the aromatic carbon bound to selenium. Correlations of a similar nature have been found in a large number of substituted selenophenes (Gronowitz *et al.*, 1975), in 79 mono- and disubstituted benzo[*b*]selenophenes **(II)** (Baiwir *et al.*, 1981), and in substituted selenoanisoles (Kalabin *et al.*, 1979a, 1980). Seleno-loisothiazoles, selenoloisoselenazoles (Onyamboko *et al.*, 1982), and nitrobenzo[*b*]selenophenes (Baiwir *et al.*, 1981) have also been studied,

II

the last providing indications of effects due to selenium d-orbital interaction. Similar results have been obtained for organic selenenyl sulfides (Luthra *et al.*, 1982b) and other compounds (Bzehzovskii *et al.*, 1981). In general, the sensitivity to changes in electronic structure seems to be about six times greater for $\delta_{77_{Se}}$ than for δ_{13_C} of a similarly located carbon.

All these observations support the simplifying notion that selenium chemical shifts are governed by the electron density at the selenium atom. Indeed, CNDO calculations of charge densities in S—Se and Se—Se isosteres of phenanthrene such as **(III)** and **(IV)** show fair correlations with $\delta_{77_{Se}}$. There is also an interesting additive effect of ring fusion in these molecules (Baiwir *et al.*, 1980).

| 500 | 583 | $\longleftarrow \delta_{77_{Se}} \longrightarrow$ | 451 |

III   **IV**

Similarly, in species such as **(V)** and **(VI)** there are excellent correlations between $\delta_{77_{Se}}$ and the total and the π-electron densities at selenium (Abronin *et al.*, 1982). Other heterocycles show similar trends (Gronowitz *et al.*, 1982; Konar and Gronowitz, 1980; Talbot *et al.*, 1981).

| 579 | $\longleftarrow \delta_{77_{Se}} \longrightarrow$ | 739 |

V   **VI**

However, notwithstanding the foregoing, there is an abundance of evidence showing that in certain circumstances other effects can dominate. In many alkyl derivatives the selenium chemical shift increases with the amount of α branching of the alkyl group; e.g., t-BuSeH has a selenium resonance 394 ppm to high frequency of MeSeH (McFarlane and Wood, 1972). This is completely analogous to the behavior of many phosphorus compounds and cannot be predicted from simple considerations of electron density. The species $H_3SeO_3^+$, H_2SeO_3, $HSeO_3^-$, and SeO_3^{2-} have chemical shifts that display no correlation at all with charge density on selenium (Kolshorn and Meier, 1977). For example, the first and last have almost identical chemical shifts. In this series, there is a linear relation between $\delta_{77_{Se}}$ and the Se—O σ-bond force constant, a stronger bond being associated with a high-frequency resonance. In species with a selenium—carbon double bond, such as $(t\text{-Bu})_2CSe$, $PhC(Se)OEt$, and the molecule **(VII)** there is an excellent correlation between $\delta_{77_{Se}}$ and λ_{max}, the wavelength of the $n \rightarrow \pi^*$ electronic transition associated with the double bond. This is certainly consistent with normal formulation of the paramagnetic contribution to the shielding as containing a $(\Delta E)^{-1}$ term (Cullen et al., 1981). The same conclusion has been reached by Gombler (1981) from a study on similar species with carbon–selenium double bonds.

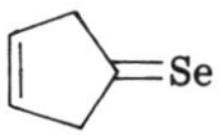

VII

Selenium chemical shifts are also sensitive to stereochemistry. Thus in **(VIII)** the selenium at A experiences enhanced deshielding owing to the proximity of the NO_2 group, whereas in **(IX)** the effect is somewhat less because the nitro group is attached to a five-membered ring and is thus further from the selenium (Granger et al., 1980).

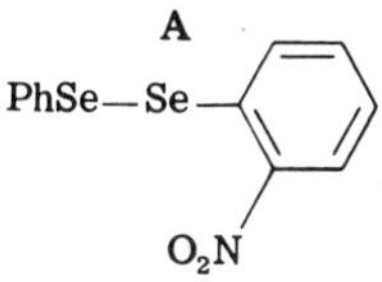

VIII 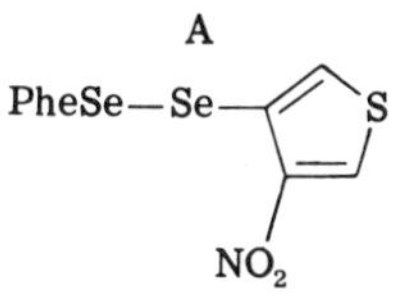   **IX**

In cyclodiphosphazene diselenides **(X)** the selenium resonates up to 100 ppm to higher frequency in the trans than in the cis isomers (Keat et al., 1979). The meso form of $[PhMeP(Se)]_2CH_2$ has $\delta_{77_{Se}}$ 17 ppm to high frequency of that in the DL form (Colquhoun and McFarlane, 1978).

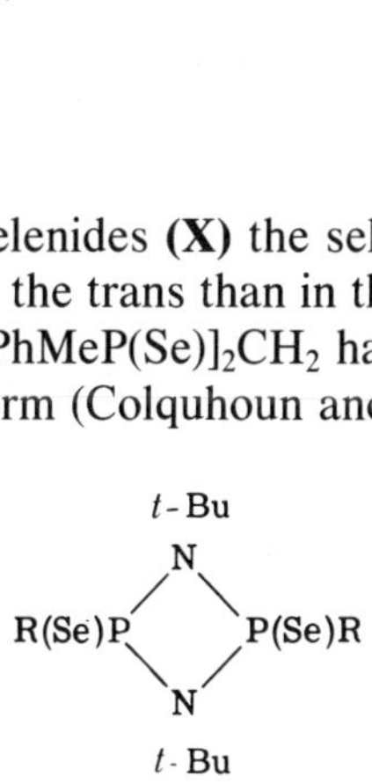

X

When selenium compounds function as ligands and form metal complexes, $\delta_{77_{Se}}$ can increase by a 100 ppm or more. This has been observed for dialkyl selenides (Anderson *et al.*, 1976; McFarlane and Wood, 1972), for selenocarbonates (Pan and Fackler, 1978), and for tertiary phosphine selenides (Colton and Dakternieks, 1980; Colquhoun and McFarlane, 1981). An interesting exception to this generalization is the complex (XI) where the selenium is part of a three-membered ring whose constraining effect leads to a remarkable shielding with $\delta_{77_{Se}} = -910$ ppm. The "ligand" (XII) with uncomplexed selenium has $\delta_{77_{Se}} = +160$ ppm (Colquhoun *et al.*, 1982).

$$\eta^5\text{-}C_5(CO)_2W\!\!-\!\!PPh_2 \qquad\qquad \eta^5\text{-}C_5H_5(CO)_3W\!\!-\!\!PPh_2$$
$$\diagdown\diagup \qquad\qquad\qquad\qquad\qquad\qquad \|$$
$$Se \qquad\qquad\qquad\qquad\qquad\qquad\qquad Se$$

$$\textbf{XI} \qquad\qquad\qquad\qquad\qquad\qquad \textbf{XII}$$

Solutions of phosphine selenides in liquid sulfur dioxide have $\delta_{77_{Se}}$ to high frequency of that in other solvents owing to complex formation (Dean, 1979), whereas complexation of Cd^{2+} by $PhSe^-$ leads to small reductions in $\delta_{77_{Se}}$ (Carson and Dean, 1982).

B. Coupling Constants

Coupling between ^{77}Se and another nucleus X manifests itself in the spectrum of X in the form of symmetrically placed satellites flanking the main peak, each normally having ~3.7% of the intensity of this peak, although this pattern may be perturbed (Fackler and Pan, 1979) when the molecule contains more than one selenium atom. When X is one of the more receptive nuclei (e.g., ^1H, ^{19}F, ^{31}P), its observation undoubtedly is the best way of determining J_{SeX}, and most of the data in the literature have been obtained in this way. It is also considerably easier to detect and assign correctly the ^{77}Se satellites in ^{13}C spectra than to attempt to observe the often complex pattern of ^{13}C satellites (of only 0.5% intensity) in ^{77}Se spectra. The shorter relaxation time of ^{199}Hg made it better to study this nucleus rather than ^{77}Se in the measurement of $^1J_{199_{Hg}77_{Se}}$ in $[Bu_3P(Se)]_2HgX_2$ (Colquhoun and McFarlane, 1981). Many of the earliest measurements of coupling between ^{77}Se and nuclei other than ^1H or ^{19}F were actually done indirectly by means of ^1H—^{77}Se spin-tickling experiments, a procedure that utilizes the high sensitivity of the proton and also gives the *sign* of the coupling in many cases (Kennedy and McFarlane, 1973; McFarlane, 1967; Pfisterer and Dreeskamp, 1969).

TABLE IV

Values of $^nJ_{(^{77}SeH)}$

n	Compound	^nJ (Hz)	Reference
1	H_2Se	65.4	Dreeskamp and Pfisterer (1968)
	MeSeH	44	McFarlane et al. (1977)
	PhSeH	56	McFarlane et al. (1977)
2	$MeSe^-K^+$	6.6	McFarlane et al. (1977)
	Me_2Se	10.5	McFarlane et al. (1977)
	Me_3Se^+	9.3	McFarlane (1967)
	Me_2SeBr_2	10.0	McFarlane et al. (1977)
	Et_2Se	10.7	Breuninger et al. (1966)
	Et_2SeBr_2	5.3	McFarlane et al. (1977)
	$MeSeO_3^-$	-8.8	McFarlane et al. (1977)
	Selenophene		Simmonnin et al. (1976)
	Seleninium$^+$	45.3	Sándor and Radics (1981)
3	Et_2Se	10.8	Breuninger et al. (1966)
	Et_3Se^+	17.4	McFarlane et al. (1977)
	$(MeO)_2SeO$	7.6	McFarlane et al. (1977)
	Selenophene	47.0	Simmonnin et al. (1976)
	Seleninium$^+$	6.2	Sándor and Radics (1981)

Indeed, even today indirect detection via portons can still be the quickest approach to measuring $J_{^{77}SeX}$ when X is of low receptivity.

(a) *Coupling to protons:* This topic was reviewed by Svanholm (1973) and by McFarlane and McFarlane (1978). Rather little in the way of fundamentally new results has appeared since then. Typical values of $^nJ_{^{77}SeH}$ are given in Table IV.

The one-bond coupling is quite sensitive to solvent effects and the coupling increases with deuterium substitution in H_2Se. There is also a *primary* isotope effect, the value of $^1J_{^{77}SeD}(\gamma_H/\gamma_D)$ being ~5% larger than $^1J_{^{77}SeH}$ (Jakobson et al., 1980). The two-bond coupling appears to require quite drastic molecular changes to be affected very much and becomes negative only when the selenium atom loses *all* its lone-pair electrons. In a series of substituted selenoanisoles, $4-XC_6H_4SeMe$, the value of $^2J_{^{77}SeH}$ was essentially invariant to the nature of X (Kalabin et al., 1979a). $^3J_{^{77}SeH}$ is sensitive to stereochemical relationships (Olsson and Almquist, 1969; Reich and Trend, 1976). Evidence for the stereochemical dependence of $^3J_{^{77}SeH}$ also comes from a study on molecules with a cyclopropyl group bound to selenium (Kalabin et al., 1980). In $(PhSe)_2SiH_2$ the geminal coupling $^2J_{^{77}SeSiH}$ has the relatively large value of 20.6 Hz (Drake and Hem-

TABLE V

$^nJ_{^{77}Se^{13}C}$ in Selected Compounds

Compound	$^nJ_{^{77}Se^{13}C}$ (Hz)		Reference
	$n = 1$	$n = 2$	
Me$_2$Se	−62	—	McFarlane (1967)
Et$_2$Se	−61.2	8.5	Breuninger et al. (1966)
EtSeO$_3^-$K$^+$	−18.5	7.1	McFarlane et al. (1977)
(MeO)$_2$SeO	—	0.9	McFarlane et al. (1977)
(MeO)$_2$SeO$_2$	−103.6	11.6	Kalabin and Kushnarev (1979)
(4-FC$_6$H$_4$)$_2$Se$_2$	−124.8	11.6	Kalabin et al. (1978)
t-Bu$_2$C=Se	−213.6	—	Cullen et al. (1981)
PhC(Se)OEt	−211.1	—	Cullen et al. (1981)

	−155.4	−13.3	Sándor and Radics (1981)
		22.9 (^3J)	

	100 (w),	12.3 (u),	Reich and Trend (1976)
	46 (y)	13.7 (x),	
		31.3 (z)	

mings, 1976). Additive effects of substituents have been found for the various selenium–proton couplings in selenophenes (Simonnin et al., 1976).

(b) *Coupling to ^{13}C:* Values are summarized in Table V. The coupling $^1J_{^{77}Se^{13}C}$ is negative (McFarlane, 1967), and its magnitude varies from 46 Hz in MeSeH to over 242 Hz in PhSeCN, there being an evident dependence on the hybridization of the carbon atom and the selenium atom (Cullen et al., 1981; Kalabin and Kushnarev, 1979), sp^3 hybridizations leading to relatively small values, whereas an increase in s-character produces an increase in the magnitude of $^1J_{^{77}Se^{13}C}$. Evidence for a dependence on selenium lone-pair orientation has been obtained, and correlations have been noted with Hammett σ constants in aromatic compounds (Kalabin and Kushnarev, 1979; Kalabin et al., 1979a, b). Exceptionally, in MeSeO$_3^-$K$^+$ the coupling is only −13 Hz; this probably represents an attempt by selenium to emulate phosphorus, because this is one of the few species in which selenium has lost all its lone-pair electrons (McFarlane et al., 1977). Longer-range ^{77}Se—^{13}C couplings are usually much smaller, so that a magnitude in excess of 45 Hz is usually diagnostic of a selenium–

carbon bond. There is evidence that these couplings depend on stereochemistry (Reich and Trend, 1976), and simple correlations with other parameters have been difficult to find. The coupling $^2J_{^{77}Se^{13}C}$ is normally positive, becoming negative only when the selenium atom loses all its lone-pair electrons (McFarlane et al., 1977). The seleninium ion in which this coupling is negative provides an exception to this generalization (Sándor and Radics, 1981).

(c) *Coupling to ^{31}P:* This coupling is negative, has a magnitude of several hundred hertz, and so is normally very easily measured in ^{31}P spectra. In polyphosphorus compounds the presence of ^{77}Se can introduce magnetic nonequivalence which may make it possible to measure an

TABLE VI

$^1J_{^{77}Se^{31}P}$ in Selected Compounds

Compound	$^1J_{^{77}Se^{31}P}$ (Hz)	Reference
Me$_2$PSeMe	−205	McFarlane and Nash (1969)
Me$_2$P(S)SeMe	−341	McFarlane and Nash (1969)
Me$_3$PSe	−684	McFarlane and Rycroft (1972, 1973)
Ph$_3$PSe	−736	McFarlane and Rycroft (1972, 1973)
(Me$_2$N)$_3$PSe	−805	McFarlane and Rycroft (1972, 1973)
(MeO)$_3$PS	−963	McFarlane and Rycroft (1972, 1973)
(EtO)$_3$PSe	−935	Stec et al. (1972)
(2-MeC$_6$H$_4$)$_3$PSe	−708	Pinnell et al. (1973)
Bu$_3$PSe	−689	Grim et al. (1978)
(Bu$_3$PSe)$_2$HgCl$_2$	−518	Grim et al. (1978)
(2-Thienyl)$_3$PSe	−757	Allen and Taylor (1982)
η^5-C$_5$H$_5$(CO)$_3$WP(Se)Ph$_2$	−621	Colquhoun et al. (1982)
η^5-C$_5$H$_5$(CO)$_2$WP(Se)Ph$_2$	−472	Colquhoun et al. (1982)
MeC(CH$_2$O)$_3$PSe	−1053	Kroschefsky et al. (1979)
MeC[CH$_2$N(Me)]$_3$PSe	−854	Kroschefsky et al. (1979)
cis-MeO(Se)P⟨N(t-Bu)⟩$_2$P(Se)OMe	−954	Keat et al. (1979)
trans-MeO(Se)P⟨N(t-Bu)⟩$_2$P(Se)OMe	−952	Keat et al. (1979)

otherwise inaccessible ^{31}P—^{31}P coupling constant (Colquhoun and Mc-Farlane, 1981; Colquhoun et al., 1979). Typical values are listed in Table VI, and there is a clear dependence on the electron-withdrawing ability of the substituents attached to phosphorus, greater charge withdrawal leading to an increase in $^1J_{31_P 77_{Se}}$. In addition, the P=Se linkage is associated with a substantially greater magnitude than the single bond. Cage compounds such as (XII) and (XIII) have magnitudes of $^1J_{77_{Se} 31_P}$, some 10% larger than $(MeO)_3PSe$ (Kroschefsky and Verkade, 1979), an effect that can be attributed to a change in the hybridization of phosphorous and consequent diversion of s-character into the P=Se bond.

$$Me—C\underset{CH_2—O}{\overset{CH_2—O}{\underset{|}{\overset{|}{-}}CH_2—O}}PSe \qquad Me—C\underset{CH_2—O}{\overset{CH_2—O}{\underset{|}{\overset{|}{-}}O}}PSe$$

XII   **XIII**

The formation of a metal complex by a tertiary phosphine selenide achieves a reduction in the magnitude of $^1J_{31_P 77_{Se}}$ of 100 Hz or more (Colquhoun and McFarlane, 1978; Colton and Dakternieks, 1980; Grim et al., 1978). Longer range couplings between selenium and phosphorus have also been reported: They are generally only a few hertz in magnitude (Colquhoun et al., 1979), although in P_4Se_3 $^2J_{77_{Se} 31_P}$ is 57 Hz (Dwek et al., 1969), and in $[PtSe_2CN-i-Bu_2)(PPh_3)_2]^+$ this coupling is 90 Hz (Fackler and Pan, 1979).

(d) *One-bond coupling to other nuclei:* Only a selection of values can be given here. Coupling $^1J_{77_{Se} 19_F}$ has been studied extensively by Seppelt (1973, 1975). The following values (all presumably negative) are typical: SeF_6, 1432; SeF_5OCl, 1350, 1453 (equatorial); SeF_5Cl, 1352, 1258 (equatorial); SeF_4, 302, 1200 Hz.

The coupling $^1J_{77_{Se} 77_{Se}}$ is small in diselenides (Granger et al., 1980), a typical value being 22 Hz in MeSeSePh (McFarlane, 1969), whereas $^1J_{77_{Se} 125_{Te}}$ is -169 Hz in MeSeTeMe (Pfisterer and Dreeskamp, 1969). This type of coupling in polycations is discussed later.

The coupling $^1J_{113_{Cd} 77_{Se}}$ changes linearly from 126 to 46 Hz as n increases from 0 to 3 in $[Cd(SPh)_n(SePh)_{4-n}]^{2-}$ (Carson and Dean, 1982). $^1J_{195_{Pt} 77_{Se}}$ is 259 and -68 Hz in $[PtCl_5SeMe_2]^-$ and $[PtI_5SeMe_2]^-$, respectively (Goggin et al., 1975). $^1J_{119_{Sn} 77_{Se}}$ increases with n in $Me_{4-n}Sn(SeMe)_n$ from 1015 to 1520 Hz (Kennedy and McFarlane, 1973); in the cyclic trimer $(Me_2SnSe)_3$ (Blecher et al., 1980) it is 1228 Hz. In metal complexes of tertiary phosphine selenides a range of values of $^1J_{M 77_{Se}}$ is found (Colquhoun and McFarlane, 1978; Colquhoun et al., 1982): ^{199}Hg, -751 Hz in

$(Bu_3PSe)_2HgCl_2$; ^{103}Rh, 20 Hz in dpmseRh(CO)Cl, where dpmse = Ph_2P-(Se)CH_2PPh_2; ^{195}Pt, -633 Hz in dpmsePtMe$_2$, and 160 Hz in dpmsePtCl$_2$.

C. Chemical Applications

Now that the ground rules of selenium NMR have been established, a number of applications are beginning to appear. Several studies on oriented molecules containing selenium have used the dipolar couplings from ^{77}Se to protons to provide additional geometric information. These include selenophene, 1,2,5-selenadiazole, 2,2′-biselenophene (Chidichimo *et al.*, 1978, 1979, 1980), selenophen-2-carbaldehyde (Bucci *et al.*, 1979), and dimethyl selenide (Diehl *et al.*, 1978). An early application was an assignment of the structure of $(EtO)_2P(S)SeEt$ on the basis of the magnitude (477 Hz) of $^1J_{^{77}Se^{31}P}$ (Stec *et al.*, 1972), and this coupling has also been used to derive conformations of cyclic phosphite derivatives (Kroshefsky and Verkade, 1979).

Selenium-77 spectra have provided information on intermolecular exchange in mercury (Colton and Dakternieks, 1980) and cadmium (Carson and Dean, 1982) complexes. The values of $\delta_{^{77}Se}$ and $^1J_{^{77}Se^{31}P}$ for solutions of tertiary phosphine selenides in liquid sulfur dioxide demonstrate the formation of complexes (Dean, 1979). The ^{77}Se spectra of **(XIV)** provided

XIV

good evidence for this structure as the third "classic" selenophene (Konar and Gronowitz, 1980), and $\delta_{^{77}Se}$ and $^1J_{^{77}Se^{13}C}$ were used to deduce the solution structures of 2-carboxyphenyl methyl selenoxide and its sodium salt as a cyclic selenurane and an acyclic selenoxide, respectively (Nakanishi *et al.*, 1981a). Selenium-77 NMR was similarly valuable in a study of **XV** and some of its derivatives (Nakanishi *et al.*, 1981b), and of intermediates in some selenoxide elimination reactions (Reich *et al.*, 1982). The reaction of selenium dioxide with aromatic ditellurides gave species with Te—Se—Te linkages whose characterization was aided by ^{77}Se (and ^{125}Te) NMR (Dereu *et al.*, 1982).

XV

An especially interesting application of selenium and tellurium NMR was a study of the mixed polyanions, $Se_2Te_2^{2+}$, etc., formed in oleum, and this is discussed in Section III. Finally, an investigation of molten selenium over the temperature range 193°C (supercooled liquid) to 1625°C (supercritical fluid) showed that δ_{77Se} and T_1 could be used to determine the degree of polymerization, which ranged from ~750 atoms per chain at 600°C to 7 atoms at 1550°C. This was possible because both the chemical shift and the relaxation time depended on the paramagnetic centers at the polymeric chain ends (Warren and Dupree, 1980).

III. Tellurium-125

Although ^{123}Te also has $I = \frac{1}{2}$, its low abundance makes it unsatisfactory for direct study, except perhaps to measure J_{TeTe} in symmetric species such as the cyclic cation Te_4^{2+} and diorganoditellurides, and even then the ^{123}Te satellites in ^{125}Te spectra would provide the required information. In fact there have been very few reports of ^{123}Te measurements (Buckler et al., 1977; Chapelle and Granger, 1981; Granger and Chapelle, 1980; Lassigne and Wells, 1978; Pfisterer and Dreeskamp, 1969), and as expected there are no significant differences between ^{125}Te and ^{123}Te chemical shifts. The first reports of tellurium NMR involved indirect detection via protons (McFarlane and McFarlane, 1973; McFarlane et al., 1976), but since then almost all work has used the FT technique (Drakenberg et al., 1976, 1979).

The theoretical maximum proton NOE is $\eta = -1.58$ for ^{125}Te, but no measurements have been published of values other than zero. A 4 M solution of K_2TeO_3 in water had $T_1 = 2.5$ s and $T_2 = 0.32$ s, the latter low value being attributed to chemical exchange of an unspecified nature (Buckler et al., 1977). In $(4\text{-MeC}_6\text{H}_4)_2\text{Te}_2$ and $(4\text{-EtOC}_6\text{H}_4)_2\text{Te}_2$ values of 1.4 and 1.7 s, respectively, were obtained for T_1 (Granger et al., 1981), and in some tetraalkoxytelluranes such as $Te(OCH_2CF_3)_4$ and (**XVI**), T_1 ranged from 1.5 to 3 s (Denney et al., 1981). A report on Me_2Te gives T_1 as 0.75 s (Granger et al., 1981a).

XVI

Detailed studies by Granger and Chapelle (1980) and by Chapelle and Granger (1981) of Me_2Te showed a magnetic field independence and also a temperature dependence consistent with domination of the relaxation by the spin–rotation mechanism. In a field of 2.1, T, T_1 varied from 0.41 s at 68°C to 1.34 s at −44°C.

A. Chemical Shifts

The overall chemical shift range probably exceeds 4000 ppm. Table VII gives a selection of values for different chemical types. As with selenium, several different reference compounds have been used for ^{125}Te NMR. Here the shifts are referred to Me_2Te which has $\Xi_{125Te} = 31.549802$ MHz; however, its shielding is temperature-dependent, and δ_{125Te} decreases by ~13 ppm over the temperature range −44 to +68°C, that is, by −0.11 ppm °C^{-1} (Granger and Chapelle, 1980). This effect is in the direction opposite that expected if changes in $(\Delta E)^{-1}$ in the paramagnetic shielding contribution were the dominating factor. However, in ditellurides there is a significant concentration dependence of both δ_{125Te} and λ_{max} in a way that is consistent with dominance by the paramagnetic team in this case (McFarlane and McFarlane, 1973). In the absence of any definite chemical interaction, solvent effects on δ_{125Te} are normally small. Studies on Me_2Te, MeTePh, and MeTeC≡CBu in a wide range of different solvents showed variations in δ_{125Te} of up to 30 ppm (Kalabin et al., 1981b).

The first study of tellurium shielding pointed out that in organic tellurium compounds the chemical shifts run remarkably parallel to selenium chemical shifts in corresponding compounds, a plot of δ_{125Te} against δ_{77Se} being linear with a slope of ~1.8 (McFarlane and McFarlane, 1973). The ratio $\langle r_{5p}^{-3}\rangle$ (Te)/$\langle r_{4p}^{-3}\rangle$ (Se) is ~1.25, so there is evidently an increased sensitivity of the tellurium atom to changes in the molecular environment. More recent results have confirmed this overall trend with high electron density producing increased shielding, but the fit to a linear plot now is not as good as it was when a smaller range of compounds was considered.

For example, in tellurophene δ_{125Te} is +782 ppm, whereas in selenophene δ_{77Se} is +605 ppm; in (t-Bu)$_3$P=Te δ_{125Te} is −80 ppm, whereas comparison with selenium analogs would have predicted a shift in the region −400 to −600 ppm. Both these results might be due to the extent of multiple bonding to tellurium in these species being significantly different from that involving selenium in the corresponding compounds. However, the value of $\delta_{125Te} = -60$ ppm (Du Mont and Kroth, 1981) in $(Me_3Si)_2Te$ is puzzling when considered alongside $\delta_{77Se} = -666$ ppm in $(SiH_3)_2Se$ and −547 ppm in $(Me_3Sn)_2Se$, and $\delta_{125Te} = -1214$ ppm in $(Me_3Sn)_2Te$.

Inorganic derivatives of tellurium, where again there may be significant differences in bonding compared with selenium, also deviate strongly

TABLE VII

Tellurium-125 Chemical Shifts in Selected Compounds

Compound	Conditions	$\delta_{125_{Te}}$ [a]	Reference
K_2TeO_3	2 M, D_2O	1732	Goodfellow (1977, private communication)
$Te(OH)_6$	H_2O	207	Tötsch et al. (1981)
$TeCl_4$	Me_2CO	1138	Goodfellow (1977, private communication)
H_2TeCl_6	2 M, acid D_2O	1403	Goodfellow (1977, private communication)
H_2TeI_6	1 M, acid D_2O	857	Goodfellow (1977, private communication)
TeF_6	BrF_5	545	Keller and Schrobilgen (1981)
$HOTeF_5$	MeCN	601	Tötsch et al. (1981)
$XeOTeF_5$	SbF_5	576	Keller and Schrobilgen (1981)
Te_4^{2+}	30% Oleum	2811	Schrobilgen (1978)
$Se_2Te_2^{2+}$	30% Oleum	3102	Schrobilgen (1978)
Te_6^{4+}	30% Oleum	152	Schrobilgen (1978)
Me_2Te	Liquid	0	McFarlane and McFarlane (1973)
$Me_3Te^+I^-$	Me_2SO	443	McFarlane and McFarlane (1973)
Me_2TeCl_2	CH_2Cl_2	749	McFarlane and McFarlane (1973)
Me_2Te_2	CH_2Cl_2	63	McFarlane and McFarlane (1973)
Ph_2Te	CH_2Cl_2	688	McFarlane and McFarlane (1973)
Ph_2TeCl_2	CH_2Cl_2	981	McFarlane et al. (1976)
Ph_2Te_2	CH_2Cl_2	422	Goodfellow (1977, private communication)
$(2\text{-Thienyl})_2Te_2$	$CDCl_3$	264	Granger et al. (1981a)
Tellurophene	Me_2CO	782	Kalabin et al. (1981b)

	$CDCL_3$	440	Lahner and Praefcke (1981a)
$(CF_3)_2Te$	?	1368	Gombler (1981)
$(CF_3)_2Te_2$	?	686	Gombler (1981)
$(Me_3Si)_2Te$	C_6D_6	-60	DuMont and Kroth (1981)
$(Me_3Sn)_2Te$	CH_2Cl_2	-1214	McFarlane and McFarlane (1973)
$(t\text{-}Bu_2P)_2Te$	C_6D_6	-174	DuMont and Kroth (1981)
$t\text{-}Bu_3PTe$	C_6D_6	-80	DuMont and Kroth (1981)
TeI_2	$CDCl_3$, DMSO	3129	Fazerkeley and Celotti (1979)
$Te(OCH_2CH_3)_4$	$CDCl_3$	1503	Denney et al. (1981)

[a] In parts per million to high frequency of Me_2Te. Accuracy is plus or minus a few parts per million.

from the previously noted generalization. Compare, for example, the chemical shifts in Tables III and V for TeF_6 and SeF_6, or $TeCl_4$ and $SeCl_4$. Stepwise replacement of OH in *o*-telluric acid $Te(OH)_6$ by fluorine to yield eventually TeF_6 produces steady decrements in $\delta_{125_{Te}}$ from 700 to 550 ppm (Tötsch et al., 1981), although this is not a change in the direction predicted on electronegativity grounds, and similar behavior is exhibited by mixed MeO/OH/F/Te(VI) species (Tötsch and Sladky, 1980). There are good parallels between the tellurium and selenium chemical shifts in the $Te_{4-n}Se_n^{2+}$ series of polycations, with the elements in the $+\frac{1}{2}$ oxidation state, and the increase in δ as n increases can be understood on the basis of the greater electronegativity of selenium. The species $Te_{6-n}Se_n^{2+}$, with $+\frac{1}{3}$ oxidation states, have lower values of δ, whereas δ_{125Te} for Te_6^{4+} (oxidation state $+\frac{2}{3}$) is very much to low frequency of any of the other species (Lassigne and Wells, 1978; Schrobilgen et al., 1978). In a range of OTeF₅ derivatives changes in δ_{125Te} were used to show that F was more electronegative than OTeF₅ (Birchall et al., 1982).

Systematic studies on tellurium chemical shifts in organotellurium compounds have been undertaken for 17 symmetric and 36 unsymmetric dialkyl ditellurides for which the effects of alkyl group substitution are found to be additive (O'Brien et al., 1982); for a selection of compounds with tellurium bound to mainly sp^2- or sp-hybridized carbon (Kalabin et al., 1981a); in a range of diaryl tellurides and ditellurides and their dichlorides, good correlations being found with Hammett σ constants (Kalabin et al., 1981c); for substituted tellurophenes with parallel results for selenium (Drakenberg et al., 1976, 1979); for benzo[(b)]tellurophenes where there are excellent correlations with selenium chemical shifts (Baiwir et al., 1982; Drakenberg et al., 1979); and for a range of tetraalkoxytelluranes which again correlated with corresponding results for selenium (Denney et al., 1981). Other compounds that have been studied include various phenyl derivatives (Lohner and Praefcke, 1981a), di-2-tellurienyl ketone (Lohner and Praefcke, 1981b), aromatic derivatives ArTeSe-Te(O)Ar (Dereu et al., 1982), miscellaneous species with the tellurium phosphorus bands (Du Mont and Kroth, 1981), diaryl ditellurides (Granger et al., 1981b), complexes of diorganotellurides for which δ_{125Te} increases by ~100 ppm on complex formation (du Mont and Nordhoff, 1980; Gysling et al., 1981); vinyl compounds (Bzehzovskii et al., 1981) and some species R_2TeX_2 (X = Cl, Br, I; R_2, cyclic) where δ_{125Te} increases with the electronegativity of X (Zumbulyadis and Gysling, 1980).

B. Coupling Constants

As for selenium, the coupling constants are often most readily determined by measuring the positions of the ^{125}Te satellites in the spectrum of the other nucleus. The existence of ^{123}Te makes it possible to measure Te—Te couplings in symmetric species. For couplings dominated by the Fermi contact interaction it is to be expected that the ratio J_{125TeX}/J_{77SeX} in analogous situations should be equal to $[\psi^2_{OTe}\gamma_{Te}]/\psi^2_{OSe}\gamma_{SE}]$ or ~ -2.0, the minus sign arising from the negative magnetogryic ratio of tellurium. Exceptions to this are likely to occur with small coupling constants, for these are generally the sum of two large terms of opposite sign, and a small change in either can produce a relatively large change in the coupling constant. In practice, tellurium coupling constants have generally been found to be two to three times those for selenium. Some typical values in hertz are as follows.

$^1J_{125_{TeH}}$: H_2Te, -59 (Glidewell et al., 1969).

$^2J_{125_{TeH}}$: Me_2Te, -20.7 and Me_3Te^+, -20.9 (McFarlane, 1967); tellurophenes, $\sim$90 (Fringuelli, et al., 1974; Lohner and Praefcke, 1981b; Martin et al., 1981); compound (XVI), $(-)40$ (Zumbulyadis and Gysling, 1980). This coupling is also thought to depend on stereochemistry (Dewan et al., 1978).

XVI

$^3J_{125_{TeH}}$: Et_2Te, -22.7 (Breuninger et al., 1966); tellurophenes, -12 Hz (Fringuelli et al., 1974; Martin et al., 1982).

$^1J_{125_{Te^{13}C}}$: Me_2Te, 162 (McFarlane 1967); $(CH_2{=}CH)_2Te$, $(+)285$; $MeTeC{\equiv}CBu$, $(+)531$ (Kalabin et al., 1981a); tellurophenes, 274–374 (Martin et al., 1982). The coupling appears to correlate well with the carbon s-character (Kalabin et al., 1981a) and also depends on the polarity of the tellurium–carbon bond (Chadha and Miller, 1982).

$^1J_{125_{Te^{19}F}}$: TeF_6, 3688 (Fraser et al., 1969; Muetterties and Phillips, 1959); $HOTeF_5$, $(+)3339$; $(HO)_5TeF$, $(+)2754$ (Tötsch et al., (1981); $XeOTeF_5$, $(+)3810$ (Keller and Schrobilgen, 1981).

$^1J_{125_{Te^{31}P}}$: $(t$-$Bu)_3PTe$, $(+)1600$; $[(t$-$Bu)_2P]_2Te$, $(+)451$; $[(t$-$Bu)_2P]_2Te$ $Cr(CO)_5$, $(+)324$ (Du Mont and Kroth, 1981).

$^1J_{125_{Te^{77}Se}}$: $MeTeSeMe$, -169 (Pfisterer and Dreeskamp, 1969); cis-$Te_2Se_2^{2+}$, ±470; trans-$Te_2Se_2^{2+}$, ±550 (Schrobilgen et al., 1978).

$^1J_{125_{Te^{119}Sn}}$: $(Me_3Sn)_2Te$, -2770; $Me_6Sn_3Te_3$, $(-)3098$ (Blecher et al., 1980; Kennedy, 1981, pers. com.).

$^1J_{125_{Te^{123}Te}}$: Te_4^{2+}, ±608; Te_6^{4+}, ±791, ±1196 (Schrobilgen et al., 1978); diaryl ditellurides, 213–269 (Granger et al., 1981b).

$^1J_{125_{Te^{195}Pt}}$:   cis-$[PtCl_2(TeCH_2CH_2Ph)_2]$,   900;   trans-$[PtCl_2)Te(CH_2$ $CH_2Ph)_2]$, 544; (Gysling et al., 1981); $[(PtCl_3)_2TeMe_2]^{2-}$, -5923 (Goggin et al., 1975.

Less is known about longer-range couplings to the majority of nuclei, but for ^{13}C they are generally a few tens of hertz, although the existence of $^2J_{129_{Xe}125_{Te}}$ of over 1000 Hz in TeF_5 derivatives (Jacob *et al.*, 1981) demonstrates that large values can occur.

C. Chemical Applications

Tellurium-125 NMR confirms the structures of the products of the reaction between selenium dioxide and aromatic ditellurides that contain Te—Se—Te linkages (Dereu *et al.*, 1982). The formation of unsymmetric ditellurides in mixtures of symmetric ones was readily demonstrated by the appearance of their ^{125}Te spectra (O'Brien *et al.*, 1982), and detailed studies have been undertaken on the exchange processes in such systems (Granger *et al.*, 1981b). The ^{125}Te spectra of the products of the solvolysis of $Te(OH)_6$ in HF show the formation of all possible octahedral species $Te(OH)_{6-n}F_n$ ($n = 0$ to 6) and their geometric isomers (Tötsch *et al.*, 1981), and similar results have been obtained for methoxy and other derivatives (Bildstein *et al.*, 1981; Tötsch and Sladky, 1980, 1982). The larger value of $^2J_{195_{Pt}125_{Te}}$ made it possible to distinquish the cis from the trans isomer of $[PtCl_2(TeCH_2CH_2Ph)_2]$ (Gysling *et al.*, 1981). The change in $^1J_{125_{Te}31_P}$ on dissolution of $(Me_2N)_3PTe$ in liquid sulfur dioxide was indicative of complex formation (Dean, 1979). Couplings to ^{129}Xe have facilitated establishment of the structures of many $OTeF_5$–Xe species (Birchell *et al.*, 1982; Jacob *et al.*, 1981; Keller and Schrobilgen, 1981; Seppelt and Rupp, 1974).

An especially elegant demonstration of the use of ^{125}Te (and ^{123}Te) NMR was provided by two studies on species obtained by dissolving selenium and tellurium in oleum (Lassigne and Wells, 1978; Schrobilgen, *et al.*, 1978). A range of polymeric cations is formed, including **(XVII–XXII)**, and the spectra of these compounds, including ^{123}Te and ^{125}Te satellites, provide unequivocal demonstrations of their structures. Interestingly, the longer range couplings in these species are often of magnitudes comparable to those of the shorter range ones, indicating that it is probably necessary to adopt a multicenter molecular orbital approach to the bonding in these compounds. A study on tellurophene oriented in a liquid crystal used ^{125}Te satellites in the proton spectrum to obtain additional geometric information (Chidichimo et al., 1981).

References

Abronin, I. A., Djumanazarova, A. Z., and Litvinov, V. P. (1982). Chem. Scr. **19**, 75–77.

Allen, D. W., and Taylor, B. F. (1982). Chem. Soc. Dalton Trans. 51–54.

Anderson, S. J., Goggin, P. L., and Goodfellow, R. J. (1976). Chem. Soc. Trans. 1959–1964.

Arnold, D. E. J., Dryburgh, J. S., Ebsworth, E. A. V., and Rankin, D. W. H. (1972). Chem. Soc. Dalton Trans. 2518–2522.

Baiwir, M., Llabrès, G., Piette, J. L., and Christiaens, L. (1980). Org. Magn. Reson. **14**, 293–295.

Baiwir, M., Llabrès, G., Christiaens, L., and Piette, J. L. (1981). Org. Magn. Reson. **16**, 14–16.

Baiwir, M., Habir, G., Christiaens, L., and Piette, J. L. (1982). Org. Magn. Reson. **18**, 33–37.

Bildstein, B., Tötsch, W., and Sladky, F. (1981). Z. Naturforsch. **36B**, 1542–1543.

Birchall, T., Gillespie, R. J., and Vekris, S. L. (1965). Can. J. Chem. **43**, 1672–1679.

Birchall, T., Myers, R. D., de Waard, H., and Schrobilgen, G. J. (1982). Inorg. Chem. **21**, 1068–1073.

Blecher, A., Mathiasch, B., and Mitchell, T. N. (1980). J. Organ. Chem. **184**, 175–180.

Breuninger, V., Dreeskamp, H., and Pfisterer, G. (1966). Ber. Bunsegesell. Phys. Chem. **70**, 613–617.

Bucci, P., Chidichimo, G., Lelj, F., Longeri, M., and Russo, N. (1979). J. Chem. Soc. Perkin Trans. **2**, 109–111.

Buckler, K. U., Kronenbitter, J., Lutz, O., and Nolle, A. (1977). Z. Naturforsch. A. **32A.**, 1263–1265.

Bzehzovskii, V. M., Kushnarev, D. F., and Efremova, G. G. (1981). Izv. Akad. Nauk. S.S.S.R. Ser. Khim. 2507–2512.

Carr, S. W., and Colton, R. (1981). Aust. J. Chem. **34**, 35–44.

Carson, G. K., and Dean, P. A. W. (1982). Inorg. Chem Acta **66**, 37–39.

Chadha, R. K., and Miller, J. M. (1982). J. Chem. Soc. Dalton Trans. 117–120.

Chapelle, S., and Granger, P. (1981). Mol. Phys. **44**, 459–467.

Chidichimo, G., Lelj, F., Barili, P. L., and Veracini, C.A. (1978). Chem. Phys. Lett. **55**(3), 519–522.

Chidichimo, G., Lelj, F., Longeri, M., Russo, N., and Veracini, C. A. (1979). Chem. Phys. Lett. **67**(2,3), 384–387.

Chidichimo, G., Lelj, F., Longeri, M., Russo, N., and Veracini, C. A. (1980). J. Magn. Reson. **41**, 35–41.

Chidichimo, G., Bucci, P., Lelj, F., and Longeri, M. (1981). Mol. Phys. **43**, 877–886.

Christiaens, L. et al. (1976). Org. Magn. Reson. **8**, 354–356.

Colquhoun, I. J., and McFarlane, W. (1978). J. Chem. Res. Synop. 368–369.

Colquhoun, I. J., and McFarlane, W. (1981). *J. Chem. Soc. Dalton Trans.* 658–660.
Colquhoun, I. J. *et al.* (1979). *Org. Magn. Reson.* **12**, 473–475.
Colquhoun, I. J., Malisch, W., Maisch, R., and McFarlane, W. (1982). *Phosphorus Sulphur* (in press).
Colton, R., and Dakternieks, D. (1980). *Aust. J. Chem.* **33**, 1463–1470.
Cradock, S., Ebsworth, E. A. V., Rankin, D. W. H., and Savage, W. J. (1976). *J. Chem. Soc. Dalton Trans.* 1661–1665.
Cullen, E. R., Guziec, F. S., Jr., Murphy, C. J., Wong, T. C., and Andersen, K. K. (1981). *J. Am. Chem. Soc.* **103**, 7055–7057.
Dawson, W. H., and Odom, J. D. (1977). *J. Am. Chem. Soc.* **99**, 8352–8354.
Dean, P. A. W. (1979). *Can. J. Chem.* **57**, 754–761.
Denney, D. B., Denney, D. Z., Hammond, P. J., and Hsu, Y. F. (1981). *J. Am. Chem. Soc.* **103**, 2340–2347.
Dereu, N. L. M., Zingaro, R. A., and Meyers, E. A. (1982). *Organ.* **1**, 111–115.
Dewan, J. C., Jennings, W. B., Silver, J., and Tolley (1978). *Org. Mag. Reson.* **11**, 449–452.
Diehl, P., Kunwar, A., and Bosiger, H. (1978). *J. Organ. Chem.* **145**, 303–306.
Drake, J. E., and Hemmings, R. T. (1976). *J. Chem. Soc. Dalton Trans.* 1730–1733.
Drakenberg, T. *et al.* (1976). *Chem. Scr.* **10**, 139–140.
Drakenberg, T., Hörnfeldt, A. B., Gronowitz, S., Talbot, J. M., and Piette, J. L. (1979). *Chem. Scr.* **13**, 152–154.
Dreeskamp, H., and Pfisterer, G. (1968). *Mol. Phys.* **14**, 295–297.
Du Mont, W. W., and Kroth, H. J. (1981). *Z. Naturforsch.* **36B**, 332–334.
Du Mont, W. W., and Nordhoff, E. (1980). *J. Organ. Chem.* **198**, C 58–60.
Dwek, R. A., Richards, R. E., Taylor, D., Penney, G. J., and Sheldrick, G. M. (1969). *J. Chem. Soc.* A 935–937.
Fackler, J. P., Jr., and Pan, W. H. (1979). *J. Am. Chem. Soc.* **101**, 1607–1608.
Fazakerley, G. V., and Celotti, M. (1979). *J. Mag. Reson.* **33**, 219–220.
Fraser, G. W., Peacock, R. D., and McFarlane, W. (1969). *Mol. Phys.* **17**, 291–297.
Fredga, A., Gronowitz, S., and Hörnfeldt, A. B. (1975). *Chem. Scr.* **8**, 15–20.
Fringuelli, F., Gronowitz, A. B., Johnson, J., and Taticchi, A. (1974). *Acta Chem. Scand. B* **28**, 175–184.
Gansow, O. A., Vernon, W. D., and Dechter, J. J. (1978). *J. Magn. Reson.* **32**, 19–21.
Glidewell, C., Rankin, D. W. A., and Sheldrick, G. M. (1969). *Trans. Farad. Soc.,* **65**, 1409–1412.
Goggin, P. L., Goodfellow, R. J., and Haddock, S. R. (1975). *Chem. Commun.* 1767–1768.
Gombler, W. (1981). *Z. Naturforsch.* **36 B**, 535–543.
Granger, P., and Chapelle, S. (1980). *J. Magn. Reson.* **39**, 329–334.
Granger, P., Chapelle, S., and Paulmier, C. (1980). *Org. Magn. Reson.* **14**, 240–243.
Granger, P., Chapelle, S., and Brevard, C. (1981a). *J. Magn. Reson.* **42**, 203–207.
Granger, P., Chapelle, S., McWinnie, W. R., and Al-Rubaie, A. (1981b). *J. Organ. Chem.* **220**, 149–158.
Grim, S. O., Walton, E. D., and Satek, L. C. (1978). *Inorg. Chim. Acta* **27**, L115–117.
Gronowitz, S., Johnson, J., and Hörnfeldt, A. B. (1973). *Chem. Scr.* **3**, 94–99.
Gronowitz, S., Johnson, J., and Hörnfeldt, A. B. (1975). *Chem. Scr.* **8**, 8–12.
Gronowitz, S., Konar, A., and Hörnfeldt, A. B. (1977). *Org. Magn. Reson.* **9**, 213–216.
Gronowitz, S., Konar, A., and Hörnfeldt, A. B. (1982). *Chem. Scr.* **19**, 5–12.
Gysling, H. J., Zumbulyadis, N., and Robertson, J. A. (1981). *J. Organ. Chem.* C **209**, 41–44.
Iwamura, H., and Nakanishi, W. (1981). *Yuki Gosei Kagaku Kyokaishi* **39**, 795–804, C.A. 96:5748s.

Jacob, E., Lentz, D., Seppelt, K., and Simon, A. (1981). *Z. Anorg. Allg. Chem.* **472**, 7–25.

Jakobsen, H. J., and Hansen, R. S. (1978). *J. Magn. Reson.* **30**, 397–399.

Jakobsen, H. J., Zozulin, A. J., Ellis, P. D., and Odom, J. D. (1980). *J. Magn. Reson.* **38**, 219–227.

Kalabin, G. A., and Kushnarev, D. F. (1979). *Zh. Strukt. Chem.* **20**, 617–621.

Kalabin, G. A., Kushnarev, D. F., Kamaeva, L. M., Kashurnikova, L. V., and Vinokurova, R. I. (1978). *Zh. Org. Khim* **14**, 2478–2479.

Kalabin, G. A., Kushnarev, D. F., Bzesovsky, V. M., and Tschumutova, G. A. (1979a). *Org. Mag. Reson.* **12**, 598–603.

Kalabin, G. A., Kushnarev, D. F., Chmutova, G. A., and Kashurnikova, L. V. (1979). *Zh. Org. Khim.* **15**, 24–31.

Kalabin, G. A., Kushnarev, D. F., and Mannafov, T. G. (1980). *Zh. Org. Khim.* **16**, 505–511.

Kalabin, G. A., Kushnarev, D. F., and Valeev, R. B. (1981a). *Zh. Org. Khim.* **17**, 1139–1142.

Kalabin, G. A., Valeev, R. B., and Kushnarev, D. F. (1981b). *Zh. Org. Khim.* **17**, 947–953.

Kalabin, G. A., Valeev, R. B., Kushnarev, D. F., Sadekov, I. D., and Minkin, V. I. (1981c). *Zh. Org. Khim.* **17**, 206–208.

Keat, R., Rycroft, D. S., and Thompson, D. G. (1979). *Org. Magn. Reson.* **12**, 391–392.

Keller, N., and Schrobilgen, G. J. (1981). *Inorg. Chem.* **20**, 2118–2129.

Kennedy, J. D., and McFarlane, W. (1973). *J. Chem. Soc. Dalton Trans.* 2134–2139.

Kennedy, J. D., and McFarlane, W. (1975). *J. Organ. Chem.* **94**, 7–11.

Koch, W., Lutz, O., and Nolle, A. (1978). *Z. Naturforsch. A* **33**, 1026–1030.

Kohne, B., Lohner, W., Jakobsen, H. J., Praefcke, K., and Villadsen, B. (1979). *J. Organ. Chem.* **166**, 373–377.

Kolshorn, H., and Meier, H. (1977). *J. Chem. Res. Synop.* 338–339.

Konar, A., and Gronowitz, S. (1980). *Tetrahedron* **36**, 3317–3323.

Kroschefsky, R. D., and Verkade, J. G. (1979). *Inorg. Chem.* **18**, 469–472.

Lardon, M. (1970). *J. Am. Chem. Soc.* **92**, 5063–5066.

Lardon, M. (1973). "Organic Selenium Compounds: Their Chemistry and Biology" (D. L. Klayman and W. H. H. Gunther, eds.), pp. 933–939. Wiley (Interscience), New York.

Lassigne, C. R., and Wells, E. J. (1978). *J. Chem. Soc. Chem. Commun.* **21**, 956–957.

Lentz, D., and Seppelt, K. (1978). *Angew. Chem. Int. Engl. Ed.* **17**, 355–361.

Llabrès, G., Baiwir, M., Piette, J. L., and Christiaens, L. (1981). *Org. Magn. Reson.* **15**, 152–154.

Lohner, W., and Praefcke, K. (1981a). *J. Organ. Chem.* **208**, 39–42.

Lohner, W., and Praefcke, K. (1981b). *J. Organ. Chem.* **208**, 43–46.

Luthra, N. P., Costello, R. C., Odom, J. D., and Dunlap, R. B. (1982a). *J. Biol. Chem.* **257**(3), 1142–1144.

Luthra, N. P., Dunlap, R. B., and Odom, J. D. (1982b). *J. Magn. Reson.* **46**, 152–157.

Martin, M. L., Trierweiler, M. Galasso, V., Fringuelli, F., and Tatichi, A. (1981). *J. Magn. Res.* **42**, 155–158.

Martin, M. L., Trierweiler, M. Galasso, V., Fringuelli, F., and Taticchi A. (1982). *J. Magn. Reson.* **47**, 504–506.

McFarlane, W. (1967). *Mol. Phys.* **12**, 243–248.

McFarlane, W. (1969). *J. Chem. Soc. A* 670–672.

McFarlane, H. C. E., and McFarlane, W. (1973). *J. Chem. Soc. Dalton Trans.* 2416–2418.

McFarlane, H. C. E., and McFarlane, W. (1978). "NMR and the Periodic Table" (R. K. Harris and B. E. Mann, eds.), pp. 402–419. Academic Press, London.

McFarlane, W., and Nash, J. A. (1969). *Chem. Commun.* 913–914.

McFarlane, W., and Rycroft, D. S. (1972). *Chem. Commun.* 902–903.

McFarlane, W., and Rycroft, D. S. (1973). *J. Chem. Soc. Dalton Trans.* 2162–2166.

McFarlane, W., and Wood, R. J. (1972). *J. Chem. Soc. Dalton Trans.* 1397–1402.

McFarlane, W., Berry, F. J., and Smith, B. C. (1976). *J. Organ. Chem.* **113,** 139–141.

McFarlane, W., Rycroft, D. S., and Turner, C. J. (1977). *Bull. Soc. Chim. Belge.* **86,** 457–463.

McFarlane, W., and Rycroft, D. S. (1973). *Chem. Commun.* 9–11.

Muetterties, E. L., and Phillips, W. D. (1959). *J. Am. Chem. Soc.* **81,** 1084–1088.

Nakanishi, W. *et al.* (1981a). *Chem. Lett.* 1353–1356.

Nakanishi, W. *et al.* (1981b). *Tetr. Letts.* **22,** 4241–4244.

O'Brien, D. H., Dereu, N., Grigsby, R. A., and Irgolic, K. J. (1982). *Organometallics,* **1,** 513–517.

Odom, J. D., Dawson, W. H., and Ellis, P. D. (1979). *J. Am. Chem. Soc.* **101,** 5815–5822.

Olsson, K., and Almquist, S. O. (1969). *Acta Chem. Scand.* **23,** 3271–3272.

Onyamboko, N. V., Renson, M., Chappell, S., and Granger, P. (1982). *Org. Magn. Reson.* **19,** 74–77.

Pan, W. H., and Fackler, J. P. (1978). *J. Am. Chem. Soc.* **100,** 5783–5789.

Pfisterer, G., and Dreeskamp, H. (1969). *Ber. Bunsengesell. Phys.Chem.* **73,** 654–661.

Pinnell, R. P., Mergerle, C. A., Mannatt, S. L., and Kroon, P. A. (1973). *J. Am. Chem. Soc.* **95,** 977–978.

Ramsey, N. F. (1950). *Phys. Rev.* **78,** 689–702.

Reich, H. J., and Trend, J. E. (1976). *Chem. Commun.* 310–311.

Reich, H. J., Hoeger, C. A., and Willis, W. W. (1982). *J. Am. Chem. Soc.* **104,** 2936–2937.

Sándor, P., and Radics, L. (1981). *Org. Magn. Reson.* **16,** 148–155.

Schrobilgen, G. J., Burns, R. C., and Granger, P. (1978). *J. Chem. Soc. Chem. Comm.* **21,** 957–960.

Seppelt, K. (1973). *Z. Anorg. Allg. Chem.* **399,** 65–69.

Seppelt, K. (1975). *Z. Anorg. Allg. Chem.* **416,** 12–15.

Seppelt, K., and Rupp, H. H. (1974). *Z. Anorg. Allg. Chem.* **409,** 338–345.

Simonnin, M. P., Pouet, M. J., Cense, J. M., and Paulmier, F. (1976). *Org. Magn. Reson.* **8,** 508–515.

Stec, W. J., Okruszek, A., Uznanski, B., and Michalski, J. (1972). *Phosphorus* **2,** 97–102.

Svanholm, V. (1973). "Organic Selenium Compounds: Their Chemistry and Biology" (D. L. Klayman and W. H. H. Ganther, eds.), pp. 903–932. Wiley (Interscience), New York.

Talbot, J. M. *et al.* (1981). *Chem. Scr.* **18,** 147–151.

Tötsch, W., and Sladky, F. (1980). *J. Chem. Soc. Chem. Commun.* **19,** 927–928.

Tötsch, W., and Sladky, F. (1982). *Chem. Ber.* **115,** 1019–1027.

Tötsch, W., Peringer, P., and Sladky, F. (1981). *J. Chem. Soc. Chem. Commun.* **16,** 841–842.

Valeev, P. B., Kalabin, G. A., and Kushnarev, D. F. (1980). *Zh. Org. Khim.* **16,** 2482–2485.

Wachli, H. E. (1953). *Phys. Rev.* **90,** 331–333.

Warren, W. W., Jr., and Dupree, R. (1980). *Phys. Rev. B* **22**(5), 2257–2275.

Zumbulyadis, N., and Gysling, H. J. (1980). *J. Organ. Chem.* **192,** 183–188.

11 Rhodium-103

B. E. Mann

The University of Sheffield
Department of Chemistry
Sheffield, England

Although ^{103}Rh is 100% abundant with $I = \frac{1}{2}$, until recently direct observation was thought to be difficult because of the low receptivity, $D_C = 0.177$, and the belief that the spin–lattice relaxation time T_1 was very long. Work has now shown that T_1 can be short and that ^{103}Rh-NMR spectra can be obtained for a 0.25 M solution of Rh(acac)(CO)$_2$ in under 1 min (Cocivera *et al.* 1982) (Fig. 1). Early measurements of ^{103}Rh chemical shifts were indirect, using INDOR, and this technique still remains a powerful method whenever $J_{^{103}Rh^1H}$ is observed, or indirectly via ^{1}H-{^{31}P, ^{103}Rh} or similar measurements. Direct observation has benefited from $J_{^{103}Rh^1H}$ coupling to enhance the sensitivity by a factor of 31.77 with proton polarization transfer using the INEPT pulse sequence; see Chapter 1, Vol. 1 (Brevard *et al.,* 1981). This use of the INEPT pulse sequence is very promising for future work using either proton or phosphorus polarization transfer. In addition to these measurements, Quadriga NMR has been used to observe slowly relaxing (Rh(acac)$_3$ (Grüninger *et al.,* 1980).

Because of the novelty of ^{103}Rh-NMR spectroscopy, reviews are limited (Goodfellow, 1978; Kidd, 1980).

I. Chemical Shifts

The choice of the reference for ^{103}Rh is still controversial, and different workers use different references. Three main references are in common

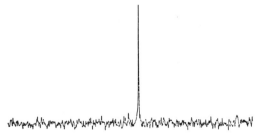

Fig. 1 Rhodium-103 NMR signal at 12.65 MHz and 25°C for a 10-mm sample tube containing 0.25 M Rh(CO)$_2$(acac) in CDCl$_3$. (Reproduced with permission from Cocivera *et al.*, 1982.)

use. Rhodium metal has been used by several workers (Brown and Green, 1970a; Gill *et al.*, 1979; Sogo and Jeffries, 1955). It has the advantage that it gives positive chemical shifts for all compounds so far investigated, but it is difficult to determine, and its value is subject to considerable error (32 ppm). In practice, the published $\Xi_{103_{Rh}}$ = 3, 155, 700 ± 100 for rhodium metal (Sogo and Jeffries, 1955) is taken as a reference, so that trimethylsilane (TMS) becomes the de facto internal reference. In view of the artificiality of this procedure and the inconvenience of using a number with five significant figures, the simpler procedure of taking $\Xi_{103_{Rh}}$ = 3.16 MHz as the reference has been adopted (Goodfellow, 1978). It is this procedure that is used in this chapter. The third reference is a saturated solution of *mer*-RhCl$_3$(SMe$_3$)$_3$ in dichloromethane. This referencing procedure suffers from the usual problems of external referencing and the temperature dependence of its chemical shift, 1.0 ± 0.1 ppm K^{-1} (McFarlane, 1976).

On account of the large chemical shift range found for ^{103}Rh, re-referencing is not simply an additive procedure but requires the use of Eq. (1) (Goodfellow, 1978). In all cases the currently approved sign convention in which high frequency is positive is used (IUPAC, 1976).

$$\delta = \delta' \frac{\Xi_{ref}'}{\Xi_{ref}} + \frac{\Xi_{ref}' - \Xi_{ref}}{\Xi_{ref}} \times 10^6 \tag{1}$$

Examination of the chemical shifts listed in Tables I to V shows that at present ^{103}Rh chemical shifts cover the range δ −1714 for one rhodium site (Fernandez and Maitlis, 1982) in Rh(C$_5$Me$_5$)H$_2$(SiMe$_3$)$_2$ to δ 9992 in [Rh(OH$_2$)$_6$]$^{3+}$ (Mann and Spencer, 1982a), although examination of the [RhF$_n$(OH$_2$)$_{6-n}$]$^{3-n}$ system would almost certainly yield higher frequency shifts. Rhodium(III) gives shifts covering the range δ 9992 to δ −231, whereas rhodium(I) gives shifts covering the much narrower range δ 2496 to δ −736, ignoring the problem of assigning oxidation states to metal carbonyl clusters. The rhodium chemical shifts can be qualitatively ex-

TABLE I

Rhodium-103 Chemical Shifts of Rhodium(III) Complexes Containing Rhodium–Carbon Bonds, Excluding Cluster Compounds

Complex	$\delta_{^{103}Rh}{}^{a}$	Reference
$[RhMe(CN\text{-}t\text{-}Bu)_5]^{2+}$	82	Taylor (1973)
trans-$[RhMe(CN\text{-}t\text{-}Bu)_4(py)]^{2+}$	526	Taylor (1973)
trans-$[RhMe(CN\text{-}t\text{-}Bu)_4(4\text{-Mepy})]^{2+}$	525	Taylor (1973)
trans-$[RhMe(CN\text{-}t\text{-}Bu)_4(PMe_3)]^{2+}$	−156	Taylor (1973)
trans-$[RhMe(CN\text{-}t\text{-}Bu)_4(PMe_2Ph)]^{2+}$	−105	Taylor (1973)
trans-$[RhMe(CN\text{-}t\text{-}Bu)_4(PPh_3)]^{2+}$	−39	Taylor (1973)
trans-$[RhMe(CN\text{-}t\text{-}Bu)_4(P(OMe)_3)]^{2+}$	−151	Taylor (1973)
trans-$[RhMe(CN\text{-}t\text{-}Bu)_4(P(OPh)_3)]^{2+}$	−95	Taylor (1973)
trans-$[RhMe(CN\text{-}t\text{-}Bu)_4(AsPh_3)]^{2+}$	−57	Taylor (1973)
trans-$[RhMe(CN\text{-}t\text{-}Bu)_4(SbPh_3)]^{2+}$	−231	Taylor (1973)
trans-$[RhMe(CN\text{-}t\text{-}Bu)_4(SMe_2)]^{2+}$	289	Taylor (1973)
trans-$[RhMe(CN\text{-}t\text{-}Bu)_4Cl]^{+}$	526	Taylor (1973)
trans-$[RhMe(CN\text{-}t\text{-}Bu)_4Br]^{+}$	367	Taylor (1973)
trans-$[RhMe(CN\text{-}t\text{-}Bu)_4I]^{+}$	107	Taylor (1973)
cis,trans-$RhCl_2Me(CO)(PMe_2Ph)_2$	499	Hyde et al. (1977)
cis,trans-$RhBr_2Me(CO)(PMe_2Ph)_2$	350	Hyde et al. (1977)
cis,trans-$RhI_2Me(CO)(PMe_2Ph)_2$	52	Hyde et al. (1977)
$RhCl{-}^{b}_{b}I{-}^{f}_{f}Me{-}^{a}_{a}(CO){-}^{d}_{d}(PMe_2Ph)_2{-}^{ce}$	194	Hyde et al. (1977)
$RhCl{-}^{b}_{b}I{-}^{f}_{f}Me{-}^{a}_{a}(CO){-}^{d}_{d}[PMe_2(C_6H_4OMe\text{-}o)]_2{-}^{ce}$	128	Hyde et al. (1977)
$RhBr{-}I{-}^{b}Me{-}^{a}(CO){-}^{d}(PMe_2Ph)_2{-}^{ce}$	268	Hyde et al. (1977)
trans-$[RhEt(CN\text{-}t\text{-}Bu)_4I]^{+}$	183	Taylor (1973)
trans-$[Rh\text{-}n\text{-}Pr(CN\text{-}t\text{-}Bu)_4I]^{+}$	174	Taylor (1973)
trans-$[Rh\text{-}n\text{-}Bu(CN\text{-}t\text{-}Bu)_4I]^{+}$	175	Taylor (1973)
trans-$[Rh(CH_2Ph)(CN\text{-}t\text{-}Bu)_4I]^{+}$	338	Taylor (1973)
$RhHCl(t\text{-}Bu_2PCH_2CH_2CHCH_2CH_2P\text{-}t\text{-}Bu_2)$	1029	Crocker et al. (1979, 1980)
$RhHCl(t\text{-}Bu_2PCH_2CH_2CMeCH_2CH_2P\text{-}t\text{-}Bu_2)$	1011	Crocker et al. (1980)
$RhCl{-}^{b}I{-}^{f}(CF_3){-}^{a}(CO){-}^{d}(PMe_2Ph)_2{-}^{ce}$	706	Hyde et al. (1977)
$RhBr{-}^{b}I{-}^{f}(CF_3){-}^{a}(CO){-}^{d}(PMe_2Ph)_2{-}^{ce}$	625	Hyde et al. (1977)
$RhI_2{-}^{bf}(CF_3){-}^{a}(CO){-}^{d}(PMe_2Ph)_2{-}^{ce}$	453	Hyde et al. (1977)
mer,trans-$RhCl_3(CO)(PMe_2Ph)_2$	1588	Hyde et al. (1977)
mer,trans-$RhBr_3(CO)(PMe_2Ph)_2$	1074	Hyde et al. (1977)
cis,trans-$RhCl_2H(CO)(PMe_2Ph)_2$	240	Hyde et al. (1977)

^{a} In parts per million to high frequency of $\Xi_{^{103}Rh}$ = 3.16 MHz.

plained using the Ramsey equation (Ramsey, 1950), by analogy to ^{59}Co chemical shifts:

$$\delta = B\langle r^{-3}\rangle_{md}/\Delta E \qquad (2)$$

where A is the diamagnetic shift, B a constant, and r the effective radius of the d electron. The value of ΔE is expected to follow the spectrochemical series

TABLE II

^{103}Rh Chemical Shifts of Rhodium(III) Complexes of the Type Rh(III)$X_n L_{6-n}$ where X = halogen and L = neutral ligand

Complex	$\delta_{103\text{Rh}}$[a]			Reference
	X = Cl	X = Br	X = I	
[RhX₆]³⁻	7975, 7986	7077	—	Mann and Spencer (1982a, b)
[RhX₅(OH₂)]²⁻	8233	—	—	Mann and Spencer (1982a)
[RhX₅(SOMe₂)]²⁻	5728	—	—	Barnes et al. (1979)
[RhX₅(SMe₂)]²⁻	6521	—	—	Anderson et al. (1978)
cis-[RhX₄(OH₂)₂]⁻	8295	—	—	Mann and Spencer (1982a)
cis-[RhX₄(SMe₂)₂]⁻	4882	4339	3070	Anderson et al. (1978)
cis-[RhX₄(SOMe₂)₂]⁻	4145	3637	—	Barnes et al. (1979)
trans-[RhX₄(NMe₃)₂]⁻	7942	—	—	Barnes et al. (1978)
trans-[RhX₄(OH₂)₂]⁻	8605	—	—	Mann and Spencer (1982a)
trans-[RhX₄(SMe₂)₂]⁻	5226	4532	2958	Anderson et al. (1978)
trans-[RhX₄(SOMe₂)₂]⁻	4506	3836	—	Barnes et al. (1979)
fac-[RhX₃(PMe₃)₃]	1437	1229	—	Taylor (1973)
fac-[RhX₃(AsMe₃)₃]	2125	1798	—	Barnes et al. (1978)
fac-[RhX₃(OH₂)₃]	8545	—	—	Mann and Spencer (1982a)
fac-[RhX₃(SMe₂)₃]	3661	—	—	Anderson et al. (1978)
fac-[RhX₃(SMePh)₃]	3658	—	—	McFarlane et al. (1976)
fac-[RhX₃(SEt₂)₃]	3904	—	—	McFarlane et al. (1976)
mer-[RhX₃(SEt₂)₃]	3599	—	—	McFarlane et al. (1976)
mer-[RhX₃(PMe₃)₃]	2202	1746	809	Barnes et al. (1978); Taylor (1973)
mer-[RhX₃(PMe₂Ph)₃]	2426	2026	1177	McFarlane et al. (1973)

mer-[RhX$_3$(P-n-Bu$_3$)$_3$]	2770	2334	—	Brown and Green (1970)
mer-[RhX$_3$(AsMe$_3$)$_3$]	2806	2276	1191	Barnes et al. (1978)
mer-[RhX$_3$(OH$_2$)$_3$]	8858	—	—	Mann and Spencer (1982a)
mer-[RhX$_3$(SMe$_2$)$_3$]	3897	3437	2448	Anderson et al. (1978)
	3896	3436	—	McFarlane et al. (1976), Hyde et al. (1977)
mer-[RhX$_3$(SMePh)$_3$]	4121	3646	—	McFarlane et al. (1976)
mer-[RhX$_3$(SeMe$_2$)$_3$]	3904	3385	—	McFarlane et al. (1976)
mer-[RhX$_3$(TeMe$_2$)$_3$]	3179	2567	1352	Anderson et al. (1978)
mer-[RhX$_3$(SOMe$_2$)$_3$]	3478	3061	—	Anderson et al. (1978)
cis-[RhX$_2$(AsMe$_3$)$_4$]$^+$	1457	1222	—	Barnes et al. (1978)
cis-[RhX$_2$(OH$_2$)$_4$]$^+$	8794	—	—	Mann and Spencer (1982a)
cis-[RhX$_2$(MeSCH$_2$CH$_2$SMe)$_2$]$^+$	2288[b]	—	—	McFarlane et al. (1976)
	2330[b]	—	—	McFarlane et al. (1976)
	2371[b]	—	—	McFarlane et al. (1976)
trans-[RhX$_2$(PMe$_3$)$_4$]$^+$	1564	1154	398	Taylor (1973), Barnes et al. (1978)
trans-[RhX$_2$(AsMe$_3$)$_4$]$^+$	1946	1486	—	Barnes et al. (1978)
trans-[RhX$_2$(OH$_2$)$_4$]$^+$	9115	—	—	Mann and Spencer (1982a)
trans-[RhX$_2$(SMe$_2$)$_4$]$^+$	3081	2756	—	Anderson et al. (1978)
trans-[RhX$_2$(MeSCH$_2$CH$_2$SMe)$_2$]$^+$	2502[b]	—	—	McFarlane et al. (1976)
	2547[b]	—	—	McFarlane et al. (1976)
	2563[b]	—	—	McFarlane et al. (1976)
	2585[b]	—	—	McFarlane et al. (1976)
[RhX(OH$_2$)$_5$]$^{2+}$	9479	—	—	Mann and Spencer (1982a)

[a] In parts per million to high frequency of $\Xi_{103\text{Rh}}$ = 3.16 MHz.
[b] Isomers, see text.

TABLE III

Rhodium-103 Chemical Shifts of Miscellaneous Rhodium(III) Complexes

Complex	$\delta_{103_{Rh}}{}^{a}$	Reference
$Rh_2H_2Cl_4[t\text{-}Bu_2P(CH_2)_5P\text{-}t\text{-}Bu_2]_2$	2736	Crocker *et al.* (1980)
$Rh_2(P\text{-}n\text{-}Bu_3)_4Cl_6$	2978	Brown and Green (1970a)
$[Rh(OH_2)_6]^{3+}$	9992	Mann and Spencer (1982a)
$Rh(acac)_3$	8358	Grüninger *et al.* (1980)
cis-$[RhCl(OSMe_2)_4(SOMe_2)]^{2+}$	6853	Barnes *et al.* (1979)
cis-$[RhBr(OSMe_2)_4(SOMe_2)]^{2+}$	6729	Barnes *et al.* (1979)
fac-$[RhCl(OSMe_2)_3(SOMe_2)_2]^{2+}$	4760	Barnes *et al.* (1979)
fac-$[RhBr(OSMe_2)_3(SOMe_2)_2]^{2+}$	4591	Barnes *et al.* (1979)
cis,cis,cis-$[RhCl_2(OSMe_2)_2(SOMe_2)_2]^+$	4661	Barnes *et al.* (1979)
cis,cis,cis-$[RhBr_2(OSMe_2)_2(SOMe_2)_2]^+$	4424	Barnes *et al.* (1979)
trans,cis,cis-$[RhCl_2(OSMe_2)_2(SOMe_2)_2]^+$	4340	Barnes *et al.* (1979)
trans,cis,cis-$[RhBr_2(OSMe_2)_2(SOMe_2)_2]^+$	4049	Barnes *et al.* (1979)
mer,cis-$RhCl_3(OSMe_2)(SOMe_2)_2$	4133	Barnes *et al.* (1979)
mer,cis-$RhBr_3(OSMe_2)(SOMe_2)_2$	3739	Barnes *et al.* (1979)
mer,trans-$RhCl_3(OSMe_2)(SOMe_2)_2$	4918	Barnes *et al.* (1979)
mer,trans-$RhBr_3(OSMe_2)(SOMe_2)_2$	4437	Barnes *et al.* (1979)
mer,cis-$RhCl_3(OH_2)(SOMe_2)_2$	3926	Barnes *et al.* (1979)
mer,cis-$RhBr_3(OH_2)(SOMe_2)_2$	3481	Barnes *et al.* (1979)
trans,mer-$RhCl_2Br(SMe_2)_3$	3776	McFarlane *et al.* (1976)
cis,mer-$RhCl_2Br(SMe_2)_3$	3732	McFarlane *et al.* (1976)
trans,mer-$RhBr_2Cl(SMe_2)_3$	3605	McFarlane *et al.* (1976)
cis,mer-$RhBr_2Cl(SMe_2)_3$	3564	McFarlane *et al.* (1976)
trans,mer-$RhCl_2Br(SMePh)_3$	3997	McFarlane *et al.* (1976)
cis,mer-$RhCl_2Br(SMePh)_3$	3840	McFarlane *et al.* (1976)
trans,mer-$RhBr_2Cl(SMePh)_3$	3821	McFarlane *et al.* (1976)
cis,mer-$RhBr_2Cl(SMePh)_3$	3779	McFarlane *et al.* (1976)
trans,mer-$RhCl_2Br(SeMe_2)_3$	3775	McFarlane *et al.* (1976)
cis,mer-$RhCl_2Br(SeMe_2)_3$	3719	McFarlane *et al.* (1976)
trans,mer-$RhBr_2Cl(SeMe_2)_3$	3578	McFarlane *et al.* (1976)
cis,mer-$RhBr_2Cl(SeMe_2)_3$	3531	McFarlane *et al.* (1976)
mer-$RhCl_3(SeMe_2)_2(SMePh)$	3943	McFarlane *et al.* (1976)
mer-$RhCl_3(SMe_2)_2(SMePh)$	4003	McFarlane *et al.* (1976)
mer-$RhCl_3(SMe_2)(SMePh)_2$	4010	McFarlane *et al.* (1976)
mer-$RhCl_3(SMe_2)(SMePh)_2$	4048	McFarlane *et al.* (1976)
$RhCl_2(SMe_2)_2(\mu\text{-}Cl)_2RhCl_2(SMe_2)_2$	4768, 5465	Anderson *et al.* (1978)
$[Rh_2Br_5(SOMe_2)(\mu\text{-}Br)_3]^{2-}$	4702	Barnes *et al.* (1979)
$[RhCl_5Br]^{3-}$	7848	Mann and Spencer (1982b)
cis-$[RhCl_4Br_2]^{3-}$	7707	Mann and Spencer (1982b)
trans-$[RhCl_4Br_2]^{3-}$	7712	Mann and Spencer (1982b)
fac-$[RhCl_3Br_3]^{3-}$	7556	Mann and Spencer (1982b)
mer-$[RhCl_3Br_3]^{3-}$	7561	Mann and Spencer (1982b)
cis-$[RhCl_2Br_4]^{3-}$	7403	Mann and Spencer (1982b)
trans-$[RhCl_2Br_4]^{3-}$	7409	Mann and Spencer (1982b)
$[RhClBr_5]^{3-}$	7243	Mann and Spencer (1982b)

[a] In parts per million to high frequency of $\Xi_{103_{Rh}} = 3.16$ MHz.

(weak) $I^- < Br^- \leqslant S^{2-} < Cl^- < NO_3^- < F^- < OH^-$
$< OH_2 < NH_3 < NO_2^- < CN^- < CO$ (strong)

whereas the effective radius of the d electron is smallest for hard ligands (e.g., F^-, OH_2) and largest for soft ligands (e.g., PR_3, CO, SMe_2) and is expected to follow the nephelauxetic series (Phillips and Williams, 1966)

(hard) $F^- < OH_2 < NH_3 < OH^- < Cl^- \lesssim CN^- < Br^- < I^-$ (soft)

Hence it is not surprising that the weak hard ligand OH_2 produces a high-frequency shift of δ 9992 for $[Rh(OH_2)_6]^{3+}$ (Mann and Spencer, 1982a). In contrast, strong soft ligands produce low-frequency shifts, e.g., δ −231 in trans-$[RhMe(CN-t-Bu)_4(SbPh_3)]^{2+}$ (Taylor, 1973). The important influence of soft ligands, in comparison with ligand strength, is clearly shown for series of compounds where covalency dominates the chemical shift order:

(a) $F^- > Cl^- > Br^- > I$, for example,

trans-$[RhCl_4(SMe_2)_2]^-$ > trans-$[RhBr_4(SMe_2)_2]^-$ > trans-$[RhI_4(SMe_2)_2]^-$
  δ 5226 δ 4532 δ 2958

and

trans-$RhF(CO)(PPh_3)_2$ > trans-$RhCl(CO)(PPh_3)_2$ > trans-$RhBr(CO)(PPh_3)_2$
  δ −148 δ −369 δ −421
                > trans-$RhI(CO)(PPh_3)_2$
                  δ −532

(b) $O > S > \simeq Se > Te$, e.g.,

mer-$RhCl_3(OH_2)_3$ > mer-$RhCl_3(SMe_2)_3$ > mer-$RhCl_3(SeMe_2)_3$ > mer-$RhCl_3(TeMe_2)_3$
  δ 8858 δ 3897 δ 3904 δ 3179

(c) $P \simeq As > Sb$, e.g.,

trans-$[RhMe(CN-t-Bu)_4(PPh_3)]^{2+}$ $\simeq$ trans-$[RhMe(CN-t-Bu)_4(AsPh_3)]^{2+}$
  δ −39 δ −57
                > trans-$[RhMe(CN-t-Bu)_4(SbPh_3)]^{2+}$
                  δ −231

but

fac-$RhCl_3(PMe_3)_3$ < fac-$RhCl_3(AsMe_3)_3$
  δ 1437 δ 2125

For the series $F^- > Cl^- > Br^- > I^-$ the chemical shift order is dominated by changes in $\langle r^{-3} \rangle$, whereas for the series $O > S \simeq Se > Te$ and P, $As >$ Sb there is a balance between $\langle r^{-3} \rangle$ and ΔE^{-1}. When ΔE^{-1} is large and the chemical shifts are to high frequency, then the ΔE^{-1} term dominates, as for fac-$RhCl_3(PMe_3)_3$ and fac-$RhCl_3(AsMe_3)_3$. But when the ΔE^{-1} term is

TABLE IV

Rhodium-103 Chemical Shifts of Rhodium(I) Complexes

Complex	$\delta_{103_{Rh}}$[a]			Reference
	X = Cl	X = Br	X = I	
trans-RhX(CO)(PMe₃)₂	−415	−448	—	Taylor (1973)
[RhX₂(CO)₂]⁻	−16.7	47	—	Brown et al. (1980)
Rh₂X₂(CO)₄	151	−1	—	Brown et al. (1980)
trans-RhX(CO)(PPh₃)₂	−369	−421	−532	Colquhoun and McFarlane (1982)
	−368	—	—	Brown and Green (1970a)
RhX(PPh₃)₃	−81	−142	−267	Brown and Green (1970a)
trans-RhX(CO)(PMe₂Ph)₂	−441	−523	—	Hyde et al. (1977)
trans-RhX(CO)[PMe₂(C₆H₄OMe-p)]₂	−407	—	—	Hyde et al. (1977)

Complex[b]	$\delta_{103_{Rh}}$	Reference
Rh(CO)₂(acac)	288	Cocivera et al. (1982)
	292	Brown et al. (1980)
Rh(dq)(Bpz₄)	2496	Cocivera et al. (1982)

Compound	Value	Reference
Rh(cod)(Bpz₄)	914	Cocivera *et al.* (1982)
Rh(nbd)(Bpz₄)	979	Cocivera *et al.* (1982)
[Rh(CO)₂]₂[(μ-R,S)-1,2-((2-C₄H₃N)CH=CHCH₂CH=N)₂cyclohexane]	46	Brevard *et al.* (1981)
RhCl(*t*-Bu₂PCH₂CH₂CH=CHCH₂P-*t*-Bu₂)	601	Crocker *et al.* (1980)
RhCl(*t*-Bu₂PCH₂CHC(=CH₂)CH₂CH₂P-*t*-Bu₂)	460	Crocker *et al.* (1980)
RhCl(*t*-Bu₂PCH₂CH₂CH=CHCH₂CH₂P-*t*-Bu₂)	243	Crocker *et al.* (1980)
[Rh(CO)(*t*-Bu₂PCH₂CH₂CH=CHCH₂CH₂P-*t*-Bu₂)]⁺	−736	Crocker *et al.* (1980)
RhCl(*t*-Bu₂PCH₂CH=CHCH₂P-*t*-Bu₂)	983	Crocker *et al.* (1980)
trans-RhF(CO)(PPh₃)₂	−148	Colquhoun and McFarlane (1982)
trans-Rh(OH)(CO)(PPh₃)₂	−300	Colquhoun and McFarlane (1982)
trans-Rh(O₂CCH₃)(CO)(PPh₃)₂	−261	Colquhoun and McFarlane (1982)
trans-Rh(O₂CCF₃)(CO)(PPh₃)₂	−276	Colquhoun and McFarlane (1982)
trans-Rh(NCS)(CO)(PPh₃)₂	−517	Colquhoun and McFarlane (1982)
trans-Rh(NO₂)(CO)(PPh₃)₂	−407	Colquhoun and McFarlane (1982)
[Rh(CO)₄]⁻	−644	Brown *et al.* (1980)
[Rh(MeCN)₂(CO)₂]⁺	−54	Brown *et al.* (1980)

[a] In parts per million to high frequency of $\Xi_{103_{Rh}}$ = 3.16 MHz.

[b] acac, Acetylacetonate; cod, 1,5-cyclooctadiene; dq, duroquinone; Bpz₄, tetrakis(1-pyrazole)borate; nbd, bicyclo[2 : 2 : 1]heptadiene.

309

TABLE V

Rhodium-103 Chemical Shifts of Compounds Containing
Rhodium–Rhodium Bonds

Complex	$\delta_{103_{Rh}}{}^a$	Reference
$(\eta^5\text{-Indenyl})_2Rh_2(\mu\text{-CH}{=}CH_2)(\mu\text{-CMe}{=}CHMe)$	$-560, -574$	Caddy et al. (1978)
$[Rh_4(CO)_{11}]^{2-}$	-431	Heaton et al. (1980a)
$Rh_4(CO)_{12}$	-426	Heaton et al. (1980a)
$[Rh_4(CO)_{11}(CO_2Me)]^-$	$-494, 202$	Heaton et al. (1980a)
$[Rh_4(\mu\text{-H})_4(C_5Me_5)_4]^{2+}$	585	Espinet et al. (1979)
$[Rh_5(CO)_{10}(\mu_2\text{-CO})_5]^-$	$-148.5, -948.5$	Fumagalli et al. (1980)
$[Rh_6C(CO)_{15}]^{2-}$	-277	Gansow et al. (1980)
	$-615, -1003$	Heaton et al. (1981)
$[Rh_6C(CO)_{13}]^{2-}$	-313	Heaton et al. (1981)
$[Rh_6N(CO)_{15}]^-$	-242	Heaton et al. (1981)
$[Rh_7(CO)_{16}]^{3-}$	$690, 483, -376$	Brown et al. (1976)
$[Rh_9P(CO)_{21}]^{2-}$	$-1051, -1222, -1423$	Gansow et al. (1980)
$[Rh_{12}(CO)_{30}]^{2-}$	$168, -322, -560$	Heaton et al. (1980b)
$[Rh_{14}(CO)_{25}]^{4-}$	$320, 200, -939, -1001$	Heaton et al. (1980a)
$[Rh_{14}H(CO)_{25}]^{3-}$	$231, 136, -840, -973$	Heaton et al. (1980a)
$[Rh_{14}H_2(CO)_{24}]^{3-}$	$4554, -408, -522$	Martinengo et al. (1977)
$[Rh_{14}H_3(CO)_{24}]^{2-}$	$3547, -532, -600$	Martinengo et al. (1977)
$[Rh_{17}S_2(CO)_{32}]^{3-}$	$879, -639, -1458$	Gansow et al. (1980)

^{a} In parts per million to high frequency of $\Xi_{103_{Rh}} = 3.16$ MHz.

small, with low-frequency shifts, $\langle r^{-3} \rangle$ is similar to the ΔE^{-1} term, as for *trans*-$[RhMe(CN\text{-}t\text{-}Bu)_4(PPh_3)]^{2+}$ and *trans*-$[RhMe(CN\text{-}t\text{-}Bu)_4(AsPh_3)]^{2+}$, the effects balance out, and the ^{103}Rh chemical shifts are similar.

The occurrence of rhodium(I) complexes in the low-frequency range can be similarly qualitatively interpreted. In order to stabilize rhodium(I) it is necessary to use a strong soft ligand, e.g., CO or PR_3, which produces small values of $\langle r^{-3} \rangle$ and ΔE^{-1}. Even the same ligands produce a greater splitting in a square-planar geometry compared with an octahedral geometry. Thus for $[Ni(H_2NCH_2CH_2NH_2)_2]^{2+}$ the first electronic absorption band is at 21,550 cm^{-1}, whereas for $[Ni(H_2NCH_2CH_2NH_2)_3]^{3+}$ it is at 11,500 cm^{-1} (Lever, 1968).

There have been attempts to determine A in Eq. (1). The ^{103}Rh chemical shift of rhodium metal has been determined as -1360 ppm (Sogo and Jeffries, 1955), but the problem is to estimate the Knight shift. Two estimates of the Knight shift have led to values of -3500 ppm (Sogo and Jeffries, 1955) and -5000 ppm (Seitchik et al., 1965). Perhaps a more reliable value of -4400 ppm comes from $RhSn_2$ where more reliable estimates are possible (Seitchik et al., 1965).

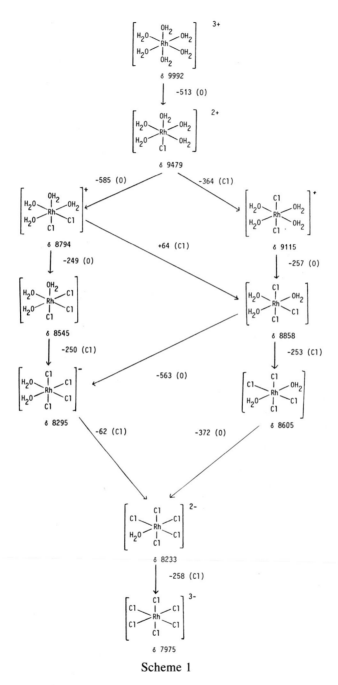

Scheme 1

Generalizations may also be drawn about the effect of stereochemistry on the ^{103}Rh chemical shift. In all cases, $\delta_{cis} < \delta_{trans}$ and $\delta_{fac} < \delta_{mer}$, providing a method for determining the structure of a compound or assigning the signals to the two isomers. In the more general case of sequential substitution, e.g., in $[Rh(OH_2)_{6-n}Cl_n]^{3-n}$, (Scheme 1), erratic stepwise substitution changes occur: The ligand atom trans to that being substituted is noted in parentheses. In this case a hard ligand, OH_2, is replaced by a weaker soft ligand, Cl^-, and the effects of the ligand on ΔE and $\langle r_d^{-3} \rangle$ in the Ramsey equation are opposed, resulting in these erratic changes. In contrast, when similar ligands are substituted, e.g., in $(RhCl_nBr_{6-n})^{3-}$ (Scheme 2), systematic stepwise substitution changes occur. Caution is necessary, however, in taking the chemical shifts of these compounds as accurate values. As noted before, the chemical shifts are dependent on temperature, and dependence on other ions present in the solution has been found. The temperature dependence of ^{103}Rh chemical shifts has been measured for a number of compounds, namely, trans-RhCl(CO)(PMe$_2$Ph)$_2$, 0.26 (Hyde et al., 1977); Rh(η-C$_5$H$_4$CO$_2$Me)(C$_2$H$_4$)$_2$, 0.25 in (CD$_3$)$_2$CO and 0.32 in CDCl$_3$ (Goodfellow, 1983); RhH(CO)(PPh$_3$)$_3$, 0.44 (Brown and Green, 1970b); mer-RhCl$_3$(SMe$_2$)$_3$, 1.0; isomers of [RhCl$_2$(MeSCH$_2$CH$_2$SMe)$_2$]Cl, 1.0 $\pm$ 0.1 (McFarlane et al., 1976); and Rh(acac)$_3$, 1.6 ppm K^{-1} (Cocivera et al., 1982).

II. Relaxation Measurements

At present spin–lattice relaxation measurements have been made on six compounds (Table VI). It is observed that T_1 varies between 0.06 and 62.8 s. The mechanism of relaxation is probably chemical shift anisotropy (CSA). For Rh(η^5-C$_5$H$_5$)(η^4-cod), the nuclear Overhauser enhancement (NOE) is zero (Mann et al., 1983), whereas for Rh(acac)(η^4-norbornadiene) the enhancement is -0.2 (Cocivera et al., 1982). As the theoretical maximum enhancement is -31.7, the dipole–dipole relaxation mechanism makes little or no contribution. Also, for these two compounds, T_1 decreases with temperature, which is inconsistent with the spin–rotation relaxation mechanism, leaving the CSA as the dominant relaxation mechanism. In the case of Rh(η^5-C$_5$H$_5$)(η^4-cod) the ^{103}Rh relaxation study was combined with a ^{13}C relaxation study to yield correlation times and hence the ^{103}Rh CSA as 524 ppm (Mann et al., 1983). Chemical shift anisotropy appears to be the dominant relaxation mechanism, it is advantageous to observe ^{103}Rh-NMR spectra at as high a magnetic field strength as possible, and less symmetric compounds, e.g., square-planar rhodium(I) com-

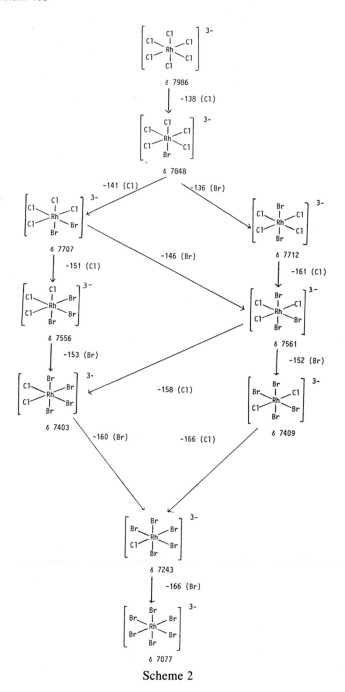

Scheme 2

TABLE VI

Spin–Lattice Relaxation Times T_1 of ^{103}Rh in Rhodium Compounds

Compound	Temperature (K)	$T_{1,^{103}Rh}$ (s)	B_0 (T)	Reference
Rh(acac)$_3$	300	62.8	2.114	Brevard *et al.* (1981)
Rh(η^5-C$_5$H$_5$)(η^4-cod)	339	12.7	9.39	Mann *et al.* (1983)
	309	8.6	9.39	Mann *et al.* (1983)
	271	5.2	9.39	Mann *et al.* (1983)
	240	2.4	9.39	Mann *et al.* (1983)
Rh(CO)$_2$(acac)	298	>30	9.39	Cocivera *et al.* (1982)
Rh(η^4-dq)(Bpz$_4$)	298	0.62 ± 0.05	9.39	Cocivera *et al.* (1982)
Rh(η^4-cod)(Bpz$_4$)	298	0.51 ± 0.05	9.39	Cocivera *et al.* (1982)
	213	0.060 ± 0.002	9.39	Cocivera *et al.* (1982)
Rh(η^4-nbd)(Bpz$_4$)	298	0.53 ± 0.07	9.39	Cocivera *et al.* (1982)

plexes, are far easier to observe than the octahedral rhodium(III) complexes.

III. Coupling Constants

There have been many observations of coupling between ^{103}Rh and the more readily observed nuclei, ^{1}H, ^{13}C, and ^{31}P. On account of the negative gyromagnetic ratio of ^{103}Rh, the one-bond coupling constants to ^{1}H and ^{31}P are negative, e.g., in *trans,trans*-RhCl$_2$H(CO)(PMe$_2$Ph)$_2$, $^1J_{^{103}Rh^1H}$ = 17.3 Hz (Hyde *et al.*, 1977), in *trans*-RhF(CO)(PPh$_3$)$_2$, $^1J_{^{103}Rh^{19}F}$ = −52.5 Hz (Colquhoun and McFarlane, 1982), and in *trans*-RhCl(CO)(PMe$_2$Ph)$_2$, $^1J_{^{103}Rh^{31}P}$ = −118 ± 1 Hz (Hyde *et al.*, 1977). In contrast, one-bond coupling constants to nuclei with negative gyromagnetic ratios are positive, e.g., in *mer*-RhCl$_3$(TeMe$_2$)$_3$, $^1J_{^{125}Te^{103}Rh}$ values are +71 Hz (^{125}Te trans to tellurium) and +94 Hz (^{125}Te trans to chlorine) (Barnes *et al.*, 1978).

There have been many observations of $^1J_{^{103}Rh^1H}$, $^1J_{^{103}Rh^{13}C}$, and $^1J_{^{103}Rh^{31}P}$, with values of 15–30, 15–100, and 80–200 Hz, respectively. Typical values of $^1J_{^{103}Rh^{13}C}$ and $^1J_{^{103}Rh^{31}P}$ have been reported in two reviews of ^{13}C- (Mann and Taylor, 1981) and ^{31}P-NMR data (Pregosin and Kunz, 1979).

The observation of ^{103}Rh coupling to itself and other metals has been limited. Some examples of $^1J_{^{103}Rh^{103}Rh}$ are 17 Hz in Rh$_2$(C$_2$H$_3$)(C$_2$Me$_3$)(η^5-idenyl)$_2$ (Caddy *et al.*, 1978), 4.2 Hz in Cp$_2$Rh$_2$(CO)$_3$, 4.4 Hz in Cp$_2$Rh$_2$(CO)$_2$(CH$_2$), and 4.4 Hz in Cp$_2$Rh$_2$(NO)$_2$ (Lawson and Shapley, 1978). Somewhat larger $^2J_{^{195}Pt^{103}Rh}$ values have been reported for

$[Rh_5Pt(CO)_{15}]^-$, where $^1J_{195_{Pt}103_{Rh}} = 24$ Hz and $^2J_{195_{Pt}103_{Rh}} = 73$ Hz, and for $[Rh_2Pt(CO)_x]_n^{n-}$, where $^1J_{195_{Pt}103_{Rh}} = 69$ Hz and $^2J_{195_{Pt}103_{Rh}} = 55$ Hz (Fumagalli *et al.*, 1978).

IV. Applications

In order to demonstrate the power of ^{103}Rh-NMR spectroscopy four different applications are discussed.

The ^{1}H-NMR spectrum of $[Rh(MeSCH_2CH_2SMe)_2Cl_2]^+$ shows a series of at least 10 signals for the methyl protons, each signal being a doublet due to $^3J_{103_{Rh}1_H}$ (McFarlane *et al.*, 1976). By use of ^{1}H—$\{^{103}$Rh$\}$ INDOR, seven different ^{103}Rh signals were identified, which fell into two groups, δ 2502 to δ 2585 and δ 2288 to δ 2371, the former group being assigned to trans isomers and the latter group to *cis* isomers. The signal at δ 2547 is associated with at least three different methyl signals and was assigned to **(I)**, the only trans isomer that gives more than one methyl signal. The other

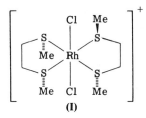

(I)

isomers were not assigned, but the differences in ^{103}Rh chemical shifts were attributed to varying amounts of internal steric strain which led to differing interbond angles at rhodium in the different isomers.

Commercial hydrated rhodium trichloride is commonly used as the starting material in the synthesis of rhodium complexes, but the yield of unstable complexes depends on the batch used. Examination of the ^{103}Rh-NMR spectra of three commercial samples shows not only several rhodium complexes present, but the proportions depend on the sample (Fig. 2). The compounds present are *cis*- and *trans*-$[RhCl_2(OH_2)_4]^+$ and *mer*- and *fac*-$RhCl_3(OH_2)_3$ (Mann and Spencer, 1982a). It is feasible that these compounds have different reactivities, and the product yield depends on the proportions of compounds present in the $[RhCl_n(OH_2)_{6-n}]^{3-n}$ mixture.

Fluxionality of the rhodium skeleton of $[Rh_9P(CO)_{21}]^{2-}$ has been demonstrated (Gansow *et al.*, 1980). At $-80°$C, three doublets, due to $^1J_{103_{Rh}31_P}$, with relative intensities $4:4:1$ were observed at δ -1052, δ -1222, and δ -1423 in agreement with the crystal structure. In contrast, at 23°C only one doublet of average chemical shift was observed, showing that fluxionality of the rhodium skeleton occurs.

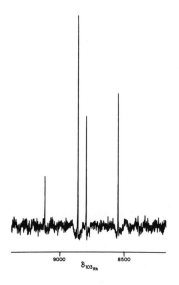

Fig. 2 Rhodium-103 NMR spectrum at 12.65 MHz for a 10-mm sample tube containing 0.8 M commercial hydrated rhodium trichloride and ~5 mg hydrated chromic chloride.

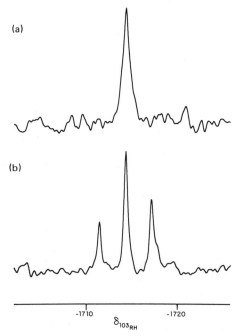

Fig. 3 Rhodium-103 NMR spectrum at 12.65 MHz for a 10-mm sample tube containing ~0.5 M $C_5Me_5RhH_2(SiMe_3)_2$ in toluene-d_8. (a) Complete broad band ^1H decoupling; (b) CW ^1H decoupling at δ ~1. (Reproduced with permission from Fernandez and Maitlis, 1982.)

A compound $(C_5Me_5)RhH_n(SiMe_3)_2$ has been shown to be a dihydride by recording the ^{103}Rh-NMR spectrum with continuous-wave (CW) ^1H decoupling at $\delta \sim 1$ to remove any coupling to the methyl groups (Fig. 3a). The ^{103}Rh-NMR spectrum is a triplet as a result of coupling to two hydrides. This coupling was removed by using noise-modulated ^1H decoupling at $\delta \sim -5$ to remove all ^1H coupling (Fig. 3b) (Fernandez and Maitlis, 1982).

V. Conclusions

Rhodium-103 NMR spectroscopy is rapidly moving from the first direct observation in 1979 (Gill *et al.*, 1979) toward becoming an important spectroscopic technique for the rhodium chemist. The present sensitivity problems are being alleviated as experience permits reliable estimates of experimental parameters. The introduction of INEPT is proving a valuable method for gaining sensitivity, but its application is limited at present by the poor 90° pulse angles for ^1H and ^{103}Rh nuclei on most commercial probes.

References

Anderson, S. J., Barnes, J. R., Goggin, P. L., and Goodfellow, R. J. (1978). *J. Chem. Res. Synop.* 286.

Barnes, J. R., Goggin, P. L., and Goodfellow, R. J. (1978). Unpublished results.

Barnes, J. R., Goggin, P. L., and Goodfellow, R. J. (1979). *J. Chem. Res. Synop.* 118.

Brevard, C., Stein, G. C., and v. Koten, G. (1981). *J. Am. Chem. Soc.* **103**, 6746.

Brown, T. H., and Green, P. J. (1970a). *J. Am. Chem. Soc.* **92**, 2359.

Brown, T. H., and Green, P. J. (1970b). *Phys. Lett.* **31A**, 148.

Brown, C. *et al.* (1976). *J. Organ. Chem.* **169**, 309.

Brown, C., Heaton, B. T., Longhetti, L., Povey, W. T., and Smith, D. O. (1980). *J. Organ. Chem.* **192**, 93.

Caddy, P., Green, M., Smart, L. E., and White, N. (1978). *J. Chem. Soc. Chem. Commun.* 839.

Cocivera, M. *et al.* (1982). *J. Magn. Reson.* **46**, 168.

Colquhoun, I. J., and McFarlane, W. (1982). *J. Magn. Reson.* **46**, 525.

Crocker, C. *et al.* (1979). *J. Chem. Soc. Commun.* 498.

Crocker, C. *et al.* (1980). *J. Am. Chem. Soc.* **102**, 4373.

Espinet, P., Bailey, P. M., Piraino, P., and Maitlis, P. M. (1979). *Inorg. Chem.* **18**, 2706.

Fernandez, M.-J., and Maitlis, P. M. (1982). *J. Chem. Soc. Chem. Commun.* 310.

Fumagalli, A. *et al.* (1980). *J. Am. Chem. Soc.* **102**, 1740.

Fumagalli, A. *et al.* (1978). *J. Chem. Soc. Chem. Commun.* 195.

Gansow, O. A. *et al.* (1980). *J. Am. Chem. Soc.* *102*, 2449.

Gill, D. S., Gansow, O. A., Bennis, F. J., and Ott, K. C. (1979). *J. Magn. Reson.* **35,** 459.

Goodfellow, R. J. (1978). "NMR and the Periodic Table" (R. K. Harris and B. E. Mann, eds.), p. 244. Academic Press, London.

Goodfellow, R. J. (1983). Unpublished results.

Grüninger, K.-D., Schwenk, A., and Mann, B. E. (1980). *J. Magn. Reson.* **41,** 354.

Heaton, B. T. *et al.* (1980a). *J. Am. Chem. Soc.* **102,** 6175.

Heaton, B. T., Strona, L., Martinengo, S., and Chini, P. (1980b). *J. Organ. Chem.* **194,** C29.

Heaton, B. T., Strona, L., and Martinengo, S. (1981). *J. Organ. Chem.* **215,** 415.

Hyde, E. M., Kennedy, J. D., Shaw, B. L., and McFarlane, W. (1977). *J. Chem. Soc. Dalton Trans.* 1571.

IUPAC. (1976). *Pure Appl. Chem.* **45,** 217.

Kidd, R. G. (1980). *Ann. Rep. Nucl. Magn. Reson. Spectrosc.* **10A,** 1.

Lawson, R. J., and Shapley, J. R. (1978). *Inorg. Chem.* **17,** 2963.

Lever, A. B. P. (1968). "Inorganic Electronic Spectroscopy", p. 305. Elsevier, Amsterdam.

McFarlane, H. C. E., McFarlane, W., and Wood, R. J. (1976). *Bull. Soc. Chim. Belge.* **85,** 864.

Mann, B. E., and Spencer, C. (1982a). *Inorg. Chim. Acta* **65,** L57.

Mann, B. E., and Spencer, C. (1982b). *Inorg. Chim. Acta* **76,** L67.

Mann, B. E., and Taylor, B. F. (1981). "^{13}C NMR Data for Organometallic Compounds." Academic Press, London and New York.

Mann, B. E., Taylor, B. F., and Yavari, P. (1983). Unpublished results.

Martinengo, S., Heaton, B. T., Goodfellow, R. J., and Chini, P. (1977). *J. Chem. Soc. Chem. Commun.* 39.

Phillips, C., and Williams, R. J. P. (1966). "Inorganic Chemistry", Vol. 2, pp. 395–402. O.U.P., London.

Pregosin, P. S., and Kunz, R. W. (1979). *In* "NMR Basic Principles and Progress," (P. Diehl, E. Fluck, and R. Kosfeld, eds.), Vol. 16. Springer-Verlag, Berlin.

Ramsey, N. F. (1950). *Phys. Rev.* **78,** 699.

Seitchik, J. A., Jaccarino, V., and Wernick, J. H. (1965). *Phys. Rev.* **138A,** 148.

Sogo, P. B., and Jeffries, C. D. (1955). *Phys. Rev.* **98,** 1316.

Taylor, B. F. (1973). Ph.D. Thesis, University of Bristol.

12 Silver-109

P. Mark Henrichs

Chemistry Division, Research Laboratories
Eastman Kodak Company
Rochester, New York

I. Introduction

Both ^{109}Ag and ^{107}Ag have spins of $\frac{1}{2}$. The NMR frequency at a magnetic field strength of 63.4 kG is 12.57 MHz for ^{109}Ag and 10.92 MHz for ^{107}Ag. Most NMR studies have involved the ^{109}Ag nucleus because of its higher frequency, even though its 49% natural abundance is slightly less than that of ^{107}Ag. Perhaps surprisingly, there have been more published reports on the NMR of silver-containing solids than silver-containing liquids. In general, the work on liquids has addressed problems more chemical in nature than has the work on solids, and this chapter is restricted to the former.

II. Experimental Aspects

A. Principal Difficulties

Experimentally, silver NMR presents severe challenges. The inherent sensitivity of ^{109}Ag is only 1.01×10^{-4} compared with an equal number of protons, and 6.54×10^{-3} compared with an equal number of ^{13}C nuclei. Furthermore, because both silver nuclei have spins of $\frac{1}{2}$, there is no quadrupolar relaxation mechanism, and dipolar relaxation is inefficient because silver ions do not normally exist in proximity to protons. Measured spin–lattice relaxation times have been in excess of 100 s (Ackerman, 1977; Kronenbitter, 1977; Kronenbitter et al., 1980). Spectral accumulation without nuclear saturation is thus difficult. To complicate things further, the gyromagnetic ratios of both ^{107}Ag and ^{109}Ag are negative. Any nuclear Overhauser enhancement (NOE) from proton irradiation is thus negative. Only a small portion of the maximum enhancement of 10.7 is found in practice, typically leading to a nulled signal (Brevard et al., 1981).

Coupling between the silver nucleus and nearby protons is normally not observed, because ligand exchange is usually fast on the NMR time scale. Signal enhancement techniques relying on polarization transfer from protons to the silver nucleus are thus possible only in a few isolated cases (Brevard et al., 1981).

B. Techniques

An early approach to the measurement of silver NMR spectra involved the use of paramagnetic additives to reduce the long relaxation times to a practical range (Brun et al., 1954). Significant chemical shifts are caused by the addition of $Fe(NO_3)_3$ to an aqueous silver nitrate solution (Burges et al., 1973), and most studies have avoided paramagnetic relaxation reagents (see also Sogo and Jeffries, 1954).

The extremely long spin–lattice relaxation times suggest that the decay of the free induction signal is probably controlled by an instrumentation effect such as magnetic field inhomogeneity. The driven equilibrium Fourier transform (DEFT) method (Becker *et al.*, 1969), which refocuses and restores longitudinal magnetization after data acquisition, appears particularly applicable to silver NMR because refocusing of the NMR signal would not be complicated by spin–spin coupling. The steady state signal, which is built up by a rapid series of 90° pulses, retains much of the enhanced intensity of the DEFT method, however, although at the expense of phase and intensity anomalies in the observed signals (Freeman and Hill, 1971). Schwenk (1971) developed the Quadriga Fourier transform (FT) procedure to eliminate these undesirable features of rapid pulsing. This method, which relies on the acquisition of spectra at four different specified reference frequencies, results in an unusual line shape, and other methods have been developed to remove the phase and intensity anomalies of rapid pulsing (Freeman and Hill, 1971; Martin *et al.*, 1980). These techniques are now possible with all commercial spectrometers.

Even with the sensitivity of modern spectrometers, silver NMR is still restricted to solutions 0.1 *M* or greater unless unusually long times for spectral averaging can be used. Interpretation of silver chemical shifts must take into account chemical effects associated with high concentrations, such as ion pairing and multiple ion formation. Meaningful results require particular care in the design of the experimental strategy.

C. Design of Experiments

One of the simplest approaches is to monitor the chemical shift of a silver salt as a function of concentration in a given solvent. As we will see, chemical shifts are strongly concentration-dependent. Plots of chemical shift versus concentration sometimes have discontinuous changes in slope, which can be related to the structures of the solvated ions present.

Studies on concentration effects on chemical shift have also shown that the counterion plays an important role in determining the chemical shift of the silver ion, presumably as a result of the existence of ion pairs in solution. Comparisons of the concentration plots for different silver salts in the same solvent are informative, and interesting additional experiments in which the concentration of the counterion was varied by the addition of another salt have been performed.

The competitive solvation of silver ions by different solvents can be studied by the examination of cosolvents of various compositions. Such experiments have led to some surprising effects which appear to require that the solvation of the counterion, as well as of the silver ion, be taken into account. Finally, the interaction of silver ions with strong complexing agents in neutral solvents has also been studied.

Ideally, for each of the types of experiments described, it would be desirable to simulate the observed chemical shift plots in terms of postulated equilibria among all the different types of solvated silver ions and ion pairs. In practice, such a detailed analysis is not usually possible (however, see Kronenbitter, 1977). In most silver solutions it is obvious that there are many species present. The chemical shifts of the silver ions in these species in the absence of chemical exchange are not usually known. Indeed, the exact number of species present is not generally known. As a result, a mathematical analysis involves many unknowns with only a few experimental data, hence the number of data points has to be rather large.

In spite of the problems in a full analysis, chemical shift data do lend themselves to semiquantitative interpretation in many cases. The following sections consider specific examples of studies on solution phenomena with silver NMR that have been especially informative.

III. Concentration Studies on Silver Ion Solvation and Ion Pairing

A. Silver Nitrate in Acetonitrile

Careful examination of the curve for acetonitrile in Fig. 1 reveals distinct features (Ackerman, 1977). In contrast to the plot for acetonitrile, the curve for water is almost featureless. The features of the acetonitrile plot are apparent if the reader sights down the length of the curve, and the derivative plot in Fig. 2 also shows these features clearly (Ackerman, 1977). Jucker *et al.* (1976) and Rahimi and Popov (1976) did not report the fine structure of the acetonitrile plot, but reexamination of the data of Jucker reveals features similar to those found by Ackerman.

Complexes containing acetonitrile and silver(I) in ratios of both 4 : 1 and 2 : 1 (as well as 1 : 1) have been reported to exist in acetonitrile solutions (Baddiel *et al.*, 1965; Balasubrahmanyam and Janz, 1970; Chang and Irish, 1974; Janz, 1971). At higher concentrations multiple-ion aggregates are postulated to exist (Janz *et al.*, 1965). The solvation of the silver ion is clearly not simple. Nevertheless, it is interesting that the breaks in the

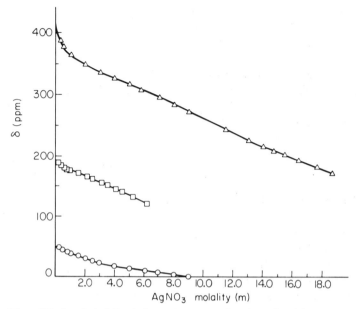

Fig. 1 Silver-109 chemical shift of silver nitrate as a function of molal concentration for three solvents, acetonitrile (△), DMSO (□), and water (○). (Reproduced by permission from Ackerman, 1977.)

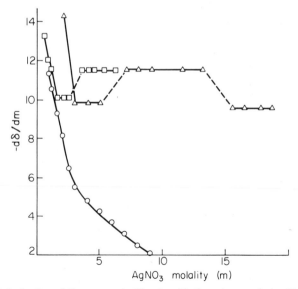

Fig. 2 First derivative of the curves in Fig. 1, with the same symbols. (Reproduced by permission from Ackerman, 1977.)

chemical shift plot occur close to the points where there are acetonitrile/ silver ratios of 3.8 (6.4 m) and 1.8 (13.4 m). Thus it is tempting (although not vigorously justified) to explain the acetonitrile results in terms of equilibria involving the 2 : 1 and 4 : 1 complexes. Ackerman (1977) also invoked equilibria involving various types of solvent-separated ion pairs. The steep portion of the curve at low salt concentrations may be a result of ion pairing, for example. Similar features of plots for silver halides have been explained by Kronenbitter *et al.* (1980) in terms of participation of the halide ion in the solvation complex.

B. Silver Nitrate in Dimethyl Sulfoxide

Ackerman (1977) proposed that in dimethyl sulfoxide (DMSO), as in acetonitrile, it is the interaction between the cation and the anion that primarily determines the silver chemical shift. This interaction is modified by the solvation of both ions, so that breaks in the chemical shift plot corresponding to specific solvated species once again occur, as shown in Figs. 1 and 2. Careful examination of the plots reveals a change in the slope of the curve at a DMSO/silver ion ratio of 4.1 : 1, corresponding to a molality of 3.1. The coordination number found by ^{1}H NMR is 4.2 (Clausen *et al.*, 1973), suggesting that the change in slope is related to the formation of this species.

C. Silver Salts in Aqueous Ethylamine

Schwenk's research group has made a number of studies on various silver salts in ethylamine mixed with water. The gaseous nature of ethylamine prevented its use as a pure solvent. The formation constants for the 1 : 1 and 2 : 1 complexes of ethylamine and Ag(I) ensure that the silver ion exists almost entirely as the 2 : 1 amine complex in this solvent, however (Kronenbitter *et al.*, 1980).

The tedious work required to obtain data at very low silver concentrations has proven worthwhile for the ethylamine system. Concentration effects on the silver chemical shifts for silver nitrate, silver bromide, and silver chloride are shown in Fig. 3 for a mixture with a 0.799 mole fraction of ethylamine. The direction of shift for the halide salts is opposite that for the nitrate. Even so, at very low concentrations the curves appear to

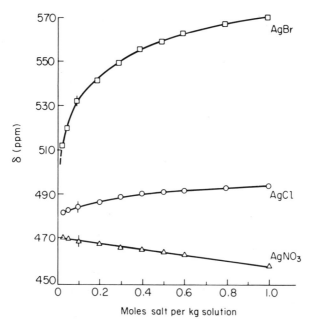

Fig. 3 Chemical shift of [109]Ag as a function of silver ion concentration in aqueous ethyl-amine solution. (Reproduced by permission from Kronenbitter, 1977.)

converge to a common point, which is the chemical shift of the pure 2 : 1 complex of ethylamine and Ag(I). The convergence is not obvious unless data points for concentrations below 0.1 m are obtained, however.

Kronenbitter *et al.* (1980) explained the steep slopes at low concentrations for the halide salts in terms of participation of the anions in the silver complex. Simulations of the observed chemical shift plots were made in terms of a postulated equilibrium between a simple complex and one containing the halide ion. The changes in chemical shift at high silver concentrations were treated phenomenologically without consideration of equilibria involving explicit chemical species. The concentration was assumed to affect the chemical shift of the pure complexed silver ion and the complex containing the halide ion equally. Such an assumption leaves open the question of the reliability of the derived equilibrium constants.

It is clear that the counterion must be considered directly in any explanation of the silver chemical shifts in all the solvents previously described. Experiments that give specific information about the nature of the cation–anion interaction are described in the next section.

IV. Counterion Effects on Silver Chemical Shifts

A. Silver Salts in Aqueous Ethylamine

Kronenbitter et al. (1980) explained the results in Fig. 3 in terms of a complex explicitly involving the halide ion. As the concentration of the silver salt is reduced, this complex breaks up and the chemical shifts of the nitrate, bromide, and chloride salts converge to a common value. Earlier work by Jucker et al. (1976) suggested that the chemical shift of silver iodide in this solvent system did not converge to this value, however. Furthermore, the direction of the shift as the salt concentration of the iodide was increased was opposite that for the other two salts. Only three data points were found for the iodide, however. Further work with this salt would be interesting to clarify the nature of the silver species involved and to determine whether an iodide ion pair is present even at very low concentrations. The direction of shift of silver fluoride is also opposite that of silver chloride and silver bromide, but the zero concentration shift does appear to converge.

B. Silver Salts in Water

Several groups have examined the silver chemical shift as a function of the concentration of silver perchlorate in water. There is some disagreement as to how much the chemical shift plot differs from that for the nitrate, however. Ackerman (1977) found a close similarity between the chemical shift plots of silver nitrate and silver perchlorate over the same concentration ranges. The total range of shifts for silver perchlorate is greater than that for silver nitrate because the solubility is much higher. Rahimi and Popov (1976) found different curves for the two salts, and the curve for silver perchlorate showed no sign of approaching an asymptotic value at high concentrations. The differences in the shape of Ackerman's curve and that of Rahimi and Popov may be a result of the fact that Rahimi and Popov expressed concentrations in molarity and Ackerman used molality. The curve of Burges et al. (1973) resembles that of Rahimi and Popov much more than that of Ackerman, however, even though units of molality were used. More work is needed to reconcile these results.

Other interesting counterion effects have been found for silver salts in water. Burges et al. (1973) observed that the silver ion becomes less shielded as the concentration of silver fluoride in water increases, whereas the direction of shift for most silver salts is in the opposite direction. The curve nevertheless extrapolates to the same value at zero concentration as the curves for silver nitrate and silver perchlorate.

C. Silver Salts in Organic Solvents

Ackerman (1977) monitored the chemical shifts of silver perchlorate in acetonitrile and DMSO as a function of silver concentration up to 1 m. The observed shifts were less than half those found in the same solvents over the same concentration range for silver nitrate. This observation may indicate a lesser importance of ion pairs for the perchlorates in organic solvents than for the nitrates. Conductance studies by Yeager and Kratochvil (1969) are consistent with such an interpretation.

D. Experiments Involving Changes in Counterion Concentration

Ackerman (1977) did a series of interesting experiments in which the ratio of the counterion concentration to the silver ion concentration was changed by the addition of other salts with the same counterion. The addition of lithium nitrate to a 1 m solution of silver nitrate in water changed the silver chemical shift by more than 50 ppm. This range is comparable to that found when the concentration of silver nitrate was varied. The addition of a tetrabutylammonium salt to silver nitrate in acetonitrile also led to shifts similar to those found when the concentration of silver nitrate itself was varied. Similar results were obtained with the addition of both ammonium nitrate and tetrabutylammonium nitrate to solutions of silver nitrate in DMSO.

These results can again be interpreted (Ackerman, 1977) in terms of the silver chemical shift being determined largely by the cation–anion interactions. The addition of excess nitrate ion facilitates the formation of ion pairs and multiple ions, regardless of whether the additional anion comes from a higher concentration of the silver salt or the presence of another salt with the nitrate ion in common.

An alternative explanation, which need not exclude the first in addition, relates to the high concentrations of the silver salts. At a 6.4 m concentration of silver nitrate in acetonitrile there are only 3.8 solvent molecules for every silver ion. At high concentrations the silver ion is preferentially solvated by as many as four acetonitrile molecules (Chang and Irish, 1974). At higher concentrations it is obvious that solvation by so much acetonitrile is impossible, especially when it is considered that the counterion will also be solvated by several solvent molecules. The addition of another salt can change the chemical shift of the silver nucleus simply by tying up additional solvent molecules.

V. Solvent Composition—Competitive Solvation Effects

A. The General Case

Under the simplest circumstances we would expect the chemical shift of silver ions in a weakly interacting solvent to change rapidly on the addition of a strongly interacting solvent, up to the point at which the solvation sphere is completely filled by the more interacting substance. Such behavior has, in fact, been found for silver nitrate and for silver perchlorate in a mixture of either DMSO and water or acetonitrile and DMSO (Ackerman, 1977). Much more interesting behavior is found when the solvent is a mixture of acetonitrile and water.

B. Acetonitrile and Water

The results for this system, studied by Rahimi and Popov (1976) and Ackerman (1977), are the most interesting and unexpected of those for any of the mixed-solvent systems. Ackerman's plot is shown in Fig. 4. The most striking feature is the maximum in the chemical shift.

Because of the indications that the interactions of the cation with the anion are important in determining the silver chemical shift, Ackerman has proposed that the unusual curve results from "heteroselective solvation" (Janz, 1971). The initial addition of acetonitrile to the aqueous solution of silver nitrate results in replacement of some of the water molecules in the immediate vicinity of the silver ions by acetonitrile molecules. This leads to strong deshielding of the silver nucleus. The nitrate ion remains largely surrounded by water molecules, however (Koepp et al., 1960; Oliver and Janz, 1970; Strehlow and Koepp, 1958; Strehlow and Schneider, 1971). At some intermediate point the silver nucleus is almost entirely solvated by acetonitrile, whereas the nitrate ion is solvated by water. Because each ion is favorably solvated, the interactions between them in the form of ion pairing are minimal. As more acetonitrile is added,

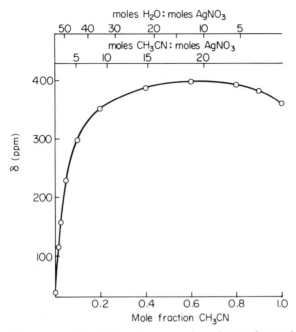

Fig. 4 Silver-109 chemical shift of 1.0 m silver nitrate in a mixture of acetonitrile and water as a function of the mole fraction of acetonitrile. (Reproduced by permission from Ackerman, 1977.)

the shell of water around the nitrate ion is disrupted, and ion pairing interactions become increasingly important again.

Some confirmation of this scenario was obtained with ^{14}N NMR of the nitrate ion (Ackerman, 1977). The total range of ^{14}N shifts is about 2 ppm for solvent mixtures varying from pure water to pure acetonitrile. There is a noticeable discontinuity in the deshielding produced by added acetonitrile at about a 0.5 mole fraction of acetonitrile, however. This is the approximate composition at which the silver chemical shift goes through its maximum and may represent the point at which the solvation sphere of the nitrate ion is changed from water to acetonitrile.

The results for silver perchlorate in mixtures of water and acetonitrile are in sharp contrast to those for silver nitrate, as shown in Fig. 5 (Ackerman, 1977). The initial rise in the curve with the addition of acetonitrile is similar, but there is no maximum. The absence of a maximum may indicate that the competition between water and acetonitrile in solvation of the perchlorate ion is much more balanced than it is for the nitrate ion.

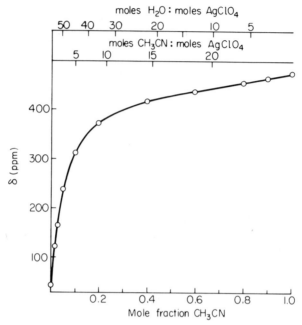

Fig. 5 Silver-109 chemical shift of 1.0 *m* silver perchlorate in a mixture of acetonitrile and water as a function of the mole fraction of acetonitrile. (Reproduced by permission from Ackerman, 1977.)

VI. Silver Complexes in Solution

This section treats cases in which the primary purpose of the research was to study the interactions of the silver ion with a complexing agent in a neutral solvent. Because of the high concentrations required for silver NMR, the complexing agent also must be present in concentrations high enough for it to play the role of a cosolvent. The distinction between the study of silver complexation and the study of mixed solvents is thus not clear, and our choice of which studies to consider in this section rather than the last admittedly is somewhat arbitrary.

A. Treatment of Chemical Shift Data

Most commonly, silver chemical shifts are monitored as a function of C_L/C_{Ag}. If the concentration of the silver ion C_{Ag} is held constant at a value that permits accumulation of the NMR spectrum in a reasonable amount of time, the concentration of the complexing agent C_L will then be the variable. Henrichs et al. (1977) have shown that plotting the silver chemical shift versus this concentration (or, more properly, the ratio of the concentration of the complexing agent to the concentration of the silver ion) leads, in the limiting case, to a series of straight-line plots that intersect at points corresponding to the composition of the various complexes formed. The limiting condition is that the formation constants for the various complexes be well separated in magnitude, so that for any given concentration of the complexing agent only two complex species are present in appreciable concentrations. The analysis further assumes that the chemical shift of the silver ion is controlled primarily by the type of complex in which it exists. The roles of the counterion and the concentration are ignored. Simple cases have actually been found in which the chemical shift plot approximates straight-line segments, but the assumptions preclude many systems from giving such plots. The model is nevertheless a useful starting point for analysis. Specific types of silver ligands are considered in the following discussion.

B. Nitrogen Ligands

The plots of chemical shift versus C_L/C_{Ag} for ligands thought to bind through nitrogen have generally been relatively simple. Henrichs et al. (1979) found that the plot for di-n-propylamine and 2-aminothiazoline consisted roughly of two straight-line segments intersecting in a rounded break at a ligand/silver ratio of about 2. The results for di-n-propylamine are shown in Fig. 6. Silver(I) typically forms binary complexes with nitrogen donors (Ahrland et al., 1958; Schwarzenbach, 1953), so that such a plot is not unreasonable. Corresponding plots of the ^{13}C chemical shifts of the ligand versus C_{Ag}/C_L also show sharp breaks corresponding to a 2:1 complex.

The curvature at the break point in the plot for the silver chemical shift illustrates the problems of the model. Furthermore, the silver chemical shift continues to change even after more than enough ligand has been added to form the 2:1 complex. It appears that concentration and ion pairing effects are important in silver plots and make analysis more difficult.

Rahimi and Popov (1979) have examined the complexation of silver(I) with pyridine in a number of solvents. Some indications of breaks in the plots of chemical shift versus C_L/C_{Ag} at the 2:1 complex were found in various solvents. The breaks were most obvious when the solvent was itself weakly interactive. Even in noninteractive solvents, however, the break points in the curves were not at all distinct.

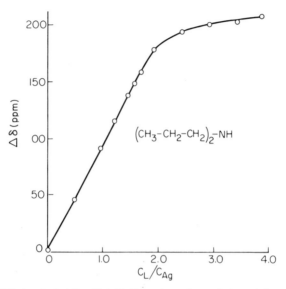

Fig. 6 Silver-109 chemical shift of 1.0 *M* silver nitrate in a solution of di-*n*-propylamine in DMSO. (Reprinted by permission from Henrichs *et al.*, copyright 1979 American Chemical Society.)

C. Sulfur Ligands

1. Pentamethylene Sulfide

The plot of the silver chemical shift for this ligand is much smoother than those found for nitrogen ligands but nevertheless suggests the formation of a 2 : 1 complex (Henrichs *et al.*, 1979). For this compound the plots of the carbon chemical shift are unusual. The plot for the four-carbon ligand shows a clear-cut break indicative of a 2 : 1 complex, but the plot for the two-carbon ligand is rounded and has no clear break at all. The plot for the three-carbon ligand shows little change in chemical shift up to a silver ion/ligand ratio of 0.5 and then a rapid change followed by leveling. Overall there is some indication of the importance of a 2 : 1 complex, but caution in interpretation is indicated.

2. Thiourea Derivatives and Related Ligands

Thiourea gave an unusual plot of the silver chemical shift versus the ligand/silver concentration ratio, as shown in Fig. 7 (Henrichs *et al.*, 1977). A simple interpretation is that 1 : 1, 2 : 1, and 3 : 1 complexes are formed sequentially with the addition of thiourea, as there appear to be breaks in the curve at or near ratios corresponding to these complexes. The authors considered the carbon results to suggest that such an explanation was incomplete, however. The plot of carbon chemical shift versus C_{Ag}/C_L showed a clear break at a ratio of 1. There was also a slight inflection in the curve between 0.0 and 0.3. The largest changes in the carbon chemical shifts occurred between the points corresponding to the 1:1 and 2:1 complexes. It was thought that the largest shifts in the carbon chemical shift should occur as the ligand passed from the free state to the complexed state, with the formation of higher-order complexes having relatively little effect on the carbon shifts. This requires that each ligand not be able to "see" the other ligands in the complex.

Higher-order complexes involving sulfur bridges between silver ions could adequately explain the carbon shifts, however. Other workers have inferred that thiourea forms polymeric silver complexes in solution (Shul'man and Savel'eva, 1970), and crystalline complexes are clearly polymeric (Udupa and Krebs, 1973; Vizzini et al., 1968). The large solvent effects on silver chemical shifts and the fact that clean breaks in the chemical shift plots are not always found even for ligands that form simple

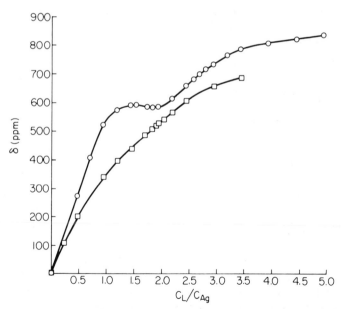

Fig. 7 Silver-109 chemical shift in solutions of 1 M silver nitrate in solutions of thiourea (○) and tetramethylthiourea (□) in DMSO. (Reprinted by permission from Henrichs *et al.*, copyright 1977 American Chemical Society.)

complexes may also explain the rounded breaks in the thiourea plot, however, without the requirement for polymeric forms.

As Fig. 7 shows, the silver chemical shift plot for tetramethylthiourea is much smoother than that for thiourea itself. Careful examination shows the changes in the slope of the curve at the same places as the changes in the slope of the thiourea plot, however. The graph of carbon chemical shifts is also similar to that for thiourea, but the inflection in the curve between the silver ion/ligand ratios of 0.3 and 0.35 is much smaller, and when the ratio is very large, the curve does not flatten out as it does for thiourea. The steric effects of the methyl groups may prevent the formation of complexes as strong as those formed by thiourea. Ethylenethiourea, in which the nitrogens are substituted, but with the attached groups held back in a ring, gives both silver and carbon results similar to those for thiourea. Any steric effect of the ring appears to be negligible.

Thiazolidine-2-thione   3-Methylthiazolidine-2-thione

The silver chemical shifts of thiazolidine-2-thione and 3-methylthiazolidine-2-thione were also measured as a function of C_L/C_{Ag} (Henrichs et al., 1979). The plot for the former is similar to that for tetramethylthiourea but appears to have breaks at ratios of 1 and 3. The carbon plot has breaks at corresponding values of C_{Ag}/C_L, but the interpretation is clouded by the possibility that the proton at the nitrogen is lost during complexation. In 3-methylthiazolidine-2-thione loss of the proton is not possible. The silver plot for this compound is featureless, but the carbon plot is similar to that for thiazolidine-2-thione.

D. Aryl Ligands

Rahimi and Popov (1979) measured the chemical shift of silver perchlorate in the presence of various amounts of benzene in a number of solvents. In weakly interacting solvents such as propylene carbonate and methanol, the silver ion was deshielded by increasing amounts of benzene, suggesting possible formation of a complex, but no sharp breaks in the chemical shift plot were found. In strongly interacting solvents such as DMSO and pyridine, the silver ion was shielded by increasing benzene concentrations. The chemical shift plots for toluene and benzene in propylene carbonate are similar, suggesting that the difference in complexing ability of the two compounds is not large.

VII. Relaxation Studies

Because of the long time required simply to acquire a single silver NMR spectrum, little work has been done to investigate the relaxation properties of the silver nucleus. Kronenbitter *et al.* (1980) have used a rapid pulse technique to measure both the spin–lattice relaxation time T_1 and the spin–spin relaxation time T_2 for several silver salts in different solvents. The values of T_2 are uniformly less than half the values of T_1.

Only small NOEs of the intensity of the silver NMR signal have been found on irradiation of the proton transitions for silver nitrate in water (Brevard *et al.*, 1981). Thus dipolar interactions of the silver nucleus with protons appear to be an insignificant source of nuclear relaxation. Chemical shift anisotropy (CSA) is a likely mechanism, and both CSA and chemical exchange may contribute to T_2. Measurement of T_1 at different magnetic field strengths would help to clarify the role of CSA.

Kronenbitter *et al.* (1980) postulate that chemical exchange in many silver systems causes T_2 to be much shorter than T_1. Occasionally the line widths of silver salt systems are excessively broad (Henrichs and Ackerman, unpublished results), suggesting that chemical exchange occurs at a rate comparable to the chemical shift range. Unfortunately, it is difficult to find solvents in which such phenomena can be observed as a function of a wide range of temperatures at the high concentrations required to obtain reasonable silver NMR intensity.

VIII. Conclusion

Silver NMR has the power to reveal many of the details of ion solvation, liquid structure, and properties of metal complexes. It is hoped that with the continued improvement of NMR spectrometers, the requirement for high concentrations will be eased and a wider range of experiments will become possible.

References

Ackerman, J. J. H. (1977). Ph.D. Thesis, Colorado State University, Fort Collins, Colorado.
Ahrland, S., Chatt, J., Davies, N. R., and Williams, A. A. (1958). *J. Chem. Soc.* 276–288.
Baddiel, C. B., Tait, M. J., and Janz, G. J. (1965). *J. Phys. Chem.* **69**, 3634–3638.
Balasubrahmanyam, K., and Janz, G. J. (1970). *J. Am. Chem. Soc.* **92**, 4189–4193.
Becker, E. D., Ferretti, J. A., and Farrar, T. C. (1969). *J. Am. Chem. Soc.* **91**, 7784–7785.

Brevard, C., van Stein, G. C., and van Koten, G. (1981). *J. Am. Chem. Soc.* **103,** 6746–6748.

Brun, E., Oeser, J., Staub, H. H., and Telschow, C. G. (1954). *Phys. Rev.* **93,** 172–173.

Burges, C.-W., Koschmieder, R., Sahm, W., and Schwenk, A. (1973). *Z. Naturforsch.* **28a,** 1753–1758.

Chang, T.-C. G., and Irish, D. E. (1974). *J. Solution Chem.* **3,** 161–174.

Clausen, A., El-Harakany, A. A., and Schneider, H. (1973). *Ber. Bunsenges.* **77,** 994–997.

Freeman, R., and Hill, H. D. W. (1971). *J. Magn. Reson.* **4,** 366–383.

Henrichs, P. M., Ackerman, J. J. H., and Maciel, G. (1977). *J. Am. Chem. Soc.* **99,** 2544–2548.

Henrichs, P. M., Sheard, S., Ackerman, J. J. H., and Maciel, G. (1979). *J. Am. Chem. Soc.* **101,** 3222–3228.

Janz, G. J. (1971). *J. Electroanal. Chem.* **29,** 107–126.

Janz, G. J., Marcinkowsky, A. E., and Ahmad, I. (1965). *J. Electrochem. Soc.* **112,** 104–107.

Jucker, K., Sahm, W., and Schwenk, A. (1976). *Z. Naturforsch.* **31a,** 1532–1538.

Koepp, H.-M., Wendt, H., and Strehlow, H. (1960. *Ber. Bunsenges.* **64,** 483–491.

Kronenbitter, J. (1977). Ph.D. Dissertation, Eberhard-Karls University at Tübingen, Germany.

Kronenbitter, J., Schweizer, U., and Schwenk, A. (1980). *Z. Naturforsch.* **35a,** 319–328.

Martin, M. L., Martin, G. J., and Delpuech, J.-J. (1980). "Practical NMR Spectroscopy", pp. 112–114. Heyden, Philadelphia.

Oliver, B. G., and Janz, G. J. (1970). *J. Phys. Chem.* **74,** 3819–3822.

Rahimi, A. K., and Popov, A. I. (1976). *Inorg. Nucl. Chem. Lett.* **12,** 703–707.

Rahimi, A. K., and Popov, A. I. (1979). *J. Magn. Reson.* **36,** 351–358.

Schwarzenbach, G. (1953). *Helv. Chim. Acta* **36,** 23–36.

Schwenk, A. (1971). *J. Magn. Reson.* **5,** 376–389.

Shul'man, V. M., and Savel'eva, Z. A. (1970). *Izv. Sib. Otd. Akad. Nauk SSSR Ser. Khim. Nauk* 124–130 (English Translation pp. 563–568).

Sogo, P. B., and Jeffries, C. D. (1954). *Phys. Rev.* **93,** 174–175.

Strehlow, H., and Koepp, H.-M. (1958). *Ber. Bunsenges.* **62,** 373–378.

Strehlow, H., and Schneider, H. (1971). *Pure Appl. Chem.* **25,** 327–344.

Udupa, M. R., and Krebs, B. (1973). *Inorg. Chim. Acta* **7,** 271–276.

Vizzini, E. A., Taylor, I. F., and Amma, E. L. (1968). *Inorg. Chem.* **7,** 1351–1357.

Yeager, H. L., and Kratochvil, B. (1969). *J. Phys. Chem.* **73,** 1963–1968.

13 Cadmium-113 NMR

Ian M. Armitage and Yvan Boulanger*

Department of Molecular Biophysics and Biochemistry
Yale University
New Haven, Connecticut

I. Cadmium-113 NMR

A. Suitability of ^{113}Cd NMR as a Structural Probe in the Study of Biological Metal Coordination Sites

A detailed understanding at the molecular level of the structural and/or functional role of the metal ion(s) in metalloproteins is essential in elucidating the mechanism of the numerous biological processes that are either dependent on or deleteriously affected by various metals. The preponderance of biological systems whose function depends on the binding of diamagnetic metal ions such as Zn^{2+}, Mg^{2+}, and Ca^{2+}, which are not directly detectable by the more conventional physical technqiues such as EPR and optical spectroscopy, have eluded the intensity of investigation afforded macromolecular complexes containing chromophoric and/or paramagnetic metal ions. These factors provided the impetus behind the efforts of our research group (Armitage *et al.*, 1976) to develop ^{113}Cd NMR as a method for studying metalloprotein complexes in which the native diamagnetic metal had been replaced by an isotopically enriched spin-$\frac{1}{2}$ ^{113}Cd nucleus. The foundation behind this development of the bio-

---

* Present address: Institut de Génie Biomédical, Université de Montréal, C.P. 6028, Succ. A, Montréal, Quebec, Canada H3C 3T8

logical applications of ^{113}Cd NMR as we know it today was established in three early papers describing the general applications of ^{113}Cd NMR (Cardin *et al.*, 1975; Kostelnik and Bothner-By, 1974; Maciel and Borzo, 1973). These papers identified many aspects pertinent to the application of this technique to biological systems, including sensitivity of the chemical shifts to subtle changes in coordination, relaxation properties, and ligand exchange dynamics. Numerous investigators have employed the ^{113}Cd-NMR technique in studying the properties of a variety of proteins and biological events in which the native Zn^{2+}, Ca^{2+}, Mg^{2+}, Mn^{2+}, Cu^{2+}, or Cd^{2+} ions were replaced by ^{113}Cd^{2+} (for a review see Armitage and Otvos, 1982). Some of the chemical and magnetic properties of the ^{113}Cd^{2+} ion to which this technique owes its success are summarized in the following list:

(a) The coordination of Cd^{2+} (ligand preferences and coordination geometry) is quite similar to that of Zn^{2+} and Ca^{2+}, and Cd^{2+} and Ca^{2+} have similar ionic radii of 0.97 and 0.99 Å, respectively.

(b) Cd^{2+} can be substituted for the native Zn^{2+} ion in all known Zn^{2+} metalloenzymes. Catalytic efficiency and specificity are frequently altered but usually not abolished. For many Cd^{2+}-substituted Zn^{2+} metalloenzymes, the pK for activity is found shifted 1–2 pH units toward the alkaline.

(c) ^{113}Cd^{2+} is diamagnetic and has a nuclear spin $I = \frac{1}{2}$ and a relative natural abundance sensitivity 7.6 times that of ^{13}C. An additional eightfold enhancement in sensitivity is possible through the use of commercially available ^{113}Cd isotopically enriched from 12 to 96 at%. Under the latter conditions, ^{113}Cd spectra of metalloproteins at millimolar concentrations can be obtained in a few hours of data accumulation.

(d) ^{113}Cd chemical shifts are extremely sensitive to the nature, number, and geometric arrangement of the ligands within the coordination sphere. This sensitivity is reflected in the chemical shift range of over 850 ppm observed for the resonances from ^{113}Cd^{2+}-substituted metalloproteins. This property is useful not only in helping to identify the ligands at a particular metal-binding site, but virtually guarantees the resoltuion of individual ^{113}Cd resonances in systems containing multiple metal ions.

We shall discuss here the basic principles and experimental methods that have played an important role in the acquisition and interpretation of biological ^{113}Cd-NMR applications. Following this, the results from one biological application of ^{113}Cd NMR in this laboratory to study the metal-binding protein metallothionein will be summarized.

B. Basic Principles

Nuclear magnetic resonance studies of the less common nuclei are potentially available to any research group having access to a modern-day multinuclear NMR spectrometer. The word "potentially" should be emphasized here because there are pertinent differences in the chemical and magnetic properties of these less common nuclei that must be understood and properly accounted for before these studies can lead to successful applications. Some of these special considerations and the manner in which they relate specifically to ^{113}Cd-NMR studies on biological systems are considered in the following discussion.

1. Sensitivity

Both naturally occurring spin-$\frac{1}{2}$ isotopes of cadmium, ^{113}Cd and ^{111}Cd, which are present at 12.3 and 12.7% natural abundance, respectively, have essentially identical NMR properties. Thus, although the following discussion deals almost exclusively with ^{113}Cd, it applies to ^{111}Cd as well (Kidd and Goodfellow, 1979). The absolute sensitivity of ^{113}Cd is 7.6 times that of ^{13}C, however, this is still insufficient to make it practical to perform typical biological Cd-NMR experiments at natural abundance levels. Instead, 96 at% isotopically enriched ^{113}Cd is employed, affording an eightfold enhancement in sensitivity. Under these conditions, typical ^{113}Cd (or ^{111}Cd) spectra can be obtained from approximately 2 ml of a 1 mM sample of Cd-substituted metalloprotein in 5–10 h of data accumulation when the T_1 values are ≤ 5 s and the line widths are ≤ 50 Hz.

2. Chemical Shifts

The greater chemical shift dispersion of ^{113}Cd in comparison to other common high-resolution nuclei (e.g., ^{1}H, ^{19}F, ^{31}P, ^{13}C, and ^{15}N) reflects the sensitivity of the shielding of the ^{113}Cd nucleus to the local environment. In the studies performed to date, the observed chemical shift dispersion covers a range of approximately 850 ppm, within which one can expect to find resonances from ^{113}Cd^{2+} coordinated to the ligands found in biological systems. This tremendous sensitivity of the ^{113}Cd chemical shift to very subtle differences in the coordination environment is the major attribute of the technique, although in some unfavorable cases this very same property can lead to difficulties in resonance detection, owing to chemical exchange broadening (Section I,B,5). An obvious benefit is the excellent spectral resolution obtained even when $^{113}Cd^{2+}$ is bound to multiple coordination sites of very similar structure (as, for example, in metallothionein; see Section II,A). The capability of monitoring separate resonances from each metal environment simultaneously makes the properties of all binding sites in a system accessible to direct investigation. Another important advantage stems from the sensitivity of the ^{113}Cd chemical shift to ligand exchange reactions occurring at a single coordination site, such as exchange of coordinated solvent with a protein ligand, enzyme substrate, or product molecule. Such processes play important roles in the catalytic mechanisms of many metalloenzymes, although in most cases the structural details and dynamics of the reactions are poorly understood. Characterization of these events has been and will continue to be one of the major goals of ^{113}Cd-NMR studies.

In even the earliest ^{113}Cd-NMR studies on small Cd^{2+} complexes in aqueous solutions, it was noted that there was a consistent correlation between the chemical shift and the identity of the ligands to the metal ion (Cardin et al., 1975; Haberkorn et al., 1976; Kostelnik and Bothner-By, 1974; Maciel and Borzo, 1973). Oxygen ligands were found to provide the greatest shielding, nitrogen and halide ligands were relatively deshielding, and sulfur ligation was by far the most deshielding. Based on these initial qualitative observations, it was hoped that once the chemical shifts of a sufficient number of Cd^{2+} complexes were measured there would emerge a reliable structure-versus-shift correlation that would allow the interpretation of ^{113}Cd chemical shifts for proteins in terms of the number, identity, and geometric arrangement of the ligands at the binding site. Unfortunately, two obstacles have so far restricted the establishment of a detailed structure-versus-shift correlation. The first problem stems from the facile chemical exchange frequently encountered with model cadmium complexes in aqueous solutions. The consequence of this is that the solution chemical shift of a cadmium compound with a known crystal structure is rarely indicative of that complex alone but instead is a weighted average of the chemical shifts of multiple species in rapid equilibrium (Ackerman et al., 1979; Kostelnik and Bothner-By, 1974). Chemical shifts for individual complexes must then be calculated from the observed shift and the known equilibrium constants. The accuracy of this method is limited by the accuracy with which the known equilibrium constants describe the system under investigation and by the possible presence of additional solution complexes unaccounted for by the known equilibrium constants. This difficulty may be overcome either by bringing the labile inorganic system into the slow-exchange regime by the use of supercooled aqueous solutions (Ackerman and Ackerman, 1980; Douzou et al., 1978; Jakobsen and Ellis, 1981) or by recording the chemical shifts

TABLE I

^{113}Cd Chemical Shifts from Dynamically Stable Cadmium Complexes in Solution or in the Solid State

Compound[a]	Chemical shift δ (ppm)[b]	Attached atoms and approximate first coordination sphere geometry	Reference
Cd(O$_2$CCHCHCO$_2$) · 2H$_2$O	12, −7	6 O	Mennitt et al. (1981)
[Cd(Ph$_2$P(O)(CH$_2$)P(O)Ph$_2$)$_3$]$^{2+}$	−7	6 O	Dean (1981)
3CdSO$_4$ · 8H$_2$O	−58, −45	6 O	Mennitt et al. (1981); Murphy and Gerstein (1981)
Cd(OH)$_2$	158	6 O	Mennitt et al. (1981)
Na$_2$Cd(EDTA)	117, 107	4 O, 2 N	Mennitt et al. (1981)
Cd-cryptate[c]	46	4 N, 2 O	Keller et al. (1983)
Im$_6$Cd(NO$_3$)$_2$	238	6 N	Mennitt et al. (1981)
Cd(en)$_3$Cl$_2$ · H$_2$O	380	6 N	Mennitt et al. (1981)
[Cd(OP(C$_6$H$_{11}$)$_3$)$_4$]$^{2+}$	94	4 O	Dean (1981)
(Me$_4$N)$_2$CdI$_4$	92	4 I	Mennitt et al. (1981)
(Me$_4$N)$_2$CdBr$_4$	391	4 Br	Mennitt et al. (1981)
(Et$_4$N)$_2$CdCl$_4$	483	4 Cl	Mennitt et al. (1981)
Cd(C$_2$H$_5$S)$_4^{2-}$	648	4 S	Carson et al. (1981)
[Cd$_{10}$(SCH$_2$CH$_2$OH)$_{16}$](ClO$_4$)$_4$	652	4 S	Murphy et al. (1981)
Cd(SCH$_2$CH$_2$S)$_2^{2-}$	829	4 S	Carson et al. (1981)

[a] Im, Imidazole; en, ethylenediamine; Me, methyl; Et, ethyl.
[b] Relative to 0.1 M Cd(ClO$_4$)$_2$.
[c] The cryptate ligand of this complex is

of solid cadmium complexes using the cross-polarization–magic angle spinning (CP-MAS) technique (Ackerman et al., 1979; Cheung et al., 1980; Mennitt et al., 1981; Murphy and Gerstein, 1981; Murphy et al., 1981; Rodesiler and Amma, 1982, and Turner et al., 1982). The latter technique is perhaps more generally applicable, and it provides an opportunity to obtain benchmark ^{113}Cd chemical shifts from structurally characterized solid species. Several structurally characterized complexes have now been studied by this method, and selected examples are given in Table I. The second limitation is the lack of benchmark chemical shift data from a representative number of suitable model compounds that mimic the distorted coordination geometries and outer-sphere interactions found at the metal-binding sites of most biological macromolecules. Whereas the data presented in Table I show the expected empirical correlation of the chemical shift with the identity of the coordinating ligands, it is apparent from these data that differences in coordination number and geometry can have a significant influence on the chemical shift. Particularly noteworthy in this regard is the range of about 200 ppm for complexes with six oxygen ligands in an octahedral arrangement. Although a quantitative understanding of these secondary contributions to the chemical shift is lacking, some general trends appear to be emerging. For example, tetrahedral coordination appears to be deshielded relative to octahedral coordination by approximately 50–100 ppm (Dean, 1981). A similar degree of deshielding between octahedral and heptacoordinated oxygen ligands has also been reported (Rodesiler and Amma, 1982). As an example of an outer-sphere effect, Carson et al. (1981) determined the effect of substitution at the α carbon in a series of related thiolate ligands to be on the order of 15 to 20 ppm per alkyl substituent and to be even smaller for substitution at the β carbon. Although these trends are useful, it is clear that additional data will be required to document the effects of coordination number, distortions in coordination geometries, and outer-sphere effects in greater detail.

Thus, although we have not yet achieved the desired level of quantitation, the existing qualitative structure-versus-shift correlation is sufficiently well established to permit its use for tentative assignment purposes in biological systems. In every case in which ^{113}Cd chemical shift determinations have been made in proteins whose binding sites had previously been characterized by X-ray diffraction, a good correspondence has been found between the observed shift and that expected on the basis of the protein ligands to the metal. This is demonstrated in Fig. 1, which shows a compilation of the solution chemical shifts of all $^{113}Cd^{2+}$-substituted metalloproteins reported to date. Resonances from cadmium bound to sites consisting exclusively of oxygen ligands are invariably found in the region of 0 to -125 ppm, those from sites containing a combination of oxygen and nitrogen ligands appear between 0 and 300 ppm, whereas resonances from sites containing two or more thiolate sulfur ligands are found in the most downfield region between 450 and 750 ppm.

The most shielded resonances, at about -100 ppm, are consistently found to arise from $^{113}Cd^{2+}$ substituted at the Ca^{2+} sites of calcium-binding proteins. The crystal structures of two of these proteins, parvalbumin and concanavalin A, indicate very similar Ca^{2+} coordination spheres consisting of six oxygen ligands in an octahedral-like arrangement. Figure 2 shows the ^{113}Cd spectra of parvalbumin, the first solid state ^{113}Cd-NMR

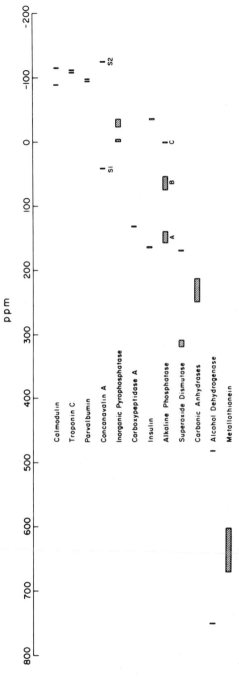

Fig. 1 Observed ^{113}Cd chemical shifts for a series of ^{113}Cd-substituted metalloproteins. All chemical shifts are reported relative to external 0.1 M Cd(ClO$_4$)$_2$.

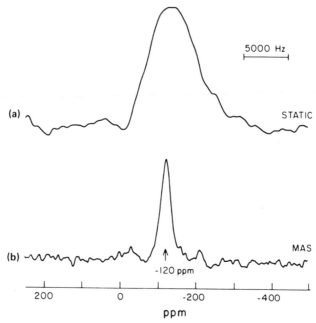

Fig. 2 High-power ^{1}H-decoupled CP-MAS ^{113}Cd-NMR spectra (44.4 MHz) of lyophilized carp ^{113}Cd$_2$-parvalbumin C3 at room temperature. (a) Static sample; (b) sample spinning at 4.2 KHz.

spectra for a ^{113}Cd^{2+}-substituted metalloprotein (Armitage *et al.,* unpublished data). The more than 20-ppm chemical shift difference between the solution (−95 ppm, see Drakenberg *et al.,* 1978) and the solid state CP-MAS spectra (−120 ppm, see Fig. 2b) is suggestive of exchange phenomena in solution or structural differences at the metal-binding site under these different conditions. The solid state spectrum obtained without MAS gives the powder pattern shown in Fig. 2a. Line shape analysis under the latter conditions can potentially provide values for the principal elements of the chemical shift tensor. These values can be directly related to the symmetry of the ligands in the metal coordination sphere, information that is lost in the MAS or solution spectra.

As oxygen ligands are replaced by nitrogen donors (usually imidazole), the ^{113}Cd chemical shifts move to a lower field by an amount that, at least qualitatively, reflects the extent of nitrogen ligation. This trend is illustrated by considering the Mn^{2+}-binding site of concanavalin A and the Zn^{2+} site of carboxypeptidase and carbonic anhydrase (Fig. 1). These sites contain one, two, and three nitrogen ligands, respectively, and give rise to ^{113}Cd chemical shifts of 46, 132, and 210–270 ppm. It should again be emphasized that differences in coordination number and geometry may exert a sizable influence on the chemical shift, and caution must therefore be exercised when using this relationship for assignment purposes. Also, as mentioned earlier, such a correlation may be complicated by limitations imposed by chemical exchange. Two cases of particular note are carbonic anhydrase and alkaline phosphatase, where the ^{113}Cd resonances can be shifted by up to 100 ppm by altering the pH or ionic composition of the medium (Coleman *et al.*, 1979; Jonsson *et al.*, 1980; Schoot Uiterkamp *et al.*, 1980). These shift changes result from either ionization of a coordinated water molecule or displacement of solvent ligands by anions in the medium and can be used as evidence that the protein-bound metal has one or more "open" coordination sites occupied by the solvent.

Cysteine thiolate sulfur ligation in metalloproteins results in large downfield chemical shifts of a magnitude consistent with that expected on the basis of studies on small inorganic cadmium–thiolate complexes (Carson *et al.*, 1981; Haberkorn *et al.*, 1976; Murphy *et al.*, 1981). For each oxygen or nitrogen ligand replaced by thiolate sulfur, deshielding by 100–200 ppm can be anticipated. A case in point is horse liver alcohol dehydrogenase, which contains two classes of Zn^{2+}-binding sites. The metal ligands at the catalytic site are two cysteinyl sulfurs, one histidyl nitrogen, and a solvent oxygen, whereas those at the structural site are four tetrahedrally disposed cysteinyl thiolates. The ^{113}Cd-substituted enzyme exhibits catalytic and structural site resonances at 483 and 751 ppm, respectively. The 751-ppm shift is considered to be close to the maximum amount of deshielding one might anticipate for ^{113}Cd^{2+} bound to a biological matrix. The cadmium ions in metallothionein, for example, are also tetracoordinated to thiolate sulfur, yet the ^{113}Cd resonances appear 80–150 ppm to higher field. This difference results from the diminished deshielding provided by thiolate ligands that bridge adjacent metal ions, as opposed to those that are nonbridging (see Section II,A; Carson *et al.*, 1981; Murphy *et al.*, 1971).

3. Coupling Constants

The observation of scalar coupling in the ^{113}Cd-NMR spectrum of biological macromolecules can provide unequivocal information about the nature of the ligands and their mode of binding. Table II shows the range of coupling constant values reported with nuclei of biological interest. Direct binding to the metal can be unambiguously concluded when one-

TABLE II

Scalar Coupling Constants of ^{113}Cd with Nuclei of Biological Interest

Coupling	^1J (Hz)	^2J (Hz)	^3J (Hz)	References
^{113}Cd—^{1}H	—	50–53	54–70	Cardin et al. (1975); Carson and Dean (1982); Turner and White (1977)
^{113}Cd—^{13}C	498–1060	3–19	5–45	Cardin et al. (1975); Jensen et al. (1981); Otvos and Armitage (1980a); Sudmeier and Bell (1977)
^{113}Cd—^{15}N	140–210	—	—	Evelhoch et al. (1981); Jakobsen and Ellis (1981)
^{113}Cd—^{31}P	1200–1710	2–30	—	Dean (1981); Mann (1971); Otvos et al. (1979); Yamasaki and Fluck (1973)
^{113}Cd—^{113}Cd	—	29–48	—	Otvos and Armitage (1980c)

bond coupling constants are observed, because their values are considerably greater than those of the corresponding multiple-bond coupling constants (Table II). The values of the two- and three-bond coupling constants, although of similar magnitude, are distinguishable in specific cases. For example, the larger value of the three-bond ^{13}C-^{113}Cd coupling constant has allowed the distinction between N-π and N-τ binding of ^{113}Cd^{2+} to ^{13}C-labeled histidine in alkaline phosphatase (Otvos and Armitage, 1980a). Three-bond coupling constants can also be potentially useful in conformational analysis, but sufficient data are presently not available to determine whether a Karplus-type relationship exists.

The usefulness of coupling constants in the study of metalloproteins has been demonstrated in at least three cases: carbonic anhydrase, alkaline phosphatase, and metallothionein. On addition of ^{13}CN$^-$ to ^{113}Cd–human carbonic anhydrase, a splitting of 1060 Hz was observed in the ^{113}Cd-NMR spectrum, in strong support of Cd—C ligation (Sudmeier and Bell, 1977). In the case of alkaline phosphatase, the ^{113}Cd—^{13}C three-bond splittings of 5 to 12 Hz were used to assign the binding histidines in the ^{13}C-NMR spectrum of the labeled protein (Otvos and Armitage, 1980a). Furthermore, when inorganic phosphate was added to this enzyme, a two-bond ^{113}Cd—^{31}P coupling of 30 Hz was measured in the ^{31}P-NMR spectrum, providing the first unequivocal evidence of a direct inner-sphere metal–phosphate interaction (Otvos et al., 1979). For metallothionein, the observation of ^{113}Cd—^{113}Cd two-bond couplings has led to elucidation of the two-cluster structure (Section II,A). Finally, it should be noted that, in addition to the structural information, the observation of scalar coupling in alkaline phosphatase and metallothionein supports the slow exchange of the metal between the aqueous solvent and the protein complex ($k_{off} \leq 10^2$ s^{-1}).

4. Relaxation Properties

The spin relaxation properties of ^{113}Cd^{2+} in a given system are of interest for two reasons. The first is a purely practical one: A working knowledge of the spin–lattice relaxation time T_1 and the nuclear Overhauser enhancement (NOE) is essential in determining the most efficient conditions for data collection. The second reason is more fundamental: By analyzing the relaxation data in terms of appropriate theoretical models one can, in principle, derive valuable information concerning motional dynamics at the metal-binding site. Unfortunately, owing to the lack of a detailed understanding regarding the several contributions to ^{113}Cd relaxation in macromolecular systems, this potential source of information remains to be exploited. In this section, we have attempted to identify those factors that are of fundamental importance in determining the spin–lattice relaxation of ^{113}Cd.

In the particular case of relaxation analysis in macromolecules, models that include internal degrees of freedom, in addition to overall molecular rotation, become increasingly important in the interpretation of spin–lattice relaxation processes. This is a consequence of the fact that common observation fields make resonance frequencies of nuclei too high for efficient exchange of energy with slow overall molecular rotations. Most models with internal modes are directed at motions that can be classified as bond isomerizations and in which nuclei relax via dipolar interactions with rigidly bound magnetic nuclei in the same rotating group, for example, ^{13}C and ^{1}H relaxation in a methyl group with rotation about its C—X bond (Woessner, 1962). Interpretation of ^{113}Cd spin–lattice relaxation behavior in a macromolecular complex presents a special problem in that there are few rigidly bound nuclei of high magnetic moment within effective relaxation distances. For ^{113}Cd at an enzyme active site, for example, protons of typical protein ligands are not very close to the metal ion (≥ 3.5 Å), whereas exchangeable protons of the hydration sphere, if present, are somewhat closer to the metal (~ 2.9 Å) and may dominate its relaxation. Present indications are that the isotropic rigid rotor model, where the motion is described completely by the rotational correlation time of the molecule τ_R, does not provide an accurate description of dipolar relaxation in macromolecular cadmium complexes. Calculated T_1 values using this model are high (≥ 10 s), whereas experimentally determined T_1 values fall in the range 0.5–5 s (Bailey *et al.*, 1980; Otvos and Armitage, 1980b). Field-dependent NOE and T_1 measurements for $^{113}\text{Cd}^{2+}$ bound to alkaline phosphatase (Otvos and Armitage, 1980b) and metallothionein (unpublished data) in H_2O and D_2O show that a dipolar modulation process at a more effective relaxation frequency than overall protein reorientation is necessary to account for the data. In these systems then, it is absolutely necessary to include the modulation of dipolar interactions with exchangeable protons by internal motions; however, it is questionable whether or not these motions should be treated as internal rotations, as is customarily done in the analysis of ^{13}C relaxation data (Doddrell *et al.*, 1972), in view of the rigidity one associates with an enzyme active site geometry and hydrogen bonding. An alternate means of modulating dipolar interactions at enzyme active sites, which was first described by Armitage *et al.* (1979) and Armitage and Prestegard (unpublished data), involves the exchange of ionizable protons on coordinated water molecules. Rates on the order of 10^8 to 10^9 s^{-1} are in fact reasonable (Coleman *et al.*, 1979). Estimates of the mean residence time of water molecules at the active sites of Mn^{2+} enzymes are reported to range from 2×10^{-8} to 5×10^{-9} s (Dwek, 1973; Hsi and Bryant, 1977). This model treats the modulation of Cd–H interaction as originating in a simple exchange of one proton for another of arbitrary spin at approximately the same site. Such a model removes some of the unnecessarily restrictive constraints the internal rotation model imposes on weakly coordinated water molecules. The protons need not move very far in this exchange. The transfer of a proton from a site where it is covalently bound to a coordinated water oxygen to a site where it is hydrogen-bonded to the same oxygen would provide a jump of sufficient magnitude to be approximated as a jump to an infinite distance. The expressions for spin–lattice relaxation and the NOE can easily be derived on the basis of such a model. The spectral density function takes the following form:

$$J(\omega) = \tau_{\text{eff}}/(1 + \omega^2 \tau_{\text{eff}}^2) \tag{1}$$

where

$$\tau_{\text{eff}}^{-1} = 1/\tau_R + 1/\tau_e \tag{2}$$

$$R_1 = \frac{\gamma_I^2 \gamma_S^2 \hbar^2 S(S+1)}{r^6} \frac{2}{15} J(\omega_I - \omega_S) + \frac{2}{5} J(\omega_I) + \frac{4}{5} J(\omega_I + \omega_S) \tag{3}$$

$$\text{NOE} = 1 + \frac{\gamma_S}{\gamma_I} \frac{\frac{2}{15}J(\omega_I - \omega_S) - \frac{4}{5}J(\omega_I + \omega_S)}{\frac{2}{15}J(\omega_I - \omega_S) + \frac{2}{5}J(\omega_I) + \frac{4}{5}J(\omega_I + \omega_S)} \tag{4}$$

where τ_R is the overall isotropic reorientation time, τ_e the time constant for the exchange of proton spins, γ_I and γ_S the gyromagnetic ratios for the

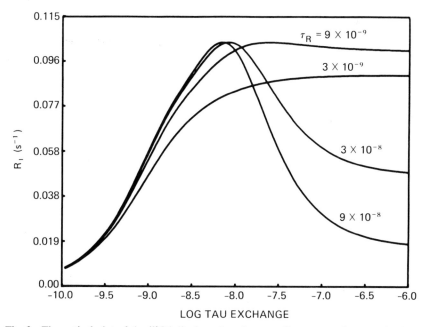

Fig. 3 Theoretical plot of the ^{113}Cd dipolar relaxation rate R_1 versus τ_e (in seconds) for a single exchanging proton 2.6 Å away and a range of τ_R values (in seconds). Results are for 2.114 T ($\omega_{Cd} = 12.54 \times 10^7$ rad s^{-1}).

observed and interacting nuclei, respectively, ω the resonance frequency in radians per second, and r the internuclear distance.

In Figs. 3 and 4, we show the values for R_1 and the NOE for a range of exchange and rotational correlation times. The values were calculated at a field strength of 2.11 T, where ^{1}H and ^{113}Cd resonate at 90.00 and 19.96 MHz, respectively. Relaxation rate plots were computed for one exchanging proton assuming dipolar relaxation via this proton ($r = 2.6$ Å) to be the only effective mechanism. When the overall rotation is sufficiently fast, $\tau_R = 3 \times 10^{-9}$ s, such that the extreme narrowing limit applies, the onset of chemical exchange decreases R_1 monotonically (Fig. 3). However, as τ_R increases ($\omega_I \tau_R \geq 1$), the onset of chemical exchange first increases R_1 until a maximum is reached (when $\tau_e \approx \omega_I^{-1}$) and then decreases asymptotically as $\tau_e \to 0$. The behavior of the NOE is shown in Fig. 4. Note that, owing to the negative magnetogyric ratio of ^{113}Cd, the NOE values ($\eta + 1$) range from -1.25 in the extreme narrowing limit to a limiting value of 0.84 for macromolecular complexes. A NOE value of zero results in the complete absence of signal intensity. When chemical exchange is very slow ($\tau_e \to \infty$), the NOE assumes a value characteristic

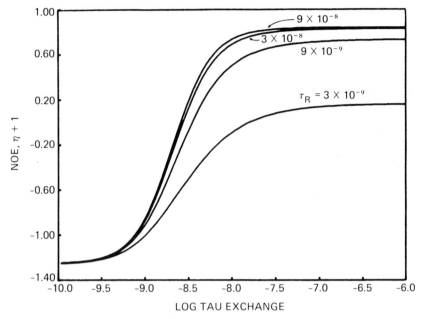

Fig. 4 Plot of the ^{113}Cd NOE versus τ_e for a range of τ_R values. See caption for Fig. 3 for other details.

of a single isotropic correlation time. The onset of chemical exchange (from right to left in Fig. 4) increases the NOE, which eventually reaches an asymptotic value equal to the maximum allowed NOE in the extreme narrowing limit.

These plots can be used to extract from relaxation data estimates of exchange rates and numbers of exchanging protons if we assume the absence of other relaxation mechanisms [e.g., chemical shift anisotropy (CSA), *vide infra*]. This is done in the following manner. From the observed NOE and literature values of τ_R a value for τ_e is read directly from the theoretical NOE plot. From the R_1 curve appropriate for τ_R, a relaxation rate due to one exchanging proton is read at the τ_e value determined from the NOE. The relaxation rate is then scaled so that it agrees with the observed rate, the scaling factor being the number of exchanging protons. The results of such an analysis on the ^{113}Cd relaxation parameters for ^{113}Cd^{2+}–carbonic anhydrase with and without I$^-$ are given in Table III. The exchange time of $10^{-8.3}$ s for carbonic anhydrase is reasonable in view of estimates of rates of water exchange for Mn^{2+} in carbonic anhydrase and other Mn^{2+} enzymes. This rate also supports the rapid proton transfer rates needed to explain the extremely high turnover numbers for carbonic

TABLE III

^{113}Cd Relaxation Parameters and Calculated Value of τ_e and the Number of Protons for ^{113}Cd^{2+}–Carbonic Anhydrase

a	R_1 (s^{-1})	NOE	τ_e (s)	Number of protons	$\tau_{R(s)}$[b]
^{113}Cd^{2+}HCAB	0.63	0.5	$10^{-8.3}$	6.3	3×10^8
^{113}Cd^{2+}HCAB + 4 equiv I$^-$	0.37	0.5	$10^{-8.3}$	3.7	3×10^8

[a] HCAB,
[b] τ_R value from Lanir and Navon, 1972.

anhydrase (Lindskog and Coleman, 1973). For carbonic anhydrase a five-coordinate site for Cd has been postulated and is supported by spectroscopic data (Bauer *et al.*, 1977). Such a site would have two water molecules in coordination, the other three sites being occupied by histidyl nitrogens. We observed six rather than four exchanging protons. At face value this supports an octahedral site, but uncertainties can be large (± 2) because of the difficulty in estimating Cd—H distances in the complex ($r = 2.6 \pm 0.2$ Å). Contributions to relaxation from coordinated nitrogens, which we have neglected, would also cause us to overestimate the number of bound water molecules, although we calculated only a 5% contribution to relaxation. I$^-$ has been shown to complex directly with the active site metal ion in both Zn^{2+}- and Cd^{2+}-substituted carbonic anhydrase. For the Cd^{2+} enzyme, I$^-$ is expected to displace one of the two water ligands from the Cd coordination sphere which remains pentacoordinate. It is therefore significant that the difference in the number of exchanging protons in the presence and absence of I$^-$ is 2.6.

It is clear that recognition of possible exchange contributions to the spin–lattice relaxation phenomenon provides a reasonable physical description of molecular details of motion and structure at the active sites of the enzymes without unnecessary restrictions of a rigid rotation geometry. Jensen *et al.* (1981) demonstrated that solvent water protons made a significant contribution to the dipolar relaxation mechanism in Cd–EDTA complexes. They interpreted this observation with a model that recognizes that the ^{113}Cd T_1 for this complex at high pH is dominated by chemical exchange processes.

Chemical shift anisotropy can also contribute to the ^{113}Cd relaxation mechanism even at relatively low magnetic field strengths (Bailey *et al.*, 1980; Otvos and Armitage, 1980b). Determination of the exact contribution that CSA makes to overall relaxation turns out to be a difficult problem, however, and is the major reason why analysis of ^{113}Cd relaxation parameters in terms of molecular motion as previously discussed has been hampered. Relaxation measurements at more than one magnetic field strength are required to determine a value for the anisotropy of the chemical shift $\Delta\sigma$, however, the accuracy of this value is critically dependent on the accuracy of the value for τ_{eff} which is required in this calculation. Line shape analysis of the powder pattern in solid state measurements can provide accurate values for the CSA $\Delta\sigma$ directly, which would greatly simplify a quantitative description of the relaxation properties of Cd in macromolecular complexes. Available solid state data report values ranging from ~100 to ~550 ppm for various ligands and coordination geometries (Cheung et al., 1980; Ellis et al., 1982; Mennitt et al., 1981; Murphy and Gerstein, 1981; Murphy et al., 1981). Unfortunately, as noted earlier, too few suitable model complexes and only one metalloprotein have been studied with this technique thus far. Nevertheless, as these data become available, theoretical plots such as those generated to account for the dipolar contribution to relaxation (Figs. 3 and 4) can be corrected for the percentage contributed by other relaxation mechanisms, thereby providing a quantitative description of the relaxation parameters in terms of structure and dynamics.

5. Chemical Exchange

The observation of ^{113}Cd-NMR resonances in biological systems can be complicated by the presence of chemical exchange. This problem is compounded for ^{113}Cd because of the wide range of chemical shifts observed and because of the molecular degrees of freedom of the central ^{113}Cd^{2+} atom or of its surrounding ligands. Both the chemical shifts and the line widths of ^{113}Cd resonances can be affected, and the extent of the effect depends on the rate of the exchange processes relative to the chemical shift difference between the exchanging species. In the fast-exchange regime, appreciable variations in the chemical shift can be obtained because the observed chemical shifts represent an average between those of the ^{113}Cd^{2+} ion in two or more states. Under intermediate-exchange conditions, problems can be more serious because of the broadening of the resonances over a fairly large range of exchange rates, which may prevent observation of the resonance.

It is possible to differentiate three major exchange processes that may exist in a macromolecular Cd^{2+} complex. These can be represented by the following equations for a pentacoordinate site with four ligands (L) contributed by the protein (P) and one by the solvent.

$$H_2O \cdot Cd^{2+}PL_4 \rightleftharpoons Cd^{2+} + PL_4 \tag{5}$$

$$H_2O \cdot Cd^{2+}PL_4 + Cl^- \rightleftharpoons Cl \cdot Cd^{2+}PL_4 + H_2O \qquad (6)$$

$$H_2O \cdot Cd^{2+}PL_4 \rightleftharpoons H_2O \cdot Cd^{2+}PL_3 + L \qquad (7)$$

Equation (5) represents the exchange of the Cd^{2+} itself between the protein and the solvent. Such an environment change could result in a difference of 50 to 700 ppm between the free and bound Cd^{2+}, depending on the protein and the buffer conditions. The chemical shift of free Cd^{2+} is strongly dependent on buffer conditions such as pH, ionic strength, ionic composition, and temperature. The large magnitude of the chemical shift differences caused by this exchange mechanism means that only relatively large off-rates ($k_{-1} \sim 10^3–10^5$ s^{-1}) affect the ^{113}Cd resonance line width of the protein-bound ^{113}Cd^{2+}. For most metalloproteins it is therefore unlikely that central metal exchange will be responsible for excessive line widths. Exceptions do exist, however, and an example is the low-affinity sites of Ca^{2+}-binding proteins whose spectra are affected by the fast exchange of the central metal ion (Forsén et al., 1979, 1980). Another example where central metal exchange occurs, but on a much slower time scale, has been reported for concanavalin A, where both free and bound ^{113}Cd^{2+} were observed (Palmer et al., 1980). A second mechanism of exchange is illustrated by Eq. (6) where a solvent ligand exchanges between the protein and the solvent. This is possible whenever the metal ion has an open coordination site that can bind a solvent molecule or an anion. The rates of solvent exhange are known to be quite rapid and well within the range that can create an intermediate-exchange situation. This exchange mechanism has been demonstrated for the various isozymes of carbonic anhydrase where ligand exchange rates vary greatly between isozymes (Jonsson et al., 1980; Schoot Uiterkamp et al., 1980). The ^{113}Cd resonances can also be affected by an exchange of the protein ligands between bound and unbound states [Eq. (7)]. The contribution of this mechanism is difficult to evaluate because of the lack of data on the rates of such processes. Under normal conditions, any of these three exchange mechanisms have the potential to cause chemical shift variations and resonance broadening and are most serious when their exchange rates are on the order of the chemical shift difference between the two exchanging species. A more complicated chemical exchange scheme, such as a three-site exchange, is possible, and a theoretical treatment for this latter case has been reported (Coleman et al., 1979). A three-site equilibrium is required to account for the ^{113}Cd-NMR spectrum observed for concanavalin A containing 2.2 equiv of ^{113}Cd (Palmer et al., 1980).

In the event that an expected resonance remains undetectable within the limits of the conventional sampling conditions, including variations in pH, temperature, and ionic conditions, alternative methods such as the use of supercooled microemulsions or solid state NMR should be considered. To our knowledge, the Cd^{2+}-glycine complex is the only reported complex undergoing fast exchange at room temperature that has been studied by ^{113}Cd NMR in supercooled microemulsions where slow-exchange conditions could be achieved (Ackerman and Ackerman, 1980; Jakobsen and Ellis, 1981). The extension of this technique to metalloproteins may, however, be complicated by the deleterious effects the experimental conditions (emulsifying agent, hydrophobic solvent, sonication) are likely to have on the stability of the protein.

The observation of solid state ^{113}Cd-NMR spectra with CP-MAS has been used to determine the chemical shifts of several model cadmium complexes, as discussed in Section I,B,2. An excellent example of this technique was provided by a ^{113}Cd-NMR study on the $[Cd_{10}(SCH_2 CH_2OH)_{16}](ClO_4)_4$ complex both in solution and in the solid state (Haberkorn et al., 1976; Murphy et al., 1981). The solid state spectrum showed three resonances for the three different Cd^{2+} sites in this complex, whereas the solution spectrum showed only two, as a result of fast exchange between two of the three sites under the latter conditions. The first application of this technique to a metalloprotein, Cd-parvalbumin, was presented in Section I,B,2 (Fig. 2). The line width of the observed resonance was approximately 1200 Hz, and preliminary experiments at low temperatures suggest that this broadening might be caused by dynamic processes with rates in the range of the CSA or ^1H-decoupling frequencies (Garroway et al., 1981). If this is the case, low-temperature CP-MAS experiments may be required to observe a high-resolution solid state ^{113}Cd-NMR spectrum for metalloproteins deleteriously affected by solution chemical exchange.

II. Biological Application of ^{113}Cd NMR

A. Metallothionein

1. Properties

Metallothioneins comprise a class of low-molecular-weight, cysteine-rich, metal-binding proteins, the first of which was isolated from equine renal cortex (Margoshes and Vallee, 1957). Since then, various investigators have characterized similar proteins from the kidney and liver tissue of other mammals, as well as from birds, fish, crustaceans, and even microorganisms (Nordberg and Kojima, 1979). The protein sequesters, through thiolate bonds, a variety of metallic cations such as cadmium, zinc, copper, and mercury (Kägi and Vallee, 1961), and its biosynthesis is closely regulated at the transcriptional level by metals (Vallee, 1979) or glucocorticoid hormones (Etzel *et al.*, 1979). The metal composition depends on the species, the tissue, the stage of development, and the exposure of the organism to metals (Ohi *et al.*, 1981; Squibb *et al.*, 1977; Vallee, 1979). It is these characteristic properties of metallothionein that are responsible for the proposal that this protein plays an important, although as yet undefined, function in the metabolism and/or homeostasis of essential heavy metals (Cherian and Goyer, 1978; Johnson and Foulkes, 1980) and also functions as a detoxifying agent by sequestering toxic metals (Winge *et al.*, 1975).

The distinguishing properties of the protein are its unusual amino acid composition and high metal content. Of the 61 amino acid residues in mammalian metallothioneins, 20 are cysteine residues and each participates in metal ion coordination via mercaptide linkages (Kägi and Vallee, 1961; Sokolowski and Weser, 1975). The other characteristic features of its amino acid composition are the complete absence of aromatic residues and histidine and the presence of a relatively large number of serine and lysine residues. The molar metal content of mammalian metallothioneins appears to be 7 g-atoms mol^{-1} for the Cd- and Zn-proteins but may be higher for Cu- and Hg-metallothionein (Winge *et al.*, 1981). Several reports of somewhat lower and variable metal stoichiometries suggest that some loss of native metal can occur during purification or storage of the protein. Most metallothioneins characterized to date are resolvable by ion exchange chromatography into two major isoprotein forms designated MT-1 and MT-2. These two isoprotein forms are usually present in comparable amounts, however, their ratio has been reported to be affected by metal induction (Suzuki and Yamamura, 1980).

Sequences of the two isoproteins from several sources indicate that the positions of the 20 cysteine residues as well as those of most of the serine and lysine residues are identical, suggesting that specific metal–thiolate complexes are essential for structural and functional viability. The sequence differences between MT-1 and MT-2 occur outside these highly conserved regions and are extensive enough to suggest that the isoproteins are coded by different cistrons. The two forms of the equine liver and kidney protein, for example, differ in 8 amino acid positions (Kojima *et al.*, 1979), whereas those from mouse liver differ in 15 positions (Huang *et al.*, 1977, 1981). Apart from these relatively minor differences in amino acid composition, it is not currently known whether significant structural or functional differences exist between the two isoproteins. Beyond this, however, additional structural characterization of the seven metal complexes and their spatial relationship with one another were unavailable prior to the [113]Cd-NMR studies considered in the following discussion. The importance of the [113]Cd-NMR studies is further amplified by the lack of a detailed X-ray crystal structure as a consequence of many unsuccessful efforts in our laboratory and others to crystallize metallothionein. The structural elucidation power of ^{113}Cd NMR in the case of metallothionein derives mainly from the sensitivity of the ^{113}Cd chemical shift to small differences in the coordination environment, which makes it possible to resolve separate resonances from ^{113}Cd^{2+} bound to each of the seven metal-binding sites. In addition, spin–spin coupling is observed between ^{113}Cd^{2+} ions located at adjacent sites of the polynuclear metal cluster in the protein. As discussed in the next section, a complete analysis of the chemical shift and spin coupling parameters has led to a detailed understanding of the structures of the metal–thiolate clusters in both mammalian and invertebrate metallothioneins.

2. Structure of the Metal Clusters in Metallothionein

A detailed structural elucidation of the individual metal coordination sites was obtained from ^{113}Cd-NMR studies on isotopically labeled ^{113}Cd-metallothionein isolated from rabbit, calf, and human livers (Armitage et al., 1982; Boulanger and Armitage, 1982; Briggs and Armitage, 1982; Otvos and Armitage, 1979, 1980c) and from Scylla serrata hepatopancreas (Otvos et al., 1982). Rabbit metallothionein, isolated from the livers of rabbits injected with 50 μmol ^{113}CdCl$_2$ per kilogram of body weight, was found to contain both ^{113}Cd^{2+} and Zn^{2+} in a molar ratio of ~2.5. The native Zn^{2+} was subsequently exchanged for ^{113}Cd^{2+} in vitro to give the homogeneous ^{113}Cd–MT (Otvos and Armitage, 1980c). Human MT, isolated from human livers obtained at autopsy, contained greater than 90% Zn^{2+} which was completely replaced in vitro by ^{113}Cd^{2+} (Boulanger and Armitage, 1982). The native calf liver MT contained both Cu$^+$ and Zn^{2+} in a Zn^{2+}/Cu$^+$ molar ratio of ~1.3. Selective in vitro replacement of the Zn^{2+} by ^{113}Cd^{2+} resulted in a protein with a ^{113}Cd^{2+}/Cu$^+$ molar ratio of 1.5 (Briggs and Armitage, 1982). The ^{113}Cd-labeled crab metallothionein was obtained from S. serrata hepatopancreas following induction by repeated injections of ^{113}CdCl$_2$ (Otvos et al., 1982). The native protein contained 6 g-atoms of Cd^{2+} per mole of protein. Figure 5 shows the representative proton-decoupled ^{113}Cd-NMR spectrum from these ^{113}Cd^{2+}-labeled metallothioneins for the MT-1 isoprotein. Several multiplets are observed in the chemical shift range 600–670 ppm. Cadmium-113 chemical shifts are very sensitive to the identity, number, and coordination geometry of the metal ligands, and the metallothionein chemical shifts are all consistent with tetrahedral sulfur coordination by comparison with the resonance for the CdS$_4$ site in the polynuclear Cd$_{10}$(SCH$_2$CH$_2$OH)$_{16}^{4+}$ complex where a chemical shift of 652 ppm has been reported (Murphy et al.,

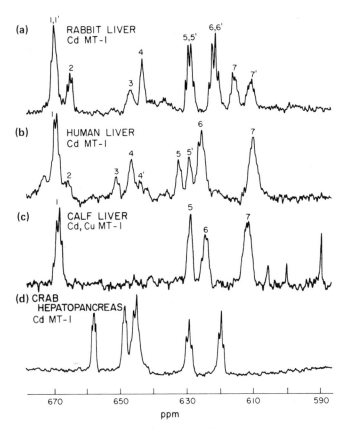

Fig. 5 Proton-decoupled ^{113}Cd-NMR spectra (44.4 MHz) of ^{113}Cd reconstituted metallothioneins in 0.01 M Tris and 0.1 N NaCl at pH 9.0. (a) Rabbit liver Cd–MT-1 ($\sim$8 mM, Cd/Zn ratio 40 : 1); (b) human liver Cd–MT-1 ($\sim$7 mM, Cd/Zn ratio 50 : 1); (c) calf liver Cd, Cu–MT-1 ($\sim$6 mM, Cd/Cu ratio 1.45 : 1); (d) $S.$ $serrata$ Cd–MT-1 ($\sim$8 mM, Cd/Zn 60 : 1).

1981). The multiple splittings of the resonances were shown to result from spin–spin coupling between adjacent ^{113}Cd^{2+} ions linked by bridging thiolate ligands. This result provided the first direct evidence for the existence of polynuclear metal clusters.

3. Metallothioneins from Scylla serrata

The structure of these clusters was determined by homonuclear decoupling experiments which are illustrated in Fig. 6b and c for crab MT-1. In Fig. 6b, the decoupling of resonance 1 causes the collapse of resonances 2 and 6 into doublets, whereas application of the decoupling pulse to the overlapping resonances 3 and 4 in Fig. 6c results in the collapse of reso-

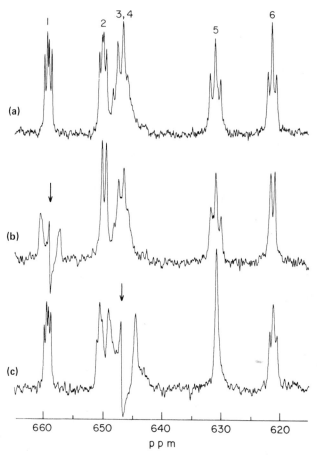

Fig. 6 ^{113}Cd-NMR spectra at 44.4 MHz of native *S. serrata* MT-1 (~8 m*M*, Cd/Zn 60 : 1) in 0.01 *M* Tris and 0.1 *N* NaCl at pH 9.0. (a) Without homonuclear decoupling; (b and c) with homonuclear decoupling pulses applied at the frequencies indicated by the arrows. (From Otvos *et al.*, 1982.)

nance 5 from a triplet to a singlet. From these two experiments alone, it may be concluded that the six metals in crab MT-1 are situated in two three-metal clusters, one containing the ^{113}Cd^{2+} ions that give rise to resonances 1, 2, and 6, and the other containing the metals corresponding to resonances 3, 4, and 5. Both clusters are arranged such that each metal is linked to the two adjacent metals in the cluster, as depicted in Fig. 7 for the type-B three-metal cluster. This structure is supported by all the ^{113}Cd-NMR data and the requirement for the participation of all 18 cysteine thiolate groups in the primary structure in metal ligation.

4. Mammalian Metallothioneins

A detailed analysis of $^{113}Cd^{2+}$-labeled mammalian metallothioneins by the same techniques has shown that the 7 g-atoms of bound Cd^{2+} ions in all mammalian metallothioneins are arranged in two separate clusters, one containing three and the other containing four metal ions (Fig. 7).

In the case of the ^{113}Cd-labeled rabbit liver metallothionein, both iso-proteins, ^{113}Cd–MT-1 and ^{113}Cd–MT-2, gave identical ^{113}Cd-NMR spectra (Fig. 5a). Homonuclear decoupling experiments showed that the cadmiums giving rise to resonances 2, 3, and 4 are connected in a three-metal cluster and those giving rise to resonances 1, 5, 6, and 7 or 1′, 5′, 6′, and 7′ are in a four-metal cluster (Fig. 5a). In the four-metal cluster, homonuclear decoupling of multiplets 6, 6′, and 7 or 7′ affected all three other multiplets assigned to this cluster, whereas homonuclear decoupling of either 1,1′ or 5,5′ affected only 6,6′ and 7,7′. The metal cluster structures consistent with all the ^{113}Cd-NMR data and the participation of all 20 cysteine thiolate groups in metal ligation are shown in Fig. 7. The three-metal cluster forms a cyclohexane-like six-membered ring requiring 9 cysteine thiolate ligands, whereas the four-metal cluster forms a bicyclo[3 : 1 : 3] structure requiring 11 cysteine thiolate ligands. In total, five metal ions are linked to 2 bridging and 2 nonbridging cysteines, and two metal ions are linked to 3 bridging and 1 nonbridging cysteine.

The observation of two sets of ^{113}Cd resonances for the four-metal cluster of rabbit liver metallothionein was originally attributed to the metal deficiency in the three-metal cluster (about 50%) (Otvos and Armitage, 1980c). However, after analyzing the ^{113}Cd-NMR spectra of metallothionein from several other species, it now appears that this results from

Cluster A   Cluster B
4-Metal Cluster   3-Metal Cluster

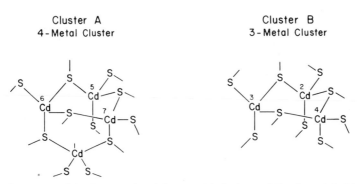

Fig. 7   Structures of the four-metal and three-metal clusters existing in metallothioneins. The Cd numbering corresponds to the respective ^{113}Cd-NMR multiplets in the spectra of mammalian metallothioneins.

a mixture of two or more proteins with limited heterogeneity in the primary structure. For example, human liver ^{113}Cd–MT-2 is also metal-deficient in the three-metal cluster, and yet it does not show duplication of the ^{113}Cd resonances assigned to the four-metal cluster (Boulanger and Armitage, 1982). On the other hand, human liver MT-1 shows heterogeneity both in the ^{113}Cd resonances of the three-metal cluster and at resonance 5 of the four-metal cluster (Fig. 5b). The primary structure of MT-2 is homogeneous, whereas MT-1 is a mixture of at least two proteins with primary structure heterogeneity in six different positions (Kissling and Kägi, 1977).

The absence of ^{113}Cd resonances in the region of the three-metal cluster in the calf liver metallothionein spectrum (Fig. 5c) has been shown to be due to the tight and specific binding of the Cu^+ ions, which were not replaced by Cd^{2+} ions under the conditions of the *in vitro* exchange, to sites in this cluster (Briggs and Armitage, 1982). Site-selective metal ion binding was also observed in studies on the ^{113}Cd-induced native ^{113}Cd^{2+},Zn^{2+}-metallothionein from rabbit liver where about 85% of the bound Zn^{2+} ions were located in the three-metal cluster. Although Cd^{2+} and Zn^{2+} both prefer tetrahedral coordination with sulfur ligands, it should be noted that the Zn—S bond length (2.34 Å) is shorter than the Cd—S bond length (2.52 Å) as a result of the smaller Zn^{2+} ionic radius (0.69 Å) relative to Cd^{2+} (1.03 Å) (Tossell and Vaughan, 1981). We attribute Cd^{2+} preferential binding (over Zn^{2+}) to the four-metal cluster to increased steric hindrance between the residues in the more contracted clusters formed by Zn^{2+}. Cu^+ usually prefers a trigonal coordination with sulfur, and its mode of binding to the three-metal cluster in calf liver metallothionein may differ from that of Cd^{2+} or Zn^{2+}. The cysteine positions allocated to the three-metal cluster may be more favorably disposed for trigonal (or other) coordination than those associated with the four-metal cluster. Whatever the correct explanation, ^{113}Cd-NMR studies have revealed significant differences in the affinity of the two mammalian clusters for different metal ions, which have important functional considerations. For the three-metal cluster, the affinity is found to decrease in the order $Cu^+ > Zn^{2+} > Cd^{2+}$, whereas the exact reverse order of affinity applies to the four-metal cluster.

Titration of a rabbit liver ^{113}Cd–MT-1 sample with EDTA at pH 9.0 has demonstrated the cooperative nature of metal dissociation from the three-metal cluster (Boulanger and Armitage, unpublished data). This is demonstrated by the simultaneous disappearance of all resonances of the three-metal cluster, as shown in Fig. 8. From these data, an equilibrium constant $K \approx 1.2$ could be calculated for EDTA and metallothionein binding, indicating a very strong binding to the three-metal cluster of

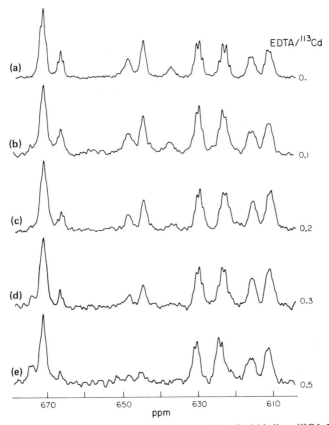

Fig. 8 Proton-decoupled ^{113}Cd-NMR spectra at 44.4 MHz of rabbit liver ^{113}Cd–MT-1 (~5 mM, ^{113}Cd/MT ratio 5.5) in 0.01 M Tris and 0.1 N NaCl at pH 9.0. The sample was titrated with EDTA in amounts corresponding to 0% (a), 10% (b), 20% (c), 30% (d), and 50% (e) of the total metal content. (From Boulanger and Armitage, unpublished data.)

metallothionein. In calf liver ^{113}Cd,Cu–MT-1 (^{113}Cd^{2+}/Cu^{+} = 9.8), a similar calculation indicated even stronger binding with $K \approx 45$. In this case, the more strongly bound Cu^{+} seems to reduce the lability of the ^{113}Cd^{2+} in the three-metal cluster. These studies also confirmed the greater affinity of the four-metal cluster for metal ions, which had previously been concluded from the metal ion deficiencies observed in the three-metal cluster of various preparations (Boulanger and Armitage, 1982; Otvos and Armitage, 1980c). Quantitative data are, however, not presently available for the four-metal cluster.

The distribution of the metal ions in two separate clusters with Cd^{2+} and presumably also Zn^{2+}, assuming tetrahedral coordination to the cys-

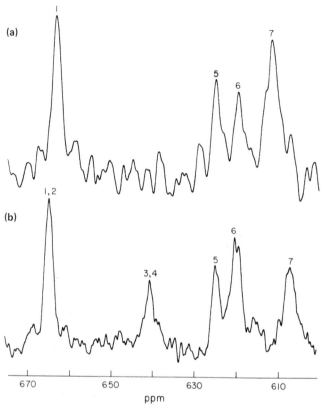

Fig. 9 Proton-decoupled ^{113}Cd-NMR spectra (44.4 MHz). (a) Rat liver Cd-α_1–MT fragment (4.2 mM). (b) human liver ^{113}Cd–Mt-2 (7.1 mM). (From Boulanger *et al.*, 1982.)

teine residues, places restrictions on the structure of metallothionein. For a protein of this small size, such an arrangement would be consistent with a random coil structure. The numerous β bends predicted by the Chou–Fasman method (Boulanger *et al.*, 1982b) do in fact generate a conformation poised to receive the proposed metal clusters. The similarities in the chemical shifts of the ^{113}Cd resonances of the four-metal cluster when the three-metal cluster is not completely filled or when it contains Cu$^+$ (Boulanger and Armitage, 1982) argue for two separate domains for the clusters, corresponding to two different portions of the amino acid chain.

The enzymatic digestion of rat liver metallothionein by subtilisin has yielded a polypeptide fragment containing 11 cysteine residues and four bound Cd^{2+} ions (Winge and Miklossy, 1982). The amino acid composition of this fragment, designated α, corresponds to the residues from positions 30–61 of the rat MT sequence. The ^{113}Cd-NMR spectrum from the Cd–α_I fragment (from MT-1) shows four resonances with the chemical shifts corresponding to the resonances assigned to the four-metal cluster in the intact protein (Fig. 9) (Boulanger *et al.*, 1982). This result provides unequivocal evidence of the existence of two separate domains in the metallothionein structure (Boulanger *et al.*, 1983).

In summary, ^{113}Cd NMR has provided unprecedented insight into the major structural features of metallothionein, especially with regard to the two-domain structure, each region containing a distinct metal cluster, and the differences in the metal binding specificities of the two clusters in mammalian metallothioneins.

References

Ackerman, M. J. B., and Ackerman, J. J. H. (1980). *J. Phys. Chem.* **84**, 3151–3153.

Ackerman, J. J. H., Orr, T. V., Bartuska, V. J., and Maciel, G. E. (1979). *J. Am. Chem. Soc.* **101**, 341–347.

Armitage, I. M., and Otvos, J. D. (1982). "Biological Magnetic Resonance" (L. J. Berliner, and J. Reuben, eds.), Vol. 4, pp. 79–144. Plenum, New York.

Armitage, I. M., and Prestegard, J. H., unpublished data.

Armitage, I. M., Pajer, R. T., Schoot Uiterkamp, A. J. M., Chlebowski, J. F., and Coleman, J. E. (1976). *J. Am. Chem. Soc.* **98**, 5710–5712.

Armitage, I. M., Otvos, J. D., Chlebowski, J. F., and Coleman, J. E. (1979). *Abst. 20th Exp. NMR Conf.* Asilomar, Calif.

Armitage, I. M., Otvos, J. D., Briggs, R. W., and Boulanger, Y. (1982). *Fed. Proc.* **41**, 2974–2980.

Armitage, I. M., Boulanger, V., Drakenberg, T., and Forsén, S., unpublished data.

Bailey, D. B., Ellis, P. D., and Fee, J. A. (1980). *Biochemistry* **19**, 591–596.

Bauer, R., Limkilde, P., and Johansen, J. T. (1977). *Carlsberg Res. Commun.* **42**, 325–339.

Boulanger, Y., and Armitage, I. M., unpublished data.

Boulanger, Y., and Armitage, I. M. (1982). *J. Inorg. Biochem.* **17**, 147–153.

Boulanger, Y., Armitage, I. M., Miklossy, K. A., and Winge, D. R. (1982). *J. Biol. Chem.* **257**, 13717–13719.

Boulanger, Y., Goodman, C. M., Forte, C. P., Fesik, S. W., and Armitage, I. M. (1983). *Proc. Natl. Acad. Sci. USA*, **80**, 1501–1505.

Briggs, R. W., and Armitage, I. M. (1982). *J. Biol. Chem.* **257**, 1259–1262.

Cardin, A. D., Ellis, P. D., Odom, J. D., and Howard, J. W. Jr., (1975). *J. Am. Chem. Soc.* **97**, 1672–1679.

Carson, G. K., and Dean, P. A. W. (1982). *Inorg. Chim. Acta* **66**, 37–39.

Carson, G. K., Dean, P. A. W., and Stillman, M. J. (1981). *Inorg. Chim. Acta* **56**, 59–71.

Cherian, M. G., and Goyer, R. (1978). *Life Sci.* **23**, 1–10.

Cheung, T. T. P., Worthington, L. E., Murphy, P. D., and Gerstein, B. C. (1980). *J. Magn. Reson.* **41**, 158–168.

Coleman, J. E., Armitage, I. M., Chlebowski, J. F., Otvos, J. D., and Schoot Uiterkamp, A. J. M. (1979). "Biological Applications of Magnetic Resonance" (R. G. Shulman, ed.), pp. 345–395. Academic Press, New York.

Dean, P. A. W. (1981). *Can. J. Chem.* **59**, 3221–3225.

Doddrell, D., Glushko, V., and Allerhand, A. (1972). *J. Chem. Phys.* **56**, 3683–3689.

Douzou, P., Balny, G., and Franks, F. (1978). *Biochimie* **60**, 151–158.

Drakenberg, T., Lindman, B., Cavé, A., and Parello, J. (1978). *FEBS Lett.* **92**, 346–350.

Dwek, R. A. (1973). "Nuclear Magnetic Resonance in Biological Systems: Applications to Enzyme Systems" Ch. 10. Clarendon, Oxford.

Ellis, P. D., Inners, R. R., and Jakobsen, H. J. (1982). *J. Phys. Chem.* **86**, 1506–1508.

Etzel, K. R., Shapiro, S. G., and Cousins, R. J. (1979). *Biochem. Biophys. Res. Commun.* **89**, 1120–1126.

Evelhoch, J. L., Bocian, D. F., and Sudmeier, J. L. (1981). *Biochemistry* **20**, 4951–4954.

Forsén, S., Thulin, E., and Lilja, H. (1979). *FEBS Lett.* **104**, 123–126.

Forsén, S., Thulin, E., Drakenberg, T., Krebs, J., and Seamon, K. (1980). *FEBS Lett.* **117**, 189–194.

Garroway, A. N., VanderHart, D. L., and Earl, W. L. (1981). *Phil. Trans. R. London A* **299**, 609–628.

Haberkorn, R. A. *et al.* (1976). *Inorg. Chem.* **15**, 2408–2414.

Hsi, E., and Bryant, R. G. (1977). *J. Phys. Chem.* **81**, 462–465.

Huang, I.-Y., Yoshida, A., Tsunoo, H., and Nakajima, H. (1977). *J. Biol. Chem.* **252**, 8217–8221.

Huang, I.-Y., Kimura, M., Hata, A., Tsunoo, H., and Yoshida, A. (1981). *J. Biochem.* **89**, 1839–1845.

Jakobsen, H. J., and Ellis, P. D. (1981). *J. Phys. Chem.* **85**, 3367–3369.

Jensen, C. F., Deskmukh, S., Jakobsen, H. J., Inners, R. R., and Ellis, P. D. (1981). *J. Am. Chem. Soc.* **103**, 3659–3666.

Johnson, D. R., and Foulkes, E. C. (1980). *Environ. Res.* **21**, 360–365.

Jonsson, N. B.-H., Tibell, L. A. E., Evelhoch, J. L., Bell, S. J. and Sudmeier, J. L. (1980). *Proc. Natl. Acad. Sci. USA* **77**, 3269–3272.

Kägi, J. H. R., and Vallee, B. L. (1961). *J. Biol. Chem.* **236**, 2435–2442.

Keller, A. D., Drakenberg, T., Freedman, T. B., and Armitage, I. M. (1983). *J. Inorg. Biochem.,* in press.

Kidd, R. G., and Goodfellow, R. J. (1979). "NMR and the Periodic Table" (R. Harris and B. Mann, eds.), pp. 261–266. Academic Press, New York.

Kissling, M. M., and Kägi, J. H. R., (1977). *FEBS Lett.* **82**, 247–250.

Kojima, Y., Berger, C., and Kägi, J. H. R. (1979). "Metallothionein" (J. H. R. Kägi and M. Nordberg, eds.), pp. 153–161. Birkhäuser, Basel.

Kostelnik, R. J., and Bothner-By, A. A. (1974). *J. Magn. Reson.* **14**, 141–151.

Lanir, A., and Navon, G. (1972). *Biochemistry* **11**, 3536–3544.

Lindskog, S., and Coleman, J. E. (1973). *Proc. Natl. Acad. Sci. USA* **70**, 2505–2508.

Maciel, G. E., and Borzo, M. (1973). *J. Chem. Soc., Chem. Commun.* 394.

Mann, B. E. (1971). *Inorg. Nucl. Chem. Lett.* **7**, 595–597.

Margoshes, M., and Vallee, B. L. (1957). *J. Am. Chem. Soc.* **79**, 4813–4814.

Mennitt, P. G., Shatlock, M. P., Bartuska, V. J., and Maciel, G. E. (1981). *J. Phys. Chem.* **85**, 2087–2091.

Murphy, P. D., and Gerstein, B. C. (1981). *J. Am. Chem. Soc.* **103**, 3282–3286.

Murphy, P. D. *et al.* (1981). *J. Am. Chem. Soc.* **103**, 4400–4405.

Nordberg, M., and Kojima, Y. (1979). "Metallothionein" (J. H. R. Kägi and M. Nordberg, eds.), pp. 41–117. Birkhäuser, Basel.

Ohi, S., Cardenosa, G., Pine, R., and Huang, P. C. (1981). *J. Bio. Chem.* **256**, 2180–2184.

Otvos, J. D., and Armitage, I. M. (1979). *J. Am. Chem. Soc.* **101**, 7734–7736.

Otvos, J. D., and Armitage, I. M. (1980a). *Biochemistry* **19**, 4021–4030.

Otvos, J. D., and Armitage, I. M. (1980b). *Biochemistry* **19**, 4031–4043.

Otvos, J. D., and Armitage, I. M. (1980c). *Proc. Natl. Acad. Sci. USA* **77**, 7094–7098.

Otvos, J. D., Alger, J. R., Coleman, J. E., and Armitage, I. M. (1979). *J. Biol. Chem.* **254**, 1778–1780.

Otvos, J. D., Olafson, R. W., and Armitage, I. M. (1982). *J. Biol. Chem.* **257**, 2427–2431.

Palmer, A. R. *et al.* (1980). *Biochemistry* **19**, 5063–5070.

Rodesiler, P. F., and Amma, E. L. (1982). *J. Chem. Soc. Chem. Commun.* 182–184.

Schoot Uiterkamp, A. J. M., Armitage, I. M., and Coleman, J. E. (1980). *J. Biol. Chem.* **255**, 3911–3917.

Sokolowski, G., and Weser, U. (1975). *Hoppe-Seyler's Z. Physiol. Chem.* **356**, 1715–1726.

Squibb, K. S., Cousins, R. J., and Feldman, S. L. (1977). *Biochem. J.* **164**, 223–228.

Sudmeier, J. L., and Bell, S. J. (1977). *J. Am. Chem. Soc.* **99**, 4499–4500.

Suzuki, K. T., and Yamamura, M. (1980). *Biochem. Pharmacol.* **29**, 2407–2412.

Tossell, J. A., and Vaughan, D. J. (1981). *Inorg. Chem.* **20**, 3333–3340.

Turner, C. J., and White, R. F. M. (1977). *J. Magn. Reson.* **26**, 1–5.

Turner, R. W., Rodesiler, P. F., and Amma, E. L. (1982). *Inorg. Chim. Acta* **66**, L13–L15.

Vallee, B. L. (1979). "Metallothionein" (J. H. R. Kägi and M. Nordberg, eds.), pp. 19–40. Birkhäuser, Basel.

Winge, D. R., and Miklossy, K. A. (1982). *J. Biol. Chem.* **257**, 3471–3476.

Winge, D. R., Premakumar, R., and Rajagopalan, K. V. (1975). *Arch. Biochem. Biophys.* **170**, 242–252.

Winge, D. R., Geller, B. L., and Garvey, J. (1981). *Arch. Biochem. Biophys.* **208**, 160–166.

Woessner, D. E. (1962). *J. Chem. Phys.* **36**, 1–4.

Yamasaki, A., and Fluck, E. (1973). *Z. Anorg. Allg. Chem.* **396**, 297–302.

14 Thallium NMR Spectroscopy

J. F. Hinton and K. R. Metz*

Department of Chemistry
University of Arkansas
Fayetteville, Arkansas

I. Introduction

Because a comprehensive presentation of thallium NMR spectroscopy has appeared recently that also contains extensive tables of data (Hinton *et al.*, 1982a), no attempt will be made in this chapter to summarize the complete thallium NMR literature or to present exhaustive data tables. Emphasis will be placed on the general characteristics of the technique. Selected applications for both the solid state and solution state will be discussed.

II. General Characteristics

A. Nuclear Properties of ^{203}Tl and ^{205}Tl

The element thallium possesses two stable isotopes, ^{203}Tl (29.5% natural abundance) and ^{205}Tl (70.5% natural abundance), that have nuclear

---

* Present address: Department of Radiology, The Milton S. Hershey Medical Center, The Pennsylvania State University, Hershey, Pennsylvania.

NMR of Newly Accessible Nuclei, Vol. 2

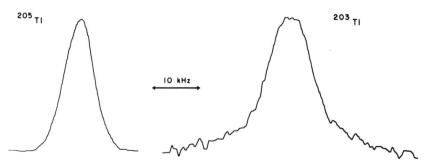

Fig. 1 ^{205}Tl and ^{203}Tl-NMR signals from TlNO$_3$ in the solid state 62°C. (Copyright 1982 American Chemical Society.)

spin $\frac{1}{2}$, and hence are NMR-active. The relative receptivities of the ^{203}Tl and ^{205}Tl nuclei are 0.055 and 0.1355, respectively, the proton having an assigned value of 1. This makes the ^{205}Tl nucleus the third most receptive spin-$\frac{1}{2}$ nuclide. The resonance frequency of the reference standard for ^{205}Tl NMR, a dilute aqueous solution of TlNO$_3$, at a field strength where the proton resonance frequency of trimethylsilane (TMS) is exactly 100 MHz, is 57.683833 MHz.

The very high receptivity of both isotopes of thallium permits one to observe them directly with relatively little difficulty, even in the solid state. This is illustrated in Fig. 1 which shows the spectrum of the simultaneous detection of both isotopes of thallium in a solid state TlNO$_3$ sample whose resonance frequencies are separated by over 500,000 Hz. The spectrum was obtained without using one of the more sophisticated techniques such as magic angle spinning (MAS), but just the normal pulsed Fourier transform (FT) technique with the solid contained in a standard 5-mm tube. The very high receptivity to detection also permits one to obtain ^{205}Tl spectra from solutions with thallium concentrations of about 5×10^{-4} M, again in 5-mm tubes.

B. Chemical Shift

The chemical shift range of the ^{205}Tl nucleus extends over 7000 ppm. Although the ^{59}Co nucleus with a shift range of 18,000 ppm appears to be more sensitive to chemical shift effects than ^{205}Tl, the chemical shift in compounds in which Co and Tl are in similar environments suggests that ^{205}Tl has the greatest NMR "shiftability" of all the elements.

The magnitude of the shiftability can be appreciated from a consideration of solvent and temperature effects on the resonance frequency. The resonance frequency of the ^{205}Tl(I) ion increases by 2000 ppm as the solvent is changed from water to liquid ammonia. The resonance frequency can also be very dependent on the temperature, as illustrated by the fact that, in liquid methylamine solutions, the resonance frequency of the Tl(I) ion changes 5 ppm deg^{-1}. As a heavy metal is involved, the origin of this large chemical shift range lies in the paramagnetic term of the nuclear screening constant. The diamagnetic contribution per valence electron is on the order of 100–250 ppm which is small compared to the observed chemical shift range of about 7000 ppm. This suggests that the paramagnetic term must be responsible for about 95% of the observed variation in the chemical shift.

Although the variety of thallium compounds is not as great as that of some other metals, such as cobalt, the available oxidation states of +1 and +3 provide an interesting array of chemical environments for thallium and concomitant chemical shift groupings. For example, the resonance frequency of thallium in the +3 oxidation state is normally higher than that of thallium in the +1 oxidation state.

As is characteristic of heavy metals, the chemical shift anisotropy (CSA) for thallium is expected to be quite large and to contribute to the spin–lattice relaxation to a significant degree. The CSA is indeed large for ^{205}Tl even in a highly ionic environment, as illustrated by the solid state ^{205}Tl spectrum of TlClO$_4$ (Fig. 2) where the CSA is 117 ppm.

C. Spin–Spin Coupling

Just as the chemical shift range of the ^{205}Tl nucleus appears to be exceptionally large, the range of the spin–spin coupling constants of this nucleus is also very large. For example, one-bond coupling constants with the ^{13}C nucleus can be larger than *10,000 Hz,* whereas a six-bond coupling constant of *66 Hz* has been observed with ^{1}H in certain organothal-

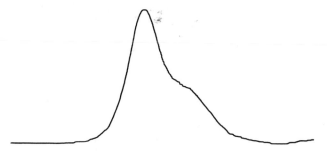

Fig. 2 Anisotropic ^{205}Tl-NMR signal from TlClO$_4$ in the solid state at 32°C.

lium(III) compounds. Because of the highly ionic nature of Tl(I) complexes, few coupling constants have been observed for this oxidation state.

The large coupling constants can in some instances create experimental problems because of an insufficiently large spectral window, aliased peaks, and computer-limited resolution. In the observation of nuclei to which thallium is coupled, it must be remembered that there are two isotopes and that decoupling of one might still leave a fairly complicated spectrum. Also, very high-power broadline irradiation might be required for efficient decoupling.

In the solid state, dipolar coupling with thallium is expected to be considerable. Consequently, very broad thallium signals are observed. The detection of any CSA, of course, depends on the relative magnitudes of the dipolar interaction and the CSA.

D. Relaxation and Line Width

The spin–lattice relaxation rate for the Tl(I) ion has been found to be extremely sensitive to environmental effects. It is quite sensitive to the presence of dissolved oxygen. In an aqueous solution the spin–lattice relaxation rate changes from 41 s^{-1} ($T_1 = 0.024$ s) at 1 atm oxygen, to 8.3 s^{-1} ($T_1 = 0.12$ s) at 0.2 atm, to 0.54 s^{-1} ($T_1 = 1.84$ s) in a degassed solution. This dependence of the spin–lattice relaxation rate on the oxygen concentration can be used to an advantage, because it normally provides a relaxation mechanism that permits faster rf pulsing and therefore more rapid spectral accumulation. The spin–lattice relaxation rate of the Tl(I) ion has also been shown to be solvent-dependent (Hinton and Briggs, 1977).

In aqueous solutions the Tl(I) spin–lattice relaxation rate is dominated by the spin–rotation mechanism (Bacon and Reeves, 1973; Chan and Reeves, 1974). However, in nonaqueous solutions the spin–lattice relaxation rate of the Tl(I) ion can be strongly dependent on the CSA mechanism (Hinton and Ladner, 1978). In the case of Tl(I)-antibiotic complexes the spin–lattice relaxation rate is determined by contributions from spin–rotation, dipolar, and CSA to varying degrees depending on the temperature and particular antibiotic complex (Briggs and Hinton, 1978b, 1979b; Briggs *et al.*, 1980). A theoretical study on the importance of the spin–rotation and CSA mechanisms in thallium relaxation has been described (Schwartz, 1976), and other mechanisms have been examined experimentally (Bangerter and Schwartz, 1974).

In aqueous solutions the ^{205}Tl(I) and ^{203}Tl(I) relaxation rates are equal and are independent of solvent isotopic substitution (H_2O, D_2O), concentration (0.03–2.0 M), anion, and resonance frequency (Bacon and Reeves, 1973; Chan and Reeves, 1974).

A number of relaxation and line width studies have been made on ^{203}Tl and ^{205}Tl nuclei in the solid state. Because of the large magnetogyric ratios of ^{203}Tl and ^{205}Tl direct dipole–dipole interaction is expected to contribute significantly to thallium line widths. The spectra of thallium in solid samples often show line broadening due to indirect spin–spin coupling. This interaction is field-independent and depends on the transmission of nuclear spin information indirectly through intervening electrons. Although the magnitude of indirect coupling may be insignificant for light nuclides, for thallium the large hyperfine interaction causes this to be a very important source of line broadening in many cases. Bloembergen and Rowland (1955) examined the ^{203}Tl and ^{205}Tl line widths of Tl_2O_3 and thallium metal as a function of the percentage of abundance of ^{205}Tl. It was shown that the ^{205}Tl line was relatively narrow in highly enriched ^{205}Tl samples, whereas the ^{203}Tl line was quite broad. The opposite effect was observed with samples enriched in ^{203}Tl. The authors concluded that spin exchange between unlike nuclei (^{203}Tl and ^{205}Tl) contributes to the second moment or line width, whereas spin exchange between like nuclei (^{205}Tl and ^{205}Tl or ^{203}Tl and ^{203}Tl) does not affect the second moment. This provides an excellent method for detecting line broadening contributions from exchange interactions in natural abundance thallium-containing samples. Since ^{203}Tl (29.5% natural abundance) is surrounded primarily by (unlike) ^{205}Tl in the solid, exchange interactions, when present, broaden this line more than the line of ^{205}Tl which is surrounded mainly by other ^{205}Tl nuclei. Equal line widths of ^{203}Tl and ^{205}Tl in natural abundance samples demonstrate the absence of exchange broadening. The favorable natural abundance and magnetogyric ratio of ^{203}Tl make this a simple and effective method for investigating exchange broadening. As shown in Fig. 1, the line width of the ^{203}Tl signal is greater than that for ^{205}Tl. This is clearly a case where exchange interactions cause considerable broadening.

Line broadening resulting mainly from exchange interactions often prevents the direct observation of CSA in the powder spectrum of thallium compounds. One method of eliminating this line broadening mechanism is by dilution in a "magnetically inert" matrix. An illustration of this technique is shown in Fig. 3 for the $TlNO_3$ salt. Because the primary source of line broadening in pure $TlNO_3$ arises from spin exchange between Tl(I) ions, dilution in a matrix of KNO_3 should produce considerable narrowing of the thallium resonance line due to the small magnetic moments of the potassium nuclides. Indeed, line narrowing occurs to the point that the CSA of the ^{205}Tl nucleus is easily observed.

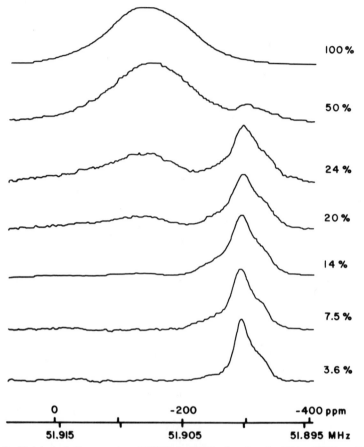

Fig. 3 Matrix isolation spectra of TlNO₃ in KNO₃ showing the CSA of ^{205}Tl at 32°C.

The effect of quadrupole coupling on the line width of a spin-$\frac{1}{2}$ nuclide in the solid state has been studied (Vanderhart *et al.*, 1967). As expected, the line width of ^{203}Tl or ^{205}Tl in the proximity of a nuclide with a quadrupole moment may dramatically increase. The line widths of ^{203}Tl and ^{205}Tl should be affected equally by quadrupole interactions.

III. Biochemical Applications of Tl NMR

A. Biological Properties of Thallium

Alkali cations participate in a variety of biological activities, including distribution of water, transmission of nerve impulses, initiation and maintenance of muscular activity, synthesis of proteins, regulation of metabolism, and activation and regulation of enzymes (Chock and Titus, 1973). Most of our knowledge about the biological chemistry of sodium and potassium has come from standard biochemical assays, electrochemical experiments, and radioactive tracer studies. Sodium is the only metal with adequate sensitivity and sufficiently high natural abundance to permit NMR studies on biological systems with relative ease. For K(I), Ca(II), and Mg(II), several probe ions with favorable spectroscopic properties can be successfully substituted for investigative purposes (Williams, 1970). A number of investigations indicate that thallium has the chemical capability of serving as a probe for potassium in biological systems. The relative receptivities to the NMR experiment for ^{39}K(I) and ^{205}Tl(I) are 0.000473 and 0.1355, respectively, which suggests that ^{205}Tl is a good spectroscopic replacement probe for potassium.

B. Biochemical Studies Using ^{205}Tl NMR

Because of its similarity to the alkali metal cations Na(I) and K(I), the ^{205}Tl(I) ion has been used as a probe in studying sodium and potassium functions in biological systems. The longitudinal relaxation rates of ^{205}Tl(I) have been used to show that the monovalent cation-binding site in Na(I)-K(I)-ATPase is very near the divalent cation-binding site (Grisham et al., 1974). The effect of the presence of pyruvate kinase and its substrates on the chemical shift, T_1, and T_2 of the ^{205}Tl(I) ion has also been studied (Ruben and Kayne, 1971). The interaction between the Tl(I) ion and antibiotic molecules, such as gramicidin, which form complexes similar to those found in membrane systems and involved in ion transport across membranes, have been investigated in nonaqueous solutions and in model membrane systems (Briggs et al., 1980; Hinton et al., 1982b,c).

There appears to be qualitative agreement between the chemical shift and the basicity of the binding functional groups of the antibiotics quite similar to that found with the Tl(I) ion in solvents of different basicity (i.e., increasing basicity produces low-field shifts). The interaction between the Tl(I) ion and gramicidin dimers incorporated into micelles has been investigated with ^{205}Tl-NMR techniques (Hinton et al., 1982b). The chemical shift data were interpreted in terms of a model in which the dimer has only one tight binding site with a binding constant of 900 M^{-1} at 30°C. The association of the Tl(I) ion with gramicidin in trifluoroethanol has been studied, and the data obtained suggest that in solution the gramicidin dimer has two identical strong binding sites (Hinton et al., 1982c). The chemical shift of the Tl(I) ion has been studied as a function of pH in phosphate solutions and in solutions of dipalmitoylphosphatidylcholine where large low-field shifts due to complexation were observed (Arnold and Scholl, 1976).

IV. Chemical Applications of Tl NMR

A. Solution Studies

Possibly the most important thing to remember about Tl NMR is that the chemical shift is extremely sensitive to solvent, concentration, and temperature. Although these features of Tl NMR make this nucleus a particularly good probe for studying a wide variety of interactions involving the Tl(I) or Tl(III) ion, this great sensitivity to the environment also means that one must be extremely aware of solvent purity, concentration accuracy, and temperature control of the NMR probe.

Conceptually, the simplest system to study in solution would be an ion interacting with only one type of solvent molecule. Here one would be interested in relating the resonance frequency of the ion to some property of the solvent, such as basicity, that reflects the strength of interaction between the ion and the solvent molecule. In practice, however, such a study can be rather difficult, because it requires one to obtain the infinite dilution resonance frequency of the ion (i.e., the resonance frequency in the absence of any effect due to ion pairing). In solvents of low polarity, ion pairing dominates the ionic equilibrium. Consequently, one must obtain the resonance frequency of the nucleus as a function of decreasing salt concentration to define the resonance frequency–salt concentration curve well enough to extrapolate to the infinite dilution resonance frequency. The extreme sensitivity of the thallium resonance frequency to its chemical environment (e.g., the difference in the thallium chemical shift for the $Tl^+-NO_3^-$ ion pair in liquid ammonia and the Tl^+ ion at infinite dilution is 66 ppm) and the nonlinear relationship between the resonance frequency and salt concentration make this extrapolation dangerous. It is advisable in such studies to use several salts whose anions are quite different to approach the infinite dilution point from more than one position on the resonance frequency–concentration plot. A computer analysis of the resonance frequency–concentration curve for each salt then yields the infinite dilution resonance frequency value for a particular solvent.

The concentration and anion dependencies of the chemical shift of thallous salts in aqueous solutions and in nonaqueous solutions have been investigated (Briggs and Hinton, 1977; Briggs et al., 1979; Dechter and Zink, 1974, 1975, 1976a, 1976b; Hinton and Briggs, 1977; Hinton and

Metz, 1980). The anion and concentration dependencies of the Tl(I) ion are in the range 10–100 ppm. A correlation between the infinite dilution resonance frequency of the Tl(I) ion and the Gutmann donor number (Gutmann, 1976), a measure of solvent basicity, has been made for a number of different solvents (Briggs and Hinton, 1977). This relationship indicates that, the more basic the solvent, the higher the resonance frequency of the Tl(I) ion. The increase in resonance frequency with increasing solvent basicity may be viewed as a measure of the strength of the interaction between the ion and solvent molecules, with the solvent acting as a Lewis base and interacting both electrostatically and covalently with the ion. Transient orbital mixing created by ion–solvent collision-induced polarization of the ion electron cloud may also contribute to the observed resonance frequency changes.

The ion pair association constant and ion pair chemical shift have been determined for a number of solvents using the method described previously (Briggs et al., 1979). As expected, K_{ip} was found to increase with decreasing dielectric constant of the solvent.

In general thallium(I) salts show low solubility in most solvents, even water. However, liquid ammonia is an excellent solvent for $TlNO_3$ and $TlClO_4$. The Tl(I) resonance frequency has been determined for a liquid ammonia solution of these salts as a function of electrolyte concentration (0.005–9.8 M) and temperature (Hinton and Metz, 1980). The dependence of the resonance frequency on concentration suggests the presence of free, fully solvated thallium ions, ion pairs, and higher order aggregates as the concentration increases. Analysis of the low-concentration data between 0 and 30°C allowed the determination of $TlNO_3$ ion pair association constants and thermodynamic parameters ($\Delta H = +6.5$ kcal mol^{-1}, $\Delta S = +36$ eu). A precipitous decrease in the resonance frequency was observed for $NH_3/TlNO_3$ molar ratios below 3:1, suggesting the formula $(NH_3)_3Tl^+NO_3^-$ for the fully solvated contact ion pair. It is interesting to note that a change of approximately 2000 ppm occurs in the Tl(I) chemical shift on going from the $TlNO_3$ melt to Tl(I) at infinite dilution for $TlNO_3$ dissolved in liquid ammonia. One word of caution seems in order when comparing data on ion pairing obtained by NMR techniques with those from conductance measurements for the same system. Conductance measurements recognize as ion pairs all species in which the component ions are not free to conduct, including contact, solvent-shared, and solvent-separated ion pairs. Nuclear magnetic resonance more easily monitors contact and solvent-separated ion pair equilibria but differentiates less distinctly between free ions and solvent-separated ion pairs. It may be that, in solvents of low dielectric constant, disparities between NMR and conductance data reflect the fact that the magnetic resonance technique monitors the equilibrium between contact and solvent-separated ion pairs, whereas conductance monitors the equilibrium between free ions and ion pairs.

The exceptional sensitivity of the thallous ion resonance frequency to its solvent environment (e.g., about a 2000 ppm change on going from water to liquid ammonia), makes this ion an excellent probe for studying ion–solvent interactions in mixed-solvent systems (Briggs and Hinton, 1977, 1978a, 1979; Dechter and Zink, 1975, 1976a). With the use of a model for preferential solvation in mixed-solvent systems based on a nonstatistical distribution of solvate species (Covington and Newman, 1976), the chemical shift for 0.005 M TlNO$_3$ in nine binary solvent systems has been analyzed (Briggs and Hinton, 1979a). The theory was used to obtain equilibrium constants and free energies of preferential solvation as one type of solvent molecule replaced another type in the solvation sphere of the ion.

Solvent isotope shifts for H$_2$O and D$_2$O have been observed to vary with concentration and anion by as much as 10 ppm and, in the case of the complexation of Tl(I) with ethylenediamine, by 100 ppm (Chan and Reeves, 1974).

The interaction between the Tl(I) ion and cryptands (Gudlin and Schneider, 1979) and crown ethers (Dechter and Zink, 1976a; Gudlin and Schneider, 1979; Srivanavit et al., 1977) has been investigated using [205]Tl-NMR spectroscopy. Thallium-205 studies were performed on Tl(I) complexes of the cryptands (2,2,1), (2,2,2), and (2,2,2B) in a number of nonaqueous solvents and in water (Gudlin and Schneider, 1979). The chemical shift of the ion complexed by a particular cryptand was found to be independent of solvent and anion, suggesting that the Tl(I) ion is completely shielded by the cryptand. It was also found that the ion resonance signal moved to high field with increasing number of oxygen atoms in the ligand. The formation constants for complexes of the crown ethers dibenzo-21-crown-7 and dibenzo-24-crown-8 with the Tl(I) ion have been determined in several nonaqueous solvents at 30°C (Shamsipur et al., 1980).

Because Tl(III) salts tend to hydrolyze in aqueous solutions, only a few investigations of the [205]Tl(III) NMR of these systems have been made (Figgis, 1959; Glaser and Henriksson, 1981; Gutowsky and McGarvey, 1953; Koppel et al., 1976; Sidei et al., 1958; Schneider and Buckingham, 1962). The concentration dependence of the chemical shift of the Tl(III) ion for Tl(NO$_3$)$_3$ in HNO$_3$ and for dilute TlCl$_3$ in varying HCl/HNO$_3$ ratios has been investigated (Gutowsky and McGarvey, 1953). Chemical shifts for the TlCl$_3$, TlBr$_3$, and Tl(NO$_3$)$_3$ salts as a function of increasing concentration of their respective acid was studied by Figgis (1959), as was the shift of $Tl(NO_3)_3$ with added H_2SO_4, HF, $HClO_4$, HBr, and HCl. The chemical shift range for the concentration, anion, and pH of the Tl(III) ion is approximately 2000 ppm.

Glaser and Henriksson (1981) have used a combination of solution and solid state ^{205}Tl-NMR experiments to study the formation and geometry of TlX_n^{3-n} complexes (X = Cl, Br). The chemical shifts for the individual species and their respective stability constants were determined for dilute (0.05 M) and concentrated (1.0–2.6 M) Tl(III) aqueous solutions. The existence of such species as $TlCl_5^{-2}$, $TlCl_6^{-3}$, and $TlBr_4^-$ was verified in these studies. This study clearly emphasizes the importance of using solid state NMR data to better understand solution state results. For example, it was found that the species $TlCl_3$ was 300–400 ppm less shielded in an aqueous solution than in the solid state, indicating a significant structural difference.

The ^{205}Tl chemical shifts have been determined for solutions of the alkylthallium(III) compounds $(CH_3)_3Tl$ (Hildenbrand and Dreeskamp, 1970; Koppel et al., 1976; Schneider and Buckingham, 1962), $(C_2H_5)_3Tl$ (Koppel et al., 1976), $(CH_3)_2Tl^+$(Koppel et al., 1976), and CH_3Tl^{+2} (Hoad et al., 1977), as well as for the complex bis(cis-1,2-dithioethane)thallate anion (Sheldrick and Yesinowski, 1975) and the ethoxide, N,N-dimethyl-amine, pyrazole, and N,N-diethyl dithiocarbamate derivates of dimethyl-thallium (Hafner and Nachtrieb, 1965; Sheldrick and Yesinowski, 1975). The temperature, concentration, anion, and solvent dependencies of the dimethyl and monomethyl cations have been investigated (Burke et al., 1981a,b; Hinton and Briggs, 1976; Hoad et al., 1977). In general, it was found that the temperature dependence of the chemical shift was greater than that resulting from changes in the anion or concentration for the compounds $(CH_3)_2TlX$ (X = NO_3, BF_4, O_2CCH_3) (Burke et al., 1981b). For several dimethylthallium(III) derivatives in a number of solvents, a linear correlation was found (Burke et al., 1981a) to exist between the ^{205}Tl chemical shift and the Drago base parameters.

The trialkyls resonate at the highest frequency, and the dialkyl and monoalkyl signals appear at about 1000–1500 and 1500–2000 ppm, respectively, to low frequency.

Solvent isotope shifts for dimethylthallium in H_2O and D_2O were found to be concentration-dependent, amounting to approximately 5 ppm (Chan and Reeves, 1974).

Thallium chemical shifts have been determined for phenylthallium(III) dichloride and its complexes with PPh_3 and dipyridine in methanol and pyridine (Koppel et al., 1976), diphenylthallium(III) chloride in liquid ammonia (Hildenbrand and Dreeskamp, 1970), diphenylthallium(III) bro-mide in dimethyl sulfoxide (DMSO) (Koppel et al., 1976), a series of substituted arylthallium(III) bis(trifluoroacetates) in a number of solvents (Hinton and Briggs, 1976), and triphenylthallium(III) in ether (Koppel *et al.*, 1976).

In general, it appears that the diaryls resonate at about 200 ppm to high frequency of the monoaryls and about 600 ppm to low frequency of the triaryls.

The general features of the relaxation of the thallium nucleus have been described in this chapter, and they may be used effectively to obtain information about the environment in which the nucleus is placed. The effect of dissolved oxygen on the spin–lattice relaxation time of the Tl(I) ion provides useful information about the solution structure of Tl(I)-antibiotic complexes. It was found that only with actin complexes was there no effect of dissolved oxygen on the spin–lattice relaxation time of Tl(I). This suggests that actins have a molecular framework that surrounds the Tl(I) ion in such a manner as to shield it from collisions with dissolved oxygen molecules. With the other antibiotics, the structures are more open and permit collisions with oxygen molecules, thereby decreasing the spin–lattice relaxation time. These solution results are consistent with the structures of the complexes determined by X-ray crystallographic techniques.

Again it is well to remember that, with heavy-metal nuclei such as thallium, CSA is frequently a dominating relaxation mechanism. The dominance of the contribution from the CSA mechanism to the spin–lattice relaxation of ^{205}Tl in dialkylthallium(III) derivatives has been demonstrated by measurements at two different magnetic field strengths and is also reflected in the line widths of the spin–spin-coupled protons (Brady *et al.*, 1981).

B. Solid State and Melt Studies

It is widely recognized that NMR spectra exhibited by solid samples are usually quite different from spectra of the same materials in solution. This is partly due to the averaging of shielding anisotropies by rapid molecular tumbling in solution. In addition, the processes that dominate relaxation are usually different in solution than in a rigid solid. Spin–spin relaxation is particularly inefficient in solutions. The result is simplification of the broad, complex spectra usually observed with solids to yield the familiar high-resolution spectra of solutions. It is important to recognize that this simplification is achieved only at a cost and that fundamentally interesting information that is lost in spectra of solutions may, at least in principle, be obtained from solid state spectra. Properties of solution spectra such as the chemical shift are strongly influenced by parameters like the CSA, which are averaged but not *eliminated* in solution.

The effect of temperature on thallium NMR shifts in solids and melts is often pronounced. The [205]Tl chemical shift may vary by as much as 5 ppm deg^{-1} [solid thallium(I) acetate] or may be temperature-independent within experimental error. It is highly significant that, among the substantial number of compounds studied, none exhibited a negative dependence of shift on temperature. At first sight this seems remarkable, because an increase in temperature is expected to weaken the thallium–anion interactions that presumably cause most salts to resonate at low fields. In an excellent study, Hafner and Nachtrieb (1964) have shown that this is probably not the dominant effect. Instead, the positive dependence of the chemical shift on temperature almost certainly arises from enhanced vibrational overlap of thallium–anion wave functions, resulting in additional paramagnetic shielding with increasing temperature.

In view of the extreme sensitivity of the thallium chemical shift to variations in the chemical environment, a change in the shift might be expected at the temperature of a phase transition. In fact, a shift discontinuity of several hundred parts per million may accompany a phase change. In most systems investigated so far, fusion results in a downfield shift of the [205]Tl resonance relative to that for the solid. It is well known that interionic distances in most salts decrease on melting, and this is consistent with observed paramagnetic thallium shifts which would result from increased cation–anion electronic overlap (Hafner and Nachtrieb, 1964).

Thallium NMR line widths are usually quite sensitive to phase changes. An excellent example is that of thallium(I) nitrate. The [205]Tl line width for this compound is essentially independent of temperature for a given phase α, β, or γ. However, a sudden narrowing of the line occurs at the transition from orthorhombic γ to hexagonal β and again at the transition from β to cubic α (Bury and McLaren, 1969; Kolditz and Wahner, 1973). The latter phase transition is accompanied by a 100-fold increase in electrical conductivity and must almost certainly mark the onset of thallium(I) diffusion in the solid (Bury and McLaren, 1969). This conclusion is supported by other excellent studies on relaxation in TlNO$_3$ (Avogadro *et al.*, 1976; Villa and Avogadro, 1976), which explain T_2 in terms of both indirect scalar exchange interactions and diffusion, T_1 via indirect coupling modulated by lattice vibrations, and T_1 by means of diffusion (α), exchange interactions (γ), or both (β).

A number of thallium NMR studies on thallium(I) halides have appeared. In addition to determinations of the [205]Tl chemical shift in TlF (Freeman *et al.*, 1959; Hafner and Nachtrieb, 1964), several studies have shown that ^{205}Tl line broadening in this compound is due mainly to direct dipolar interactions between ^{205}Tl and ^{19}F (Blicharski, 1972; Saito, 1958, 1966). Contributions from indirect spin–spin interactions are approximately four times smaller than those from direct dipolar interactions in this compound (Blicharski, 1972). Thallium-205 chemical shifts have been determined for molten Tl_2Cl_4 and Tl_2Br_2 (Hafner and Nachtrieb, 1965; Rowland and Bromberg, 1958), and in both cases, two signals have been observed that have been attributed to equal quantities of TlX and TlX_3.

Two ^{205}Tl-NMR studies on mixed-salt solids and melts have appeared. Thallium-205 shifts in melts containing various proportions of TlX and MX (X = Cl, Br, I; M = Li, Na, K, Rb, Cs, Ag) were found to be linearly dependent on the mole fraction of MX (Hafner and Nachtrieb, 1965). The direction of the shift depended on M, and for a given mole fraction the shift was linear with the radius of M. These results were interpreted as effects of MX on the TlX covalency. A similar study on TlX–CsX mixtures (X = Cl, Br) revealed a linear dependence of chemical shift on mole fraction in the melt but not in the solid (Hafner, 1966). These effects were again attributed to the influence of CsX on the TlX covalency.

Thallium(I) perchlorate displays very interesting NMR properties. The isotropic ^{205}Tl chemical shift is quite far upfield, showing that it is a highly ionic salt. Nevertheless, polycrystalline $TlClO_4$ presents a classic powder spectrum characteristic of axial symmetry, with a CSA of $\sigma_\perp - \sigma_\parallel = 117$ ppm. That a salt with such ionic character can still exhibit a shift anisotropy of this magnitude is remarkable and testifies to the extreme sensitivity of the thallium chemical shift to small electronic effects. In a single crystal, the ^{205}Tl shift depends on the orientation θ as $1 - 3\cos^2\theta$, and the line width is also quite orientation-dependent. Freeman et al. (1959) have pointed out that the ^{205}Tl line width is approximately that predicted from direct dipolar interactions, and the orientation dependence in a single crystal verifies that this is the primary source of line broadening. Possible contributions from thallium–thallium spin exchange are eliminated by the virtual equality of the ^{205}Tl and ^{203}Tl line widths and second moments in this compound (Freeman et al., 1959).

Thallium-205 spectra of two solid Tl(I)-antibiotic complexes have been studied. From the spectrum of the valinomycin complex, a thallium CSA of $\sigma_\parallel - \sigma_\perp = 150$ ppm was found (Hinton et al., 1981). The observed isotropic shift of -545 ppm is upfield from any other known Tl(I) system and attests to the highly ionic character of thallium in this complex. The solid state ^{205}Tl chemical shift in the Tl(I)–gramicidin A complex, with acetate as the counteranion, is approximately -160 ppm. This shift suggests somewhat greater thallium covalency in gramicidin than in valinomycin, and it may also reflect solid state interactions with the acetate anion.

A variety of thallium(III) salts have been investigated as solids or melts, but few of them have been studied in detail. The lack of published information is demonstrated by the fact that ^{205}Tl-NMR data for two-thirds of all the thallium(III) salts ever investigated have recently been reported in a single paper by Glaser and Henriksson (1981). The chemical shift of $Tl(ClO_4)_3$ in both the solid and the melt has been studied as a function of temperature (Hafner and Nachtrieb, 1964). This salt exhibits the only known negative change in shift on fusion, implying decreased covalency in the melt. The solid hydrate $Tl(ClO_4)_3 \cdot 6H_2O$ has also been studied (Glaser and Henriksson, 1981). The ^{205}Tl chemical shift of $TlCl_3$ in both the solid and the melt (Hafner and Nachtrieb, 1964; Schneider and Buckingham, 1962) has been determined, and both the shift and the line width of $TlCl_3 \cdot 4H_2O$ are known (Glaser and Henriksson, 1981). A previous study on the tetrahydrate revealed a complex ^{205}Tl spectrum of at least three overlapping lines. The shift of $TlBr_3 \cdot 4H_2O$ is not greatly different from that of $TlCl_3 \cdot 4H_2O$, but the line width of 14.0 kHz is nearly twice as large.

Spectra for several solids containing TlX_4^- and TlX_5^{2-} have been determined. These include $Zn(TlCl_4)_2$ (Freeman et al., 1959) and $KTlCl_4$, $KTlBr_4 \cdot 2H_2O$, and NBu_4TlI_4 (Glaser and Henriksson, 1981). In this case the ^{205}Tl line width is greatest for the iodide and smallest for the chloride. NBu_4TlI_4 exhibits a remarkable chemical shift of -1560 ppm, approximately 1000 ppm *upfield* from that of the nearest thallium(I) salt. The ^{205}Tl chemical shifts and line widths of $Na_2TlCl_5 \cdot 4H_2O$ and $Cs_2TlCl_5 \cdot H_2O$ are similar (Glaser and Henriksson, 1981).

Compounds containing TlX_6^{3-} and $Tl_2X_9^{3-}$ have been studied. The hydrates $Na_3TlCl_6 \cdot 12H_2O$ and $K_3TlCl_6 \cdot 2H_2O$ are similar in both shift (1972 and 2007 ppm) and line width (Glaser and Henriksson, 1981). The anhydrous salts K_3TlCl_6 and $(NH_4)_3TlCl_6$ exhibit identical shifts of 2220 ppm (Freeman et al., 1959), but the difference between this value and the more recent figures for the above salts may not be experimentally significant. The shift and line width of $Co(NH_3)_6TlCl_6$ are 2019 ppm and 3.3 kHz, respectively, whereas for $Co(NH_3)_6TlBr_6$ they are -1194 ppm and 19.0 kHz (Glaser and Henriksson, 1981). A similar result is obtained with $Cs_2Tl_2Cl_9$ ($\delta = +1926$ ppm, line width 9.0 kHz) and $Cs_3Tl_2Br_9$ ($\delta = -1194$ ppm, line width 12.0 kHz).

The results reported so far suggest that the thallium chemical shift in a thallium(III) salt depends largely on the particular halogen involved. In every series where data are available, $\delta_{Cl} > \delta_{Br} > \delta_I$. The reason for this apparent trend is unclear, and additional compounds must be studied before such a trend is definitely established. Certain salts would be especially interesting. For example, thallium shifts are known for the compounds $MTlX \cdot nH_2O$ (M = NBu_4, K; X = I, Br, Cl; n = 0, 2), so the shift for an analog with X = F would be very interesting. Likewise, it would be extremely useful to determine the shifts in $Co(NH_3)_6TlF_6$ and $Co(NH_3)_6TlI_6$. The extraordinary chemical shifts determined for $(NBu_4)_3$ $TlI_4{}^-$, $Co(NH_3)_6TlBr_6$, and $Cs_3Tl_2Br_9$ merit additional study. Elucidation of the mechanism that shifts the thallium resonance in these salts over 3000 ppm to high field of the usual thallium(III) chemical shift range would be of great interest.

The thallium NMR properties of only a few solid or liquid organothallium compounds have been determined. Despite a large number of studies with various solvents, the ^{205}Tl spectrum in solid dimethylthallium(I) bromide has only recently been observed (Hinton and Metz, unpublished results). The chemical shift is anisotropic, with $\sigma_{\parallel} - \sigma_{\perp}$ = 485 ppm, and the isotropic shift is 5590 ppm. By comparison, Tl_2O_3 exhibits an isotropic shift of 5500 ppm but a shift anisotropy of 1870 ppm. The large isotropic shift of Me_2TlBr presumably results from covalent thallium–methyl interactions in this compound, yet the shift anisotropy is smaller than might have been predicted. This probably reflects covalent interactions between thallium and surrounding bromine atoms, yielding greater than expected electronic symmetry about the thallium ion. In addition to Me_2TlBr, chemical shifts have been reported for thallium ethoxide and a pyrazole derivative (Koppel et al., 1976; Schneider and Buckingham, 1962).

The favorable NMR properties of thallium have permitted several investigations of thallium ions adsorbed on surfaces and thallium contained in zeolites. These studies, along with those on the NMR properties of thallium in solution, have been reviewed fairly recently (Morariu and Chifu, 1979).

The first published report in this area concerned thallium shifts and spin–spin relaxation times in the zeolite $Tl_3HGe_7O_{16} \cdot nH_2O$ (Wada and Cohen-Addad, 1969). It was found that thallium diffusion was rapid in the completely hydrated zeolite. In fact, the rate of thallium diffusion is essentially the same as that of water at room temperature. The thallium chemical shift is highly dependent on the amount of water present, suggesting major changes in the thallium environment on hydration. These same conclusions were reached in another study on this system (Bittner and Hauser, 1970), which also found more rapid thallium diffusion near room temperature than at liquid nitrogen temperatures.

Thallium-205 NMR properties in silicon–aluminum zeolites have been investigated (Morariu, 1978). Evidence for the existence of several thallium sites was found. In addition, the determination of correlation times allowed a detailed description of thallium motion as a function of temperature and water content.

The chemical shift and spin–lattice relaxation time of ^{205}Tl(I) ions adsorbed on hydrated silica have been studied (Morariu, 1978). A prolonged equilibration period was required before NMR properties stabilized. The chemical shift of adsorbed Tl(I) was found to be about 80 ppm upfield from aqueous thallium(I) acetate in the bulk solution [about -40 ppm with respect to infinitely dilute aqueous Tl(I) at zero]. The shift to high field was attributed to the dissociation of ion pairs with acetate on binding to the surface. Inversion recovery T_1 measurements suggested the dominance of the dipolar relaxation mechanism for adsorbed ^{205}Tl(I).

References

Arnold, K., and Scholl, R. (1976). *Stud. Biophys.* **59**, 47.
Avogadro, A., Villa, M., and Chiodelli, G. (1976). *Gazz. Chim. Ital.* **106**, 413.
Bacon, M., and Reeves, L. W. (1973). *J. Am. Chem. Soc.* **95**, 272.
Bangerter, B. W., and Schwartz, R. N. (1974). *J. Chem. Phys.* **60**, 33.
Bittner, H., and Hauser, E. (1970). *Monatsh. Chem.* **101**, 1471.
Blicharski, J. S. (1972). *Z. Naturforsch. A* **27**, 1355.
Bloembergen, N., and Rowland, T. J. (1955). *Phys. Rev.* **97**, 1679.
Brady, F., Matthews, R. W., Forster, M. J., and Gillies, D. G. (1981). *Inorg. Nucl. Chem. Letts.* **17**, 155.
Briggs, R. W., and Hinton, J. F. (1977). *J. Solution Chem.* **6**, 827.
Briggs, R. W., and Hinton, J. F. (1978a). *J. Solution Chem.* **7**, 1.
Briggs, R. W., and Hinton, J. F. (1978b). *J. Magn. Reson.* **32**, 155.
Briggs, R. W., and Hinton, J. F. (1979a). *J. Solution Chem.* **8**, 519.
Briggs, R. W., and Hinton, J. F. (1979b). *J. Magn. Reson.* **33**, 363.
Briggs, R. W., Metz, K. R., and Hinton, J. F. (1979). *J. Solution Chem.* **8**, 479.
Briggs, R. W., Etzkorn, F. A., and Hinton, J. F. (1980). *J. Magn. Reson.* **37**, 523.
Burke, P. J., Gilies, D. G., and Matthews, R. W. (1981a). *J. Chem. Res.* **5**, 124.
Burke, P. J., Matthews, R. W., Cresshull, I. D., and Gillies, D. G. (1981b). *J. Chem. Soc. Dalton Trans.* **1**, 132.
Bury, P. C., and McLaren, A. C. (1969). *Phys. Status Solidi* **31**, K5.
Chan, S. O., and Reeves, L. W. (1974). *J. Am. Chem. Soc.* **96**, 404.
Chock, P. B., and Titus, E. O. (1973). "Progress in Inorganic Chemistry" (S. J. Lippard, ed.), Vol. 18, pp. 287–382. Wiley, New York.
Covington, A. K., and Newman, K. E. (1976). *Adv. Chem. Ser.* **155**, 153.
Dechter, J. J., and Zink, J. I. (1974). *J. Chem. Soc. Chem. Commun.* **96**.
Dechter, J. J., and Zink, J. I. (1975). *J. Am. Chem. Soc.* **97**, 2937.
Dechter, J. J., and Zink, J. I. (1976a). *Inorg. Chem.* **15**, 1690.
Dechter, J. J., and Zink, J. I. (1976b). *J. Am. Chem. Soc.* **98**, 845.
Figgis, B. N. (1959). *Trans. Faraday Soc.* **55**, 1075.
Freeman, R., Gasser, R. P. H., and Richards, R. E. (1959). *Mol. Phys.* **2**, 301.

Glaser, J., and Henriksson, U. (1981). *J. Am. Chem. Soc.* **103**, 6642.

Grisham, C. M., Gupta, R. K., Barnett, R. E., and Mildvan, A. S. (1974). *J. Biol. Chem.* **249**, 6738.

Gudlin, D., and Schneider, H. (1979). *Inorg. Chim. Acta* **33**, 205.

Gutman, V. (1976). *Coord. Chem. Rev.* **18**, 225.

Gutowsky, H. S., and McGarvey, B. R. (1953). *Phys. Rev.* **91**, 81.

Hafner, S. (1966). *J. Phys. Chem. Solids* **27**, 1881.

Hafner, S., and Nachtrieb, N. H. (1964). *J. Chem. Phys.* **40**, 2891.

Hafner, S., and Nachtrieb, N. H. (1965). *J. Chem. Phys.* **42**, 631.

Hildenbrand, K., and Dreeskamp (1970). *Z. Phys. Chem.* (*N.F.*) **69**, 171.

Hinton, J. F., and Briggs, R. W. (1976). *J. Magn. Reson.* **22**, 447.

Hinton, J. F., and Briggs, R. W. (1977). *J. Magn. Reson.* **25**, 379.

Hinton, J. F., and Ladner, K. H. (1978). *J. Magn. Reson.* **32**, 303.

Hinton, J. F., and Metz, K. R. (1980). *J. Solution Chem.* **9**, 197.

Hinton, J. F., Metz, K. R., and Millett, F. S. (1981). *J. Magn. Reson.* **44**, 217.

Hinton, J. F., Metz, K. R., and Briggs, R. W. (1982a). "Annual Reports on NMR Spectroscopy", Vol. 13A. Academic Press, London.

Hinton, J. F., Young, G., and Millett, F. S. (1982b). *Biochemistry* **21**, 651.

Hinton, J. F., Turner, G. L., and Millett, F. S. (1982c). *Biochemistry* **21**, 646.

Hoad, C. S., Matthews, R. W., Thakur, M. M., and Gillies, D. G. (1977). *Organ. Chem.* **124**, C31.

Kolditz, L., and Wahner, E. (1973). *Z. Anorg. Allg. Chem.* **400**, 161.

Koppel, H., Dallorso, J., Hoffman, G., and Walther, B. (1976). *Z. Anorg. Allg. Chem.* **427**, 24.

Marariu, V. V., and Chifu, A. (1979). *Stud. Cercet. Fiziol.* **31**, 391.

Metz, K. R., and Hinton, J. F. (1982). *J. Am. Chem. Soc.* **104**, 6206.

Morariu, V. V. (1978). *Chem. Phys. Lett.* **56**, 272.

Reuben, J., and Kayne, F. J. (1971). *J. Biol. Chem.* **246**, 6227.

Rowland, T. J., and Bromberg, J. P. (1958). *J. Chem. Phys.* **29**, 626.

Saito, Y. (1958). *J. Phys. Soc. Jpn* **13**, 72.

Saito, Y. (1966). *J. Phys. Soc. Jpn* **21**, 1072.

Schneider, W. G., and Buckingham, A. D. (1962). *Discuss. Faraday Soc.* **34**, 147.

Schwartz, R. N. (1976). *J. Magn. Reson.* **24**, 205.

Shamsipur, M., Rounaghi, G., and Popov, A. (1980). *J. Solution Chem.* **9**, 701.

Sheldrick, G. M., and Yesinowski, J. P. (1975). *J. Chem. Soc. Dalton Trans.* 870.

Sidei, T., Yano, S., and Sasaki, S. (1958). *Oyo Butsuri* **27**, 513.

Srivanavit, C., Zink, J. I., and Dechter, J. J. (1977). *J. Am. Chem. Soc.* **99**, 5876.

Vanderhart, D. L., Gutowsky, H. S., and Farrar, T. C. (1967). *J. Am. Chem. Soc.* **89**, 5056.

Villa, M., and Avogadro, A. (1976). *Phys. Status Solidi B* **75**, 179.

Wada, T., and Cohen-Addad, J. P. (1969). *Bull. Soc. Fr. Mineral. Cristallogr.* **92**, 238.

Williams, R. J. P. (1970). *Q. Rev.* **24**, 331.

15 NMR of Less Common Nuclei

P. Granger

I.U.T. de Rouen
Mont Saint Aignan, France

I. Introduction

This chapter is limited to a consideration of some of the less common nuclei and discusses only their high-resolution NMR in solution. We shall present an overview for each nucleus with details on factors such as temperature, solvent, concentration effects, and relaxation on which successful observations depend.

Because this chapter is designed as a starting point for the study of these nuclei, we have included a reference list as complete as possible, except for platinum and tin for which there are far too many references.

II. Beryllium ($Z = 4$)

The only stable isotope of beryllium is ^{9}Be. It has a spin of $\frac{3}{2}$, and its resonance frequency in a field where tetramethylsilane (TMS) resonates at 100 MHz is $\Xi = 14.0518$ MHz for the solvated ion $Be(H_2O)_4^{2+}$ (Brevard and Granger, 1981a; Harris and Mann, 1978). This isotope is easy to observe because its receptivity compared to ^{13}C is 78.8, but its quadrupole moment of 5.1×10^{-30} m^2 leads to broad bands and precludes observation in dissymetric environments. Its magnetic moment is negative, resulting in a negative nuclear Overhauser enhancement (NOE) effect with a maximum magnitude of -3.557. The chemistry of beryllium is well known, but NMR studies on this nucleus are limited.

A. Chemical Shifts

Most of the known chemical shifts are listed in Gaines *et al.* (1981) or can be found elsewhere in the literature (Atam *et al.*, 1972; Barabas, 1972; Delpuech *et al.*, 1977; Drew and Morgan, 1977; Grigoreo *et al.*, 1971; Kovar and Morgan, 1970; Kotz *et al.*, 1967; Morgan and McVicker, 1968). They range from -22 to 21 ppm with respect to $Be(H_2O_4^{2-}$. From the limited amount of available data certain trends appear: When the substituent electronegativity increases, a deshielding effect is observed; however, when the coordination number increases, the beryllium atom is shielded (Kovar and Morgan, 1970), but the range of the tricoordinate species overlaps that of the tetracoordinate species. The chemical shift is governed essentially by the diamagnetic term. Beryllium-9 NMR has been used to study the different species formed by the dissolution $BeCl_2$ in CH_3CN (Wehrli and Wehrli, 1982).

B. Coupling Constants

The quadrupole moment often prevents measurement of the coupling constants which are all small (>40 Hz) because of the large ionic character of the bonds to the beryllium atom. The value of $^1J_{BeF}$ ranges from 28 to 33 Hz (Kotz *et al.*, 1967; Kovar and Morgan, 1970; Wehrli, 1978), and that of $^1J_{BeP}$ from 4 to 6.2 Hz (Delpuech *et al.*, 1977); $^1J_{BeB} = 3.6$ Hz (Gaines *et al.*, 1981), and $^1J_{BeC} = 1.1$ Hz (obtained from ^{13}C spectra) (Fischer *et al.*, 1976). Only one two-bond coupling constant is known: $^2J_{BeH} = 10.2$ Hz (Gaines *et al.*, 1981).

C. Relaxation

Wehrli has investigated in detail the different relaxation mechanisms of ^9Be (Wehrli, 1976, 1978). When the symmetry is low, the quadrupolar relaxation is dominant, as observed for $Be(acac)_2$. With tetrahedral symmetry dipolar relaxation is dominant for BeF_4^{2-} but is only part of the relaxation mechanism at low temperatures for $Be(H_2O)_4^{2+}$. At higher temperatures this mechanism decreases in importance and is replaced by spin–rotation relaxation. At all temperatures quadrupolar relaxation plays a role. The theory of Hertz seems unable to interpret these results (Lindman *et al.*, 1977).

III. Sulfur (*Z* = 16)

Sulfur is the only element of group VI with a poorly developed NMR because it has several unfavorable characteristics: low natural abundance (0.76% for the only active isotope ^{33}S), a low resonance frequency ($\Xi = 7.6760$ MHz), and a quadrupole moment of 5×10^{-30} m^2 equivalent to that of chlorine. Hence observation is difficult.

A. Chemical Shifts

After initial measurements of the magnetic moment of ^{33}S in 1951 (Dharmatti and Weaver, 1951) it was not until 1968 that sulfur NMR was utilized in chemical applications (Karr and Schultz, 1968; Lee, 1968). More systematic studies were undertaken by Retcofsky *et al.* (Retcofsky and Friedel, 1970, 1972) to test the use of ^{33}S NMR in the analysis of sulfur derivatives in coals and oil. Despite the difficulties, other authors have extended our knowledge of this nucleus (Faure *et al.*, 1981; Kroneck *et al.*, 1980b; Lutz *et al.*, 1973, 1976b; Retcofsky and Friedel, 1972; Schutz *et al.*, 1971). The usual reference is SO_4^{2-} associated with an alkali metal counterion (Lutz *et al.*, 1973), which was claimed to be independent of the concentration (Lutz *et al.*, 1973). However, this anion gives a variation of 94 ppm from concentrated sulfuric acid to a 10 *N* solution (Schultz *et al.*, 1971). Usually the tetrahedral environment gives rise to relatively narrow lines, but it has been observed (Faure *et al.*, 1981) with sulfones and sulfonic acids that the lines broaden when the substituent increases in size, to the extent that the resonance cannot be observed for *n*-butylsulfone. It has also been noted that, even for small molecules such as Na_2S, $Na_2S_2O_4$, $(NH_4)_2S_2O_8$, and Na_2SO_3, no signal is observed (Lutz *et al.*, 1976b).

B. Coupling Constants

The only known value for a coupling constant is $^1J_{^{33}S^{19}F} = 251$ Hz in SF_6 (Inglefield and Reeves, 1964). Broad lines prevent measurement of the fine structure, and the correspondingly very short relaxation times prevent measurement of the coupling constant from the observation of nuclei coupled to sulfur.

C. Relaxation

The relaxation mechanism arises exclusively from the quadrupole moment. Line widths are used to measure relaxation times; they range from 10 Hz in $(NH_4)_2SO_4$ (Lutz *et al.*, 1976b) to 9000 Hz in methyl-3-thiophene

(Retcofsky and Friedel, 1972). The relaxation of CS_2 has been studied in detail (Ancian *et al.*, 1979; Vold *et al.*, 1978).

IV. Vanadium ($Z = 23$)

Vanadium has two NMR-active isotopes, and their characteristics are listed in the accompanying table.

Isotope	Spin	Natural abundance (%)	Ξ (MHz)	Q (10^{-28} m^2)	Receptivity referred to ^{13}C
^{50}V	6	0.24	9.9703	+0.21	0.755
^{51}V	$\frac{7}{2}$	99.76	26.3029	−0.052	2150

Although the parameters of the usual isotope ^{51}V are well known (Brevard and Granger, 1982a), the values of the magnetic moment and the quadrupole moment of ^{50}V are still under investigation (Habayeb and Hileman, 1980a; Howarth and Richards, 1964; Lutz *et al.*, 1981). Many authors have measured the quadrupole coupling constant (QCC) in the solid state (Allerhand, 1970; Basler *et al.*, 1981; Baugher *et al.*, 1969; Dehmelt, 1953; Gornostansky and Stazer, 1967; Gossard, 1966; Narita *et al.*, 1966; Umeda *et al.*, 1965; Whitesides and Mitchell, 1969) or in nematic phases (Paulsen and Rehder, 1982; Rehder *et al.*, 1978). This parameter varies according to the chemical environment, but its value is usually less than 10 MHz.

A. Chemical Shifts

The most well-documented review on chemical shifts is by Rehder (1983). The reference is neat $VOCl_3$, and the chemical shift range depends on the oxidation state of vanadium: 450 ppm $> \delta > -900$ ppm for V(V), -450 ppm $> \delta > -1100$ ppm for V(III), -500 ppm $> \delta > -1700$ ppm for V(I), and -1650 ppm $> \delta > -1950$ ppm for V($-$I); but, because they present some overlap, these chemical shifts cannot be used unambiguously to determine the oxidation state.

Most of the work has been done by the Hamburg group (Basler *et al.*, 1981; Baumgarten *et al.*, 1979; Borowski *et al.*, 1981; Mühlbach *et al.*, 1981; Müller and Redher, 1977; Näumann and Rehder, 1981; Paulsen and Rehder, 1982; Paulsen *et al.*, 1978; Puttfarcken and Rehder, 1980; Rehder, 1972, 1976, 1977a, b, c, 1980, 1983; Rehder and Dorn, 1976; Rehder and Schmidt, 1974, 1977; Rehder and Wieghardt, 1981; Rehder *et* al., 1976a, b, 1978, 1980b, 1982; Schmidt and Rehder, 1980; Talay and Rehder, 1978, 1981) who studied the ^{51}V resonance for a wide variety of ligands. The equilibrium of vanadate ions and peroxo complexes has been studied as a function of pH (Flynn *et al.*, 1974; Habayeb and Hileman, 1980b; Heath and Howarth, 1981; Howarth and Hunt, 1979; Howarth and Jarrold, 1978; Howarth and Richards, 1965; Kazanskii and Spitsyn, 1975; Kazanskii *et al.*, 1975; O'Donnell and Pope, 1976; Tarasov *et al.*, 1981). Fluorinated and carbonylated derivatives have also been investigated (Hatton *et al.*, 1965; Howell and Moss, 1971a; Nakano, 1977). A detailed theoretical study on $V(CO)_6^-$ is presented in Nakano (1977). The variation in the chemical shift with temperature is about -0.5 ppm K^{-1} (Paulsen *et al.*, 1978).

High-resolution NMR in the solid state has also been proved possible (Meadows *et al.*, 1982) and probably will open up new areas in chemistry. The NMR of vanadium is now a valuable tool for structural determinations.

B. Coupling Constants

The coupling constants are all below 1 KHz. Only one value for $^1J_{51VH}$ (20.3 Hz) is known (Puttfarcken and Rehder, 1980), and several $^1J_{51VF}$ values have been measured and range from 88 to 140 Hz (Hatton *et al.*, 1965, Howell and Moss, 1971a). These values are temperature-dependent. Scalar coupling sometimes disappears at room temperature when exchange occurs, however, a large number of one-bond coupling constants are known with ^{31}P. Their values range from 35 to 660 Hz (Rehder, 1983). Several $^1J_{51V17O}$ values are known (Heath and Howarth, 1981; Lutz *et al.*, 1976c), and they are between 30 and 62 Hz. Only $V(CO)_6^-$ allows the determination of a coupling constant with ^{13}C (Bodner and Todd, 1974; Lauterbur and King, 1966), which has a value of 116 Hz.

C. Relaxation

The relaxation mechanism is purely quadrupolar, and the relaxation times are short. Most of the determined values are obtained from line widths. This parameter is often used to study the exchange processes or the chemical evolution of mixtures (Borowski *et al.*, 1981; Flynn *et al.*, 1974; Howarth and Jarrold, 1978; Paulsen *et al.*, 1978). Most T_1 measurements were performed to compare the quadrupole moment of both isotopes (Habayeb and Hileman, 1980a; Lutz *et al.*, 1981), and only one study monitored evolution of the transverse relaxation with temperature (Gillen and Noggle, 1970).

V. Chromium (Z = 24)

The NMR of the only active isotope ^{53}Cr, with a natural abundance of 9.55%, is severely restricted by two factors. The first is the presence of a quadrupole moment of 3×10^{-30} m^2, and the second is that the only diamagnetic molecules are Cr(0) and Cr(VI) species. The spin of ^{53}Cr is $\frac{3}{2}$, and its Ξ value is 5.6524 MHz (Brevard and Granger, 1981, p. 118).

The earliest study (Egozy and Lowenstein, 1969) was devoted to the equilibrium $CrO_4^{2-} + Cr_2O_7^{2-} \rightleftharpoons Cr_2O_7^{2-} + CrO_4^{2-}$ in H$_2$O using line widths as a function of concentration. Since then the only extensive research (Epperlein *et al.*, 1975) has involved the chemical shift of CrO_4^{2-} in relation to the nature of the counterion. When the salt concentration increases, shielding increases, and this effect is more pronounced for large cations than for smaller ones. A large isotopic effect of -1.3 ppm is also observed when H$_2$O is replaced by D$_2$O. In the same study a value of -1795 ppm was reported for Cr(CO)$_6$, which agrees with other results on transition metals.

The only measured coupling constant is $^1J_{^{53}Cr^{17}O} = 10$ Hz in CrO_4^{2-} (Lutz *et al.*, 1976c). Except for the line width, no relaxation study has been done on chromium, but the quadrupolar mechanism is dominant. The quadrupole moment is known only approximately.

VI. Iron (Z = 26)

The only iron isotope active in NMR, ^{57}Fe, has a low natural abundance, 2.19%, a low resonance frequency, $\Xi = 3.2377$ MHz (Brevard and Granger, 1981a, p. 122), and usually very long relaxation times. Before the advent of Fourier transform (FT) NMR and high-field instruments, only special techniques allowed observation of ^{57}Fe (Schwenk, 1968, 1971). Only Fe(0) and Fe(II) are observable, because the other oxidation states are paramagnetic. The INEPT sequence has been used successfully to increase the intensity of the signals (Brevard and Schimpf, 1982).

A. Chemical Shifts

The resonances observed so far range over about 3000 ppm, but comparison with other transition metals such as ruthenium (Brevard and Granger, 1983) shows that this range is probably greater. Two references are used, Fe(CO)$_5$ and Fe(C$_5$H$_5$)$_2$ (Haslinger *et al.*, 1981; Jenny *et al.*,

1981) The relationship between the two scales is complex, because $Fe(C_5H_5)_2$ = 1535.2 ppm in tetrahydrofuran (THF), 1560.1 ppm in CS_2, 1532 in C_6H_6, and 1542.8 in $CDCl_3$, with neat iron pentacarbonyl as the reference (Haslinger et al., 1981; Jenny et al., 1981; Sahm and Schwenk, 1974). These results show the solvent effect on iron resonances, but no systematic study has been published. On the contrary, the effect of the concentration seems to be less than the experimental error of ±1 ppm (Haslinger et al., 1981). The effect of the temperature on $Fe(CO)_5$ is relatively small for a transition metal: 0.5 ppm K^{-1} (Schwenk, 1970). A similar effect is observed for $Fe(C_5H_5)_2$ in CS_2 (Haslinger et al., 1981), but the influence of the temperature must be great for other compounds such as $K_4Fe(CN)_6$ in water and may reach 1.9 ppm K^{-1} (Haslinger et al., 1981).

The earliest publications contain only some chemical shift values (Koridze et al., 1975a, b, 1978; Sahm and Schwenk, 1974; Schwenk, 1970, 1971). Only two current works (Haslinger et al., 1981; Jenny et al., 1981) have dealt with chemical applications of iron NMR. Most of the studied species are ferrocene (Haslinger et al., 1981) or carbonyl derivatives (Jenny et al., 1981). For the latter, the exchange of one or more carbonyl groups for other ligands does not significantly affect the chemical shifts which remain in the region from 0 to 300 ppm, but if the new ligand is a strong electron acceptor, the deshielding is very large (about 1600 ppm). Iron in the ferrocene series remains between 1535 and 2100 ppm (Jenny et al., 1981; Koridze et al., 1978), but the ferrocenyl ion and its derivatives experience a deshielding effect of 800 to 2000 ppm (Jenny et al., 1981).

B. Coupling Constants

The coupling constants are small because the magnetic moment of ^{57}Fe is small, and they are difficult to measure because the natural abundance is low; in most cases they are measured for enriched samples or from observation of the coupled nuclei. The coupling constant with ^{13}C in carbonyl derivatives ranges from 23 to 32 Hz (Jenny et al., 1981; Mann, 1971; Nielsen et al., 1976), and couplings in ferrocene compounds are small, 2–7 Hz (Koridze et al., 1975a,b; Mann, 1971; Nielsen et al., 1976). The value of $^1J_{^{57}Fe^{31}P}$ is about 26 Hz, and $^1J_{^{57}Fe^{15}N}$ = 8 Hz (Morishima et al., 1978). The highest values of J are characteristic of σ bonds, and the lowest are for π bonds. Coupling constants are not sensitive to the substituents but to the position of the ligands (Jenny et al., 1981), as is often observed for transition metals.

The only chemical problem solved using $^1J_{FeC}$ involves the structure of the ferrocenyl carbenium ion (Koridze *et al.*, 1975b), and the only theoretical study on $^1J_{FeC}$ (Nielsen *et al.*, 1976) agrees with experiments, but no experimental sign determinations are known.

C. Relaxation

As previously mentioned, the relaxation times of ^{57}Fe are very long, probably greater than 10 s, but no measurements are found in the literature. The small magnetic moment reduces the effect of all the relaxation mechanisms. The dipole–dipole interactions and the NOE effects are negligible because, protons are usually far from iron. This effect may perhaps occur in phosphorus derivatives (Mann, 1971). Sometimes the exchange phenomenon reduces the value of T_2 (Jenny *et al.*, 1981; Schwenk, 1971).

VII. Copper (Z = 29)

Only copper(I) is diamagnetic and gives observable signals. Two isotopes are active in NMR, and both have a spin of $\frac{3}{2}$. Their resonance frequencies are on the same order: $\Xi^{63}Cu = 26.5281$ MHz and $\Xi^{65}Cu = 28.4172$ MHz. The first has a natural abundance of 69.09%, and the second an abundance of 30.91%. Their quadrupole moments are similar, -0.211×10^{-28} m^2 for ^{63}Cu and -0.195×10^{-28} m^2 for ^{65}Cu. Both isotopes are easily observed, but ^{63}Cu is the most used because its receptivity is slightly greater.

A. Chemical Shifts

The first observation of copper NMR in solution was in the study of the disproportionation reaction of copper(I) in hydrochloric acid (McConnell and Berger, 1957; McConnell and Weaver, 1956). From the variation in the line width the exchange rate constant for the reaction $Cu_a^{2+} + Cu_b^+ \rightleftharpoons Cu_a^+ + Cu_b^{2+}$ is $k = 0.5 \times 10^8$ liter mol^{-1} s^{-1}.

Only after 1970 did chemists start to use copper NMR. Yamamoto *et al.* (1970) investigated solutions containing cyanocuprate(I) ions and showed that several species were present, the concentration of which depended on the ratio [CN$^-$]/[Cu$^+$], but no chemical shift differences were detected. Ellis *et al.* (1973) have studied solutions of $K_3Cu(CN)_4$.

Several references have been proposed: the Tübingen group (Lutz *et al.*, 1978a, b) used $Cu(CH_3CN)_4BF_4$ because its line width was at the time the narrowest known—540 Hz—and several investigators have done likewise (Harris and Mann, 1978, p. 258; Brevard and Granger, 1981a, p. 129; Kidd, 1980, p. 51). Yet, sharper lines have been discovered (Kroneck *et al.*, 1980a, 1982) with $Cu[P(C_2H_5)_3]_4^+ClO_4^-$ in $CDCl_3$. The influence of different ligands containing phosphorus has been investigated in detail (Kroneck *et al.*, 1980a). Phosphites have resonances between 73 and 91 ppm, but phosphines are more deshielded and their resonances range between 134 and 274 ppm. It has been shown (McFarlane and Rycroft, 1976) that there is no concentration effect on $Cu[P(OCH_3)_3]_4$ and that the counterion has no influence. On the contrary, an increase in temperature has a variable shielding effect from -0.06 ppm K^{-1} (McFarlane and Rycroft, 1976, Marker and Gunter, 1982) to -0.75 ppm K^{-1} (Kroneck *et al.*, 1982).

The chemical shift range of Cu(I) is about 500 ppm, including complexes with CH_3CN or pyridine (Ochsenbein and Schläpfer, 1980) and some derivatives of Cu(I) (Kroneck *et al.*, 1982), but this range is probably greater. No isotopic effect is observed with ^{63}Cu or ^{65}Cu (Lutz *et al.*, 1978a).

B. Coupling Constants

Coupling constants are difficult to measure because copper line widths are large. Only $^1J_{63Cu31P}$ is known and ranges from 1100 to 1460 Hz (Kroneck *et al.*, 1980a; Lutz *et al.*, 1978a; Marker and Gunter, 1982; McFarlane and Rycroft, 1976). These copper–phosphorus coupling constants are larger for phosphites than for phosphines, and their behavior is similar to that of tungsten.

C. Relaxation

The relaxation time measurement of ^{63}Cu and ^{31}P for Cu-$[P(OCH_3)_3]_4^+BF_4^-$ and $Cu[P(OCH_2CH_3)_3]_4^+BF_4^-$ (Marker and Gunter, 1982) has been reported as a function of temperature. Above $-30°C$ an exchange process occurs. Both T_1 and T_2 have been determined for $Cu(CH_3CN)_4^+$ (Ochsenbein and Schläpfer, 1980). Many chemical studies, however, use the line width (Kroneck *et al.*, 1980a; Lutz *et al.*, 1978b; McConnell and Weaver, 1956; Ochsenbein and Schläpfer, 1980; Yamamoto *et al.*, 1970), which can vary from 115 to 4250 Hz or more.

VIII. Niobium ($Z = 41$)

Niobium-93, with a natural abundance of 100% and a spin of $\frac{9}{2}$, has a receptivity of 2740 relative to ^{13}C. It is one of the most sensitive nuclei and is easily detected, but its quadrupole moment of -0.2×10^{-28} m^2 broadens the lines despite the favorable factor $(2I + 3)/I^2(2I - 1) = 0.074$ (Brevard and Granger, 1981a, p. 20). Its resonance frequency is $\Xi = 24.4761$ MHz.

A. Chemical Shifts

Several references are used in the literature: $NbCl_6^-$ in CH_3CN (Brevard and Granger, 1981a, p. 151; Harris and Mann, 1978; Kidd and Spinney, 1973, 1981), NbF_6^- (Harris and Mann, 1978, p. 208), $NbOCl_3$ (Bechthold and Rehder, 1981; Rehder et al., 1982), and $NbCl_5$ in CH_3CN (Bechthold and Rehder, 1981) or diglyme (Rehder et al., 1980). The relationship among all these scales is not always specified, and because the solvent effect is not known, it is not easy to compare results from the literature using different references. Nevertheless we give the following approximate conversion factors: $NbCl_5$ (CH_3CN or diglyme), 450 ppm compared to $NbOCl_3$ (Rehder et al., 1980, 1982); $NbCl_6^-$ (CH_3CN), 2227 ppm relative to NbF_6 (CH_3CN) (Harris and Mann, 1978, p. 210); and $NbOCl_3$ (CH_3CN), 987 ppm referred to NbF_6^- (CH_3CN) (Harris and Mann, 1978, p. 210).

The solvent effect is not clearly established for NbF_6^- (Buslaev et al., 1973; Packer and Muetterties, 1963), but Herrmann et al. (1981) have observed a difference of 10 ppm in the resonance of niobium in $(\eta\text{-}C_5H_5)_3$ $Nb_3(CO)_2$ in THF and in CH_2Cl_2. The chemical shift range is about 2500 ppm, and the temperature effect varies between -0.2 ppm K^{-1} (Tarasov et al., 1978a) and -0.4 ppm K^{-1} (Herrmann et al., 1981).

Niobium(V) is the most investigated oxidation state, and in halogenated derivatives the interhalogen chemical exchange has been extensively studied (Buslaev et al., 1971, 1973; Hatton et al., 1965; Kidd and Spinney, 1973; Packer and Muetterties, 1963; Tarasov et al., 1978a, b). The pairwise additivity model allows identification of all the detected species. The iodine derivative presents a deshielding as compared to NbF_6^-, and this behavior is different from that of most of the transition metals. Other halogenated derivatives with —O—, —SCN$^-$, NCO$^-$, S, and Se as ligands have been also studied (Buslaev et al., 1972, 1979; Tarasov et al., 1980). The observation of compounds of valence + or − has been very limited (Bechthold and Rehder, 1981; Herrmann *et al.*, 1981; Rehder *et al.*, 1980, 1982). Their chemical shifts are toward the lower frequencies.

Some theoretical interpretations have been made (Rehder *et al.*, 1982), but more experimental results are needed and calculations must be improved. Meadows *et al.* (1982) have shown that high-resolution solid state NMR is possible for this nucleus.

B. Coupling Constants

As with many quadrupolar nuclei, few coupling constants are known. Only couplings with ^{19}F and ^{31}P have been reported: $334 \leq {}^1J_{96_{Nb}19_F} \leq 345$ Hz in NbF_6^- (Buslaev *et al.*, 1973; Hatton *et al.*, 1965; Packer and Muetterties, 1963), $^2J_{96_{Nb}19_F} = 55$Hz in $Nb(PF_3)_6^-$ (Rehder *et al.*, 1980), and $^1J_{96_{Nb}31_P} = 1050$ Hz in the same compound.

C. Relaxation

The only relaxation mechanism is the quadrupolar one, even for symmetric molecules. Sometimes this mechanism is less efficient than relaxation from chemical exchange (Tarasov *et al.*, 1978a). The narrowest line is observed for $NbCl_6^-$ (30 Hz), but it increases rapidly to values of about 500 Hz for $NbCl_5$ or $NbBr_5$ in CH_3CN (Kidd and Spinney, 1973) and may reach 6000 Hz for $(\eta-C_5H_5)Nb(CO)_4$ substituted with phosphines (Bechthold and Rehder, 1981). The line width is sometimes temperature-dependent (Herrmann *et al.*, 1981; Howell and Moss, 1971b).

IX. Molybdenum ($Z = 42$)

This element has two NMR-active isotopes with a spin of $\frac{5}{2}$. Molybdenum-95 has a natural abundance of 15.72% and a resonance frequency $\Xi = 6.5169$ MHz; the same parameters for ^{97}Mo are 9.46% and $\Xi = 6.6536$ MHz. The quadrupole moments are 0.12×10^{-28} m^2 and 1.1×10^{-28} m^2, respectively. Hence the best isotope is ^{95}Mo. Several works are devoted to determination of the quadrupole moment (Kaufmann, 1964; Kaufmann *et al.*, 1975; Krueger *et al.*, 1973; Vold and Vold, 1975).

A. Chemical Shifts

The accepted reference seems to be MoO_4^- in a basic solution, however, Bailey *et al.* (1982) used $Mo(CO)_6$. Both reference compounds give narrow lines, but the second is difficult to handle. On the other hand, $Mo(CO)_6$ resonates at high field, and almost all the chemical shifts are positive. MoO_4^- appears in the middle of the range and leads to both positive and negative values.

The effect of the concentration and of the associated cation has been studied in detail for MoO_4^- (Kautt et al., 1976). Its chemical shift varies from 10 to −35 ppm relative to infinite dilution. Consequently, when MoO_4^- is used as a reference, the exact experimental conditions must be given. Solvent effects are sometimes very large. For instance, for MoS_4^- the chemical shift changes from 2176 ppm in dimethyl sulfoxide (DMSO) to 2254 ppm in H_2O (Gheller et al., 1981), whereas this effect is limited to a few parts per million with $[C_5H_5Mo(CO)_3]^-$ (Masters et al., 1981).

The literature does not mention the temperature effect, but an isotopic effect of 0.25 ppm is observed when ^{16}O is replaced by ^{18}O (Buckler et al., 1977), and a similar variation of 0.09 ppm is obtained for ^{32}S and ^{34}S (Lutz et al., 1977a). A larger effect of about 1 ppm is observed for H_2O and D_2O solutions.

The known chemical shift range is about 5500 ppm, with the lowest oxidation states at high field (Alyea et al., 1982; Le Gall et al., 1981). But most of the work done on this element is devoted to the oxidation states 0 and VI. The first is found between −1000 and −2120 ppm (relative to MoO_4^-), and the second leads to resonances between 3350 and −700 ppm. These two ranges do not overlap, but Mo(II), for instance, is in the range −2072 to −154 ppm.

The NMR of molybdenum has been used to detect species in equilibrium that cannot be separated and to study the numerous complexes of this element (Alyea et al., 1982; Bailey et al., 1982; Christensen et al., 1981; Dysart et al., 1981; Freeman et al., 1982; Gheller et al., 1981; Lutz et al., 1976a, 1977b; Masters et al., 1980, 1981). Often the resonances are very sensitive to the change in ligands. For instance, the change from oxygen to sulfur leads to a deshielding of 500 ppm (Lutz et al., 1977b) or of 188 ppm between oxygenated and sulfur ligands (Christensen et al., 1981). But sometimes this effect is small, as proved by Bailey et al. (1982). As for many transition metals, the resonance of molybdenum may be used to distinguish between isomers (Minelli et al., 1981).

B. Coupling Constants

The presence of a quadrupole moment limits the observability of fine structure, and it is impossible to make any measurements for ^{97}Mo (Andrews and McFarlane, 1978). The best studied coupling constants are those with phosphorus, their values ranging from 123 to 290 Hz (Alyea *et al.*, 1982; Andrews and McFarlane, 1978; Bailey *et al.*, 1982; Lutz *et al.*, 1976c; Masters *et al.*, 1980; Milbrath *et al.*, 1976). Coupling with other nuclei is very rare. We have found only $^1J_{95_{Mo}17_O} = 40.3$ Hz (Lutz *et al.*, 1976c; Vold and Vold, 1974), $^1J_{95_{Mo}13_C} = 68$ Hz (Mann, 1973), $^1J_{95_{Mo}1_H} = 15$ Hz (ref 38 in Masters *et al.*, 1981), and $^1J_{95_{Mo}19_F} = 48$ Hz (Muetterties and Phillips, 1959). The only two-bond coupling constant observed is $^2J_{95_{Mo}19_F} = 14$ Hz (Bailey *et al.*, 1982).

C. Relaxation

The quadrupolar mechanism is always dominant even in symmetric molecules in which the line width is only a few hertz. But when dissymmetry occurs, lines broaden out and disappear, as observed for MoO_4^- when the pH decreases (Kautt *et al.*, 1976; Vold and Vold, 1975). For ^{97}Mo the line widths are 130 times greater, which rapidly leads to unobservable lines (Kroneck *et al.*, 1980b). Measurements of T_1 were used to determine the ratio between the quadrupole moments of the two isotopes (Kaufmann *et al.*, 1975; Vold and Vold, 1974, 1975), and one study interpreted the results using molecular dynamics (Vold and Vold, 1974). The only measurement of the quadrupolar coupling constant for solid state Na_2MoO_4 leads to a surprising zero value for both ^{95}Mo and ^{97}Mo (Lynch and Segel, 1972).

X. Ruthenium (Z = 44)

The two NMR-active isotopes of ruthenium, ^{99}Ru and ^{101}Ru, both have a spin of $\frac{5}{2}$ and similar natural abundances, 12.72 and 17.07%, respectively, and their resonance frequencies are $\Xi = 4.6140$ and 5.1713 MHz (Brevard and Granger, 1981a, p. 156). Their relative receptivities compared to ^{13}C are on the same order, 0.83 and 1.56, but their quadrupole moments are different, 0.076×10^{-28} and 0.44×10^{-28} m^2.

These characteristics are not too unfavorable for observation, but it is only recently that two laboratories have simultaneously detected the resonance of both isotopes (Brevard and Granger, 1981b; Dykstra and Harrison, 1981). Despite its greater receptivity, the ^{101}Ru isotope should not be used, because its quadrupole moment is larger than that of ^{99}Ru. Only the diamagnetic oxidation states Ru(VIII), Ru(II), and Ru(0) can be studied.

A. Chemical Shifts

Two references have been proposed: $Ru(CN)_6^{4-}$ in D_2O (Dykstra and Harrison, 1982) and RuO_4 in CCl_4 (Brevard and Granger, 1981b). The first is better because it can be easily handled. The known chemical shift range is about 8300 ppm. Potassium salts should be avoided because the resonance of K^+ appears at low field of the ruthenium range at about 11,340 ppm. Chemical shifts are a function of concentration (Dykstra and Harrison, 1982) and temperature: 1 ppm K^{-1} (Brevard and Granger, 1983; Dykstra and Harrison, 1982). An isotopic effect of 0.37 ppm is observed between ^{12}C and ^{13}C, and one of 0.02 ppm between ^{16}O and ^{17}O (Brevard and Granger, 1981b). As usually observed for transition metals, low oxidation states appear at high field. The resonances are sensitive to the chemical environment and allow a distinction between isomers (Brevard and Granger, 1983; Dykstra and Harrison, 1982). The only published chemical application involves the exchange of chlorine with water in $Ru(CO)_3Cl_3^-$ (Brevard and Granger, 1983).

B. Coupling Constants

Three coupling constants are known, $^1J_{^{99}Ru^{17}O} = 23.4$ Hz in RuO_4 (Brevard and Granger, 1981b), $^1J_{^{99}Ru^{13}C} = 44.8$ Hz in $Ru(CN)_6^-$, and $^1J_{^{99}Ru^{119}Sn} = 846$ Hz in $Ru(SnCl_3)_5Cl^{4-}$ (Brevard and Granger, 1983). The reduced coupling constant with ^{17}O, which is 31.15×10^{20} NA^{-2} m^{-3} is similar to the coupling between molybdenum and oxygen with $K = 38.0 \times 10^{20}$ NA^{-2} m^{-3} in MoO_4^{2-} (Vold and Vold, 1974), which shows the similarity of the coupling mechanisms.

C. Relaxation

The line width of ^{99}Ru may be small—0.8 Hz (Brevard and Granger, 1981b)—and often on the order of a few hertz in symmetric environments (Dykstra and Harrison, 1982). The line width of ^{101}Ru is 32 times greater. The only T_1 measurement for RuO_4 leads to 0.98 s for ^{99}Ru and 0.033 s for ^{101}Ru (Brevard and Granger, 1981b); these values are compared to MoO_4^{2-} and OsO_4 and allow determination of the ratio of the quadrupole moments of the two ruthenium isotopes.

XI. Tin (Z = 50)

Tin is the only element with three NMR-active isotopes of spin $\frac{1}{2}$. It has 10 stable isotopes, and their characteristics are listed in the accompanying table.

Isotope	Natural abundance (%)	Receptivity referred to ^{13}C	Ξ (MHz)
^{115}Sn	0.35	0.7	32.6998
^{117}Sn	7.61	19.5	35.6322
^{119}Sn	8.58	25.2	37.2906

Usually ^{119}Sn is observed, but ^{117}Sn may also be used without too much loss of sensitivity.

The gyromagnetic ratios of all three isotopes are negative, and the NOE effect is then negative with a maximum magnitude of -1.34 for ^{119}Sn. But often this effect is small or nonexistent, and decoupling may be used without appreciable loss of intensity. The primary isotope effect has been the subject of some controversy (Harris and Mann, 1978, p. 245), but studies (Lyčka et al., 1981; McFarlane et al., 1979) have shown that this effect on the three isotopes is below the experimental error (0.1 ppm). This result agrees with previous studies on ^{10}B and ^{11}B and on ^{14}N and ^{15}N.

It is impossible to consider here in detail the NMR of this element, which is essentially devoted to ^{119}Sn. This nucleus was widely observed in early NMR studies, and Emsley, Feeney, and Sutcliffe discussed this element in 1966. Since this first review, other general papers have appeared (Harris and Mann, 1978, p. 342; Kennedy and McFarlane, 1973; Pereyre et al., 1982; Petrosyan, 1977; Smith and Smith, 1973; Smith and Tupcianskas, 1978). We shall limit our discussion to a general overview.

A. Chemical Shifts

The reference is $Sn(CH_3)_4$. When solvents such as C_6D_6, CCl_4, and $CDCl_3$ are used, the solvent effect is small as long as tin is not bonded to electronegative elements. In the latter case, or if tin or the solvent contains donor atoms, solvation equilibria occur and the solvent effect may be very large—on the order of several hundred parts per million. This is, for instance, the case with dimethyl sulfoxide (DMSO) and hexamethylphosphoramide (HMPT). The chemical shifts depend on the temperature.

This influence leads to an upfield or a downfield shift of 0.05 to 0.3 ppm K^{-1}. The secondary isotope effect may be large; for instance, the chemical shift difference between $Sn(CH_3)_4$ and $Sn(CD_3)_4$ is -2.86 ppm (Lassigne and Wells, 1978).

The chemical shift range of tin is about 2700 ppm. Collected data are found in Harris and Mann (1978, p. 350) and in Smith and Tupcianskas (1978). A very small number of Sn(II) compounds have been studied (Burke and Lauterbur, 1961), and they range between -700 and -2200 ppm. Most of the results are for Sn(IV) and organic derivatives. Chemical shifts depend on several factors:

(a) The coordination number: Tetracoordinated tin compounds range between -1700 and 500 ppm, pentacoordinated tin between -200 and 200 ppm, and hexacoordinated tin between -900 and -500 ppm. As the coordination number increases, the chemical shift decreases, as observed for other nuclei such as aluminum.

(b) The electronegativity of the substituents

(c) The geometry of the bonds near the tin atom

(d) The π-bonding effect

(e) The bulky atom effect

Tin chemical shifts are now well known, and the NMR of this element may be used like ^{13}C- or ^{31}P-NMR to solve chemical problems. A recent review develops this possibility (Pereyre et al., 1982). Even chemically induced dynamic nuclear polarization (CIDNP) effects have been observed with tin (Lehnig, 1981; Lehnig et al., 1978).

B. Coupling Constants

A large range of values are possible, and they may reach more than 15,000 Hz, for instance, for two directly bonded tin atoms (Mathiasch and Mitchell, 1980). Coupling constants of tin with many different atoms are known. An unusual feature is encountered with protons where $^3J_{SnH} > {}^2J_{SnH}$, as observed also for lead. Coupling constants with ^{13}C range between 290 and 970 Hz, and the sign is usually negative but may become positive when the bond between tin and some elements has a large ionic character. This is, for instance, the case with $(CH_3)_3SnLi$, where $^1J_{SnC}$ ranges from 155 to 220 Hz depending on the solvent. As with protons, $^3J_{SnC}$ is often greater than $^2J_{SnC}$. Coupling with elements such as ^{19}F, ^{31}P, ^{11}B, ^{15}N, ^{77}Se, ^{207}Pb (Kennedy et al., 1976), ^{29}Si, ^{183}W (Kennedy et al., 1975), ^{17}O, ^{125}Te, ^{203}Tl, ^{103}Rh (Moriyama et al., 1981), ^{35}Cl, ^{79}Br, ^{127}I, and ^{119}Sn is known. These values are usually stereospecific and may be used to solve structural problems (Burke and Lauterbur, 1961).

C. Relaxation

The value of T_1 is below 2 s, and often $T_2 < T_1$. This allows for rapid accumulation without a waiting time. In organic molecules spin–rotation is the most important relaxation mechanism (Lassigne and Wells, 1977). Halogens bonded to tin may lead to a possible scalar relaxation mechanism which allows measurement of coupling constants. In some cases a dipole–dipole interaction is observed (Mitchell, 1976), but none has mentioned an effect of the shielding anisotropy.

XII. Tungsten ($Z = 74$)

The resonance frequency of the only NMR-active isotope of tungsten, ^{183}W, is $\Xi = 4.1663$ MHz. With a spin of $\frac{1}{2}$ and a natural abundance of 14.28%, its receptivity compared to ^{13}C is 0.059. This is a difficult nucleus to observe, and only in 1974 was it first detected in solution (Sahm and Schwenk, 1974) for WO_4^{2-}. Since the extension of FT NMR spectrometers and high-field instruments, chemists have observed this element more extensively. The advent of enhancement techniques such as SPI and INEPT (Brevard and Schimpf, 1982) has increased the observability of ^{183}W. Its large NOE effect, which can reach a value of 12, may also be used whenever possible by the experimentalist.

A. Chemical Shifts

Several studies have taken WF_6 as a reference (Harris and Mann, 1978, p. 216), but in addition to the fact that this compound is difficult to handle, a large solvent effect of about 12 ppm is observed (McFarlane et al., 1971). Most recent studies have preferred WO_4^{2-} (Acerete et al., 1979a, b; Brevard and Granger, 1981a, p. 193; Gansow et al., 1980; Jeannin and Martin-Frère, 1981; Lefebvre et al., 1981) because the solvent effect is limited to several parts per million (Banck and Schwenk, 1975). The solvent effect depends on the compounds studied. For instance, for α-[η-$C_5H_5Ti(PW_{11}O_{39})]^{4-}$ it reaches 12 ppm between CD_3CN and DMF (Gansow et al., 1980), but as seen for WO_4^{2-} it may be also very small (Banck and Schwenk, 1975). The temperature effect is itself dependent on the solution; the reference WO_4^{2-} has a coefficient of 0.16 ppm K^{-1} (Banck and Schwenk, 1975), but $(\eta$-$C_5H_5)WCO_2H(PX_3)$ has a larger one, 1.5 ppm K^{-1} (McFarlane et al., 1976). Other intermediate values are also found (Banck and Schwenk, 1975; Malish et al., 1981).

As of the time of writing the chemical shift range is about 6900 ppm. There is a wide gap of 1200 ppm between oxidation states VI and 0. The latter are found at high field, beyond -1200 ppm.

The chemical shifts are very sensitive to the chemical environment, distinguishing between different structures or isomers (Acerete et al., 1979a, b; Finke et al., 1981; Gansow et al., 1980; Green and Brown, 1971; Jeannin and Martin-Frère, 1981; Lefebvre et al., 1981; McFarlane et al., 1971). For instance, the chemical shift difference between cis- and trans-$(CO)_4W[P(C_4H_9)_3]_2$ is 56 ppm, the trans isomer being at higher field. The NMR of tungsten is now the most powerful tool for structure determination of heteropolytungstates (Gansow et al., 1980; Lefebvre et al., 1981). Low oxidation states are less studied (McFarlane et al., 1976).

B. Coupling Constants

Owing to the difficulty of observation, the chemical shifts are not extensively studied, however, coupling constants for this nucleus have been known for a long time. This situation arises from the fact that, except for platinum and rhodium, this is the only transition metal with spin $\frac{1}{2}$, iron having too low a natural abundance. This allows the measurement of coupling constants in all cases from satellites of more sensitive nuclei coupled to it.

Coupling with phosphorus is the most studied, with values between 15 and 485 Hz for one bond (Abd-El-Mottaleb, 1976; Biffar et al., 1981; Fischer et al., 1972; Grim and Wheatland, 1968, 1969; Grim et al., 1967, 1969; Keiter and Verkade, 1969; McFarlane and Rycroft, 1973; Mc-Farlane et al., 1976; Verkade, 1972). Some relationships exist between $^1J_{WP}$ and other parameters such as force constants (Grim et al., 1967, 1969; Keiter and Verkade, 1969) and the electronegativity of substituents (Fischer et al., 1972; McFarlane et al., 1976). More interesting is the influence of isomerism. For instance, in the series $W(CO)_4(PZ_3)_2$ one has $265 < {^1J_{WP(trans)}} < 275$ Hz and $220 < {^1J_{WP(cis)}} < 230$ Hz (Green et al., 1971; Grim and Wheatland, 1968, 1969). The two-bond tungsten–phosphorus coupling constants are below 2 Hz and are difficult to measure (Acerete et al., 1979b; Finke et al., 1981; Gansow et al., 1980). The values of $^1J_{WF}$ are smaller and range from 12 to 70 Hz (Banck and Schwenk, 1975; Keiter and Verkade, 1969; Lefebvre et al., 1981; McFarlane et al., 1970; Muetterties and Phillips, 1959) and seem to depend on the isomerism (Mc-Farlane et al., 1971). One-bond coupling constants with protons are on the same order: 28–80 Hz (Davisson et al., 1963; Deubzer and Kaesz, 1968; Jesson, 1971). Although the two-bond couplings are small, 1–3 Hz

(Kennedy et al., 1975; McFarlane et al., 1976; Shortland and Wilkinson, 1972), they sometimes have larger values such as 19.4 Hz in cyclopentadienyl derivatives (Deubzer and Kaesz, 1968). Some $^3J_{WH}$ values are known (Kennedy et al., 1975; McFarlane and Rycroft, 1973; McFarlane et al., 1976). One-bond ^{13}C coupling constants with tungsten lie between 124 and 197 Hz (Ashworth et al., 1981; Köhler et al., 1976; Mann, 1973, 1974) and are a function of the hybridization of the atoms (Köhler et al., 1976). Except for these common nuclei, couplings with other elements are limited. We have found only $^1J_{WO} < 10$ Hz (Banck and Schwenk, 1975), $^1J_{WSn} = -150$ Hz (Kennedy et al., 1975; McFarlane et al., 1976), $^2J_{WSn} = 16$ Hz (Biffar et al., 1981), $^1J_{WPt} = 177$ Hz (Ashworth et al., 1981), and several values of $^2J_{WW}$ ranging between 19 and 22 Hz or from 5 to 8 Hz according to the W—O—W bond angle (Lefebvre et al., 1981), which are of great interest in structural determinations.

C. Relaxation

The values of T_1 and T_2 have been measured only for WF_6 (Banck and Schwenk, 1974; Kronenbitter and Schwenk, 1977). Such measurements are difficult because of the low sensitivity of this nucleus, but values of $T_1 = 4.2$ s and $T_2 = 2.1$ s have been reported. The difference between T_1 and T_2, which could have been expected to be equal, are interpreted as arising from diffusion processes.

XIII. Platinum (Z = 78)

Platinum-195, with a spin of $\frac{1}{2}$, is the only NMR-active platinum isotope. With a natural abundance of 33.8% and a resonance frequency of $\Xi = 21.4617$ MHz, it has a receptivity of 19.1 compared to ^{13}C and is an easily detected nucleus. The literature is abundant, and it is outside the scope of this chapter to present all references and a detailed discussion. For more complete coverage the reader is referred to the review by Pregosin (1982).

A. Chemical Shifts

The most common reference is $PtCl_6^{2-}$ (Kerrison and Sadler, 1978; Pregosin, 1982). But this ion does not give a narrow line because of the isotope effect of chlorine, which increases the line width at higher fields. A large temperature effect is also observed. For these reasons it has been proposed (Harris and Mann, 1978, p. 250) that an arbitrary frequency reference be used, $\Xi = 21.4$ MHz. This results in a precise reference but requires magnetic susceptibility corrections (Brevard and Granger, 1981a,

p. 40) and eventually lock corrections (Brevard and Granger, 1981a, p. 43). The correspondence between the two scales is not straightforward: $\delta_{PtCl_6{}^{2-}} = 4522$ ppm as measured on the second scale (Harris and Mann, 1978, p. 251) versus a reported value of $\delta_{PtCl_6{}^{2-}} = 4533$ ppm (Pregosin, 1982).

The effect of concentration has not been extensively studied. A variation of 24 ppm is observed for $PtCl_4{}^{2-}$ from 0.065 M to 2.3 M in H_2O (Freeman et al., 1976), and a small effect on $PtCl_6{}^{2-}$ (Freeman et al., 1976). The largest detected effect on $PtCl_4{}^{2-}$ seems to have its origin in a hydrolysis equilibrium (Freeman et al., 1976). The solvent effect has been well investigated, and is very large. From H_2O to DMSO changes of -400 ppm for $PtCl_4{}^{2-}$ and -242 ppm for $PtCl_6{}^{2-}$ are observed (Pesek and Mason, 1977). This influence may be larger than the effect of chemical substitution on the platinum atom. But solvent effects are not always so large (Appleton and Hall, 1980; Balt and Meuldijk, 1981). A shift of 11 ppm between H_2O and D_2O solutions has also been observed (Pesek and Mason, 1977).

Because of the sensitivity of platinum to the change in its chemical environment, it is not surprising to find large secondary isotope effects. The effect of chlorine on $PtCl_6{}^{2-}$ has been previously mentioned, but the same effect is observed on $PtBr_6{}^{2-}$ (Ismail et al., 1980). For the former ion a shift of 0.167 ppm is observed, and an effect of 0.028 ppm for the latter. A larger perturbation of 0.63 to 1 ppm is detected when ^{16}O is replaced by ^{18}O (Groning et al., 1982). The effect of temperature may range from 0.1 ppm K^{-1} (Pregosin, 1982) to 1 or 1.5 ppm K^{-1} for $PtCl_6{}^{2-}$ and $PtCl_4{}^{2-}$ (Freeman et al., 1976).

The chemical shift range is about 15,000 ppm. Platinum(IV) appears at a higher frequency than platinum(II) or platinum(0). But all the ranges overlap (Kidd, 1980). Many chemical problems have been solved using platinum NMR, and the previously cited review (Pregosin, 1982) contains a good presentation of all the possibilities. Platinum NMR is especially useful in the study of complexes and allows determination of the different isomers. For instance, it has been shown by this method that some complexes such as $Pt(CH_3)_2Z$ are tetrameric (Appleton and Hall, 1980). The influence of paramagnetic reagents has been studied as the exchange of ligands between platinum and the paramagnetic nucleus (Hirayama and Sasaki, 1982; Zharkova et al., 1982).

B. Coupling Constants

Many coupling constants are known, and they are of great interest in structure elucidation. Because of the natural abundance of ^{195}Pt, care is necessary in the measurement of coupling constants (Brevard and Granger, 1981a, p. 6). Besides mercury, platinum is one of the nuclides that leads to very large couplings which may reach about 30 KHz (Ostoja-Starzewski and Pregosin, 1980). Several reviews have appeared on coupling constants with ^1H, ^{13}C, and ^{31}P (Appleton et al., 1973; Chisholm and Godleski, 1976; Mann, 1974; Nixon and Pidcock, 1969; Pregosin, 1981; Pregosin and Kunz, 1979). All these couplings are structurally dependent. Coupling constants are known with many other nuclei, ^{11}B, ^{14}N, ^{15}N, ^{19}F, ^{77}Se, ^{103}Rh, ^{119}Sn, ^{125}Te, ^{183}W, ^{207}Pb, and ^{195}Pt itself (Carr et al., 1982; Chikuma and Pollock, 1982; Crocker and Goodfellow, 1981; Harris and Mann, 1978, p. 256; Pregosin, 1982). At the present time, the structural interpretation of J_{PtPt} is not well understood.

C. Relaxation

Several works have been devoted to T_1 and T_2 measurements. The value of T_1 is below 1.7 s (Freeman et al., 1976; Lallemand et al., 1980; Pesek and Mason, 1977); T_2 is usually smaller than T_1 and no greater than 5×10^{-2} s. The difference between T_1 and T_2 is explained by the large chemical shift anisotropy (CSA) mechanism often met with platinum (Lallemand et al., 1980). This effect explains the line broadening with increasing field, which often leads to a loss of information on small coupling constants or small chemical shifts. A good example has been reported recently (Lallemand et al., 1980). Relaxation times are also solvent-dependent, for instance, $T_1 = 0.89$ s for $PtCl_6^{2-}$ in CH_2Cl_2 and 1.44 sec in CH_3CN (Pesek and Mason, 1977).

XIV. Lead (Z = 82)

The 22.6% natural abundance of ^{207}Pb with a resonance frequency of $\Xi = 20.8581$ MHz and a spin of $\frac{1}{2}$ leads to a receptivity of 11.8 compared to ^{13}C. A positive NOE effect, with a maximum of 2.4, is expected, but it is usually zero because the bond lengths are too great.

The NMR of lead was not well developed when last reviewed (Harris and Mann, 1978, p. 366), but numerous works have appeared since then.

A. Chemical Shifts

In addition to the initial works (Clawley and Danyluk, 1964; Schneider and Buckingham, 1962), the usual reference is neat $Pb(CH_3)_4$. Like most of the heavier nuclei, lead is sensitive to all factors: concentration, solvent, temperature. The first has been only briefly studied (Hawk and

Sharp, 1974b; Maciel and Dallas, 1973); the concentration effect is large for Pb^{2+} and may reach 35 ppm mol^{-1}. On the contrary, organometallic derivatives seem to experience a smaller effect of about several parts per million except when chemical reactions occur, as with $(CH_3)_3PbCl$ or (phenyl)$_3PbCl$ (Lucchini and Wells, 1980).

The solvent effect has been investigated more extensively. This influence is very large. For instance, $(Et)_3PbCl$ is measured at 472 ppm in $CHCl_3$ and at 305 ppm in pyridine, corresponding to a shielding of 167 ppm (Mitchell et al., 1978). Similar effects are observed on other systems (Cooper et al., 1974; Cox, 1979; Kennedy et al., 1977). It is not surprising then to observe a shielding of 31 ppm for $Pb(NO_3)_2$ between solutions in H_2O and D_2O (Lutz and Stricker, 1971). Organometallic derivatives of Pb(IV) lead to limited effects of several parts per million or less (Bakhbukh et al., 1979; Cooper et al., 1974; Cox, 1979; Kennedy et al., 1977), which, however, are not negligible for the reference $Pb(CH_3)_4$ (Bakhbukh et al., 1979; Cox, 1979). Only one report has mentioned the temperature effect (Maciel and Dallas, 1973) on $(iso\text{-}C_4H_9)_3PbOCOCH_3$ in acetic acid, for which a shielding of 0.5 ppm K^{-1} is observed. This is not an unusual value for a heavy nucleus, but larger effects may be expected.

The known chemical shift range for lead in solution is about 3500 ppm. Organometallic compounds are between -1870 and 480 ppm, and Pb^{2+} derivatives range between -1300 and -2980 ppm (Lutz and Stricker, 1971; Maciel and Dallas, 1973). Most of the literature deals with substituent effects on $PbRR'R''R'''$ molecules (Banney et al., 1970; Biffar et al., 1981; Clark et al., 1969; Cooper et al., 1974; Cox, 1979; Hays et al., 1981; Kennedy et al., 1976a, 1977; Kim and Bray, 1974; Kusaba et al., 1982; Maciel and Dallas, 1973; Mitchell and Marsmann, 1981; Mitchell et al., 1978; Otera et al., 1982; Van Beelen and Wolters, 1980; Van Beelen et al., 1979; Wrackmeyer, 1979). Some authors have found additivity rules in well-defined series (Biffar et al., 1981; Mitchell and Marsmann, 1981). It seems that a linear relation exists between the chemical shifts of lead and tin: $\delta_{Pb} \simeq 3\delta_{Sn}$ (Biffar et al., 1981; Kennedy et al., 1977). Only two studies have used NMR to solve chemical problems. The first dealt with the equilibrium $(CH_3)_3PbCl \rightleftharpoons (CH_3)_3Pb^+(solvent) + Cl^-$ (Lucchini and Wells, 1980), and the second demonstrated the existence of poly-condensated ions of the form $(Pb_9)^{4-}$ and $(Pb_{9-x}Sn_x)^{4-}$ (Rudolph et al., 1978). The NMR of this nucleus is a valuable tool for chemical studies but it is sometimes limited by the low solubility of lead derivatives.

B. Coupling Constants

As for platinum, many coupling constants were measured before the advent of ^{207}Pb NMR for satellites of more common nuclei. Only two values of $^1J_{PbH}$ are known, one for $(CH_3)_3PbH$ with $^1J_{PbH}$ = 2379 Hz (Flitcroft and Kaesz, 1963), and one for PbH_4 with $^1J_{PbH}$ = 2456 Hz (Schumann and Dreeskamp, 1970). Most of the values are two- or three-bond coupling constants between lead and protons (Clark et al., 1969; Cooper et al., 1974; De Vos, 1976; De Vos et al., 1979b; Flitcroft and Kaesz, 1963; Fritz and Schwartzhaus, 1964a,b; Hildenbrand and Dreeskamp, 1970; Kawasaki and Aritoni, 1976; Kennedy et al., 1976a, 1977, 1980; Majima and Kawasaki, 1979; McFarlane, 1976; Narasimhan and Rogers, 1961; Shier and Drago, 1966; Singh, 1968, 1975; Van Beelen and Wolters, 1980b; Van Beelen et al., 1976), but six-bond coupling is also known (Kitching et al., 1967). A feature of the proton–lead coupling constants is that $^3J_{PbH}$ is often greater than $^2J_{PbH}$, as observed for tin. A theoretical explanation has been given by De Vos et al. (1980). For eth-ylenic derivatives $^3J_{PbH(trans)}$ is greater than $^3J_{PbH(cis)}$, and $^2J_{PbH(gem)}$ is inter-mediate. Their signs are identical to those of the proton–proton coupling constants (Clawley and Danyluk, 1964; Krebs and Dreeskamp, 1969; Mitchell and Marsmann, 1981). Allenic derivatives (Simonnin et al., 1970) and heterocyclic compounds have also been studied (Barbieri and Taddei, 1972; Bulman, 1969). These couplings are sometimes solvent-dependent (De Vos et al., 1979a; Kawasaki and Aritoni, 1976; Majima and Kawasaki, 1979) and J_{PbH} increases with the donor strength of the sol-vent. Sometimes this parameter is concentration-dependent (Kawasaki and Aritoni, 1976).

The coupling constants of lead with ^{13}C are widely studied (Adcock et al., 1974, 1977; Bullpitt et al., 1976; Clark et al., 1969; De Vos, 1976; De Vos and Wolters, 1978; De Vos et al., 1979a; Doddrell et al., 1974; Grün-ing et al., 1977; Hildenbrand and Dreeskamp, 1970; McFarlane, 1967, 1976; Mitchell and Marsmann, 1981; Mitchell et al., 1978; Singh, 1975; Vaickus and Anderson, 1980; Van Beelen and Wolters, 1980b; Van Beelen et al., 1976, 1980; Wrackmeyer, 1979, 1981) and have been ob-served up to five bonds. Several relationships have been established, for instance, for a phenyl ring, $^1J_{PbC} \gg ^3J_{PbC} > ^2J_{PbC} > ^4J_{PbC}$ (De Vos and Wolters, 1978), and often for aliphatic compounds, $^3J_{PbC} > ^2J_{PbC}$ (Clark et al., 1969; McFarlane, 1976; Singh, 1975). As with protons, solvent effects are large (Mitchell et al., 1978; Wrackmeyer, 1981); for instance, $^1J_{PbC}$ ranges from 205 to 268 Hz for $(Et)_3PbCl$ in $CHCl_3$ and DMSO (Mitchell et al., 1978). A Karplus curve has been observed for $^2J_{PbC}$ (Kitching et al., 1976). Theoretical interpretations are given for simple molecules (De Vos et al., 1980).

Coupling constants of lead with many other elements have been mea-sured: $^4J_{PbF}$ = 40 Hz (Van Beelen et al., 1979) and $^5J_{PbF}$ = 18 to 23 Hz (Cooper et al., 1974; Van Beelen et al., 1979); $^1J_{Pb^{14}N}$ = 170 Hz (Kennedy *et al.*, 1980) and $^1J_{Pb^{15}N} = 261$ Hz (Kennedy *et al.*, 1976a). It has been also observed that the coupling with nitrogen is stereospecific (Alcock *et al.*, 1979); $^1J_{Pb^{15}N(trans)} = 207.5$ Hz, and $^1J_{Pb^{15}N(cis)} = 19.8$ Hz. Only one value is known for the coupling to phosphorus, $^1J_{PbP} = -1335$ Hz (Kennedy *et al.*, 1976a), to chlorine, $^1J_{PbCl} = 705$ Hz (Hawk and Sharp, 1974a), and to boron, $^1J_{Pb^{11}B} = 1330$ (Kennedy *et al.*, 1976a). Coupling with tin has been more extensively studied and is known up to three bonds (Biffar *et al.*, 1981; Kennedy *et al.*, 1976a; Wrackmeyer, 1979, 1981). The value of $^1J_{PbSe}$ is -1170 Hz (Kennedy *et al.*, 1976a), and $^1J_{PbPt} = 18,375$ Hz (Carr *et al.*, 1982). Only one homonuclear coupling constant is found in the literature (Kennedy and McFarlane, 1974), which has been measured indirectly; its value is $^1J_{PbPb} = 290$ Hz.

C. Relaxation

Several workers have studied lead relaxation. For symmetric environments the main mechanism is spin–rotation (Hays *et al.*, 1981; Hawk and Sharp, 1973, 1974b), but in one case a scalar relaxation mechanism occurs (Hawk and Sharp, 1974a) where $T_2 < T_1$, which leads to the determination of $^1J_{PbCl}$, as previously mentioned. The maximum value of T_1 measured is 1.6 s (Hays *et al.*, 1981; Maciel and Dallas, 1973). When molecules are disymmetric, the CSA mechanism appears and predominates at high field or at low temperatures (Hays *et al.*, 1981). The CSA is known in the solid state for several compounds (Lauterbur and Burke, 1965; Lutz and Nolle, 1980; Nolle, 1977). The differences $\sigma_\parallel - \sigma_\perp$ range between 50 and 880 ppm. In some cases a dipole–dipole mechanism is observed with the solvent molecules in very concentrated solutions (Hawk and Sharp, 1974b).

XV. Bismuth ($Z = 83$)

Bismuth-209, with a natural abundance of 100%, is the last stable isotope in the periodic table. It has a spin of $\frac{9}{2}$, very good receptivity of 777 compared to ^{13}C, and a resonance frequency $\Xi = 16.0692$ MHz. But its quadrupole moment of -0.4×10^{-28} m^2 limits its observation to very symmetric environments.

Since the discovery of the resonance of ^{209}Bi (Proctor and Yu, 1950, 1951) for $Bi(NO_3)_3$ in D_2O, which is the only known spectra in solution (Brevard and Granger, 1981a), only Fukushima and Mastin (1969) and Fukushima (1971) have observed $KBiF_6$ in the solid state, and they have obtained $^1J_{BiF} = 2700$ Hz and a quadrupolar coupling constant of 132 MHz. Only recently have mixtures of Bi and BiBr$_3$ been studied in the liquid state above the melting point (Dupree and Gardner, 1980). The chemical shift decreases rapidly, and the relaxation increases when BiBr$_3$ is added to the melt. These results are interpreted as proof of a short-lived species that is BiBr$_3$ dissolved in the solution.

The relaxation mechanism is only quadrupolar. But considering that the quadrupole moment of bismuth is the same as that of ^{59}Co, knowledge of the NMR of this nucleus may increase rapidly.

Acknowledgments

We are indebted to Professors D. Rehder and P. S. Pregosin for the communication of information prior to publication and to Professor B. Ancian for the bibliography on tin.

References

Abd-El-Mottaleb, M. S. A. (1976). *J. Mol. Struct.* **32**, 203.

Acerete, R., Harmalker, S., Hammer, C. F., Pope, M. T., and Baker, L. C. W. (1979a). *J. Chem. Soc. Chem. Commun.* 777.

Acerete, R., Hammer, C. F., and Baker, L. C. W. (1979b). *J. Am. Chem. Soc.* **101**, 267.

Adcock, W., Gupta, B. D., Kitching, W., Doddrell, D., and Geckle, M. (1974). *J. Am. Chem. Soc.* **96**, 7360.

Adcock, W., Cox, D. P., and Kitching, W. (1977). *J. Organomet. Chem.* **133**, 393.

Alcock, N. W., Herron, N., and Moore, P. (1979). *J. Chem. Soc. Dalton Trans.* 1486.

Allerhand, A. (1970). *J. Chem. Phys.* **52**, 2162.

Alyea, E. C., Lenkinski, R. E., and Somogyvari, A. (1982). *Polyhedron* **1**, 130.

Ancian, B., Tiffon, B., and Dubois, J. E. (1979). *Chem. Phys. Letters* **65**, 281.

Andrews, G. T., and McFarlane, W. (1978). *Inorg. Nucl. Chem. Lett.* **14**, 215.

Appleton, T. G., Clark, H. C., and Manzer, L. E. (1973). *Coord. Chem. Rev.* **10**, 335.

Appleton, T. G., and Hall, J. R. (1980). *Aust. J. Chem.* **33**, 2387.

Ashworth, T. V. *et al.* (1981). *J. Chem. Soc. Dalton Trans.* 763.

Atam, N., Mueller, H., and Dehnicke, K. (1972). *J. Organomet. Chem.* **37**, 15.

Bailey, J. T., Clark, R. J., and Levy, G. C. (1982). *Inorg. Chem.* **21**, 2085.

Bakhbukh, M., Grishin, Y. K., Ustynyuk, Y. A., and Zemlyanskii, N. N. (1979). *Vestn. Mosk. Univ. Ser. 2 Khim.* **20**, 366.

Balt, S., and Meuldijk, J. (1981). *Inorg. Chim. Acta* **47**, 217.

Banck, J., and Schwenk, A. (1975). *Z. Phys.* **20B**, 75.

Banney, P. J., McWilliam, D. C., and Wells, P. R. (1970). *J. Magn. Reson.* **2**, 235.

Barabas, A. (1972). *Rev. Roum. Chim.* **17**, 1997.

Barbieri, G., and Taddei, F. (1972). *J. Chem. Soc. Perkin Trans.* **2**, 262.

Basler, W., Lechert, H., Paulsen, K., and Rehder, D. (1981). *J. Magn. Reson.* **45**, 170.

Baugher, J. F., Taylor, P. C., Oja, T., and Bray, P. J. (1969). *J. Chem. Phys.* **50**, 4914.

Baumgarten, H., Johannsen, H., and Rehder, D. (1979). *Chem. Ber.* **112**, 2650.

Bechthold, H. C., and Rehder, D. (1981). *J. Organomet. Chem.* **206**, 305.

Biffar, W., Gasparis-Ebeling, T., Noth, H., Storch, W., and Wrackmeyer, B. (1981). *J. Magn. Reson.* **44**, 54.

Bodner, G. R., and Todd, L. J. (1974). *Inorg. Chem.* **13**, 1335.

Borowski, R., Rehder, D., and Von Deuten, K. (1981). *J. Organomet. Chem.* **220**, 45.

Brevard, C., and Granger, P. (1981a). "Handbook of High Resolution Multinuclear NMR." Wiley, New York.

Brevard, C., and Granger, P. (1981b). *J. Chem. Phys.* **75**, 4175.

Brevard, C., and Granger, P. (1983). *Inorg. Chem.* **22**, 532.

Brevard, C., and Schimpf R. (1982). *J. Magn. Reson.* **47**, 528.

Buckler, K. U., Haase, A. R., Lutz, O., Mueller, M., and Nolle, A. (1977). *Z. Naturforsh.* **32A**, 126.

Bullpitt, M., Kitching, W., Adcock, W., and Doddrell, D. (1976). *J. Organomet. Chem.* **116**, 161.

Bulman, M. J. (1969). *Tetrahedron* **25**, 1433.

Burke, J. J., and Lauterbur, P. C. (1961). *J. Am. Chem. Soc.* **83**, 328.

Buslaev, Y. A., Kopanev, V. O., and Tarasov, V. P. (1971). *J. Chem. Soc. D Chem. Commun.*, 1175.

Buslaev, Y. A., Il'In, E. G., Kopanev, V. D., and Tarasov, V. P. (1972). *Zh. Strukt. Khim.* **13**, 930.

Buslaev, Y. A., Kopanev, V. D., Sinitsyna, S. M., and Khlebodarov, V. G. (1973). *Zh. Neorg. Khim.* **18**, 2567.

Buslaev, Y. A., Tarasov, V. P., Sinitsyna, S. M., Khlebodarov, V. G., and Kopanev, V. D. (1979). *Koord. Khim.* **5**, 189.

Carr, S., Colton, R., and Dakternieks, D. (1982). *J. Magn. Reson.* **47**, 156.

Chikuma, M., and Pollock, R. J. (1982). *J. Magn. Reson.* **47**, 324.

Chisholm, M. H., and Godleski, S. (1976). "Progress in Inorganic Chemistry", Vol. 20, p. 299. Wiley, New York.

Christensen, K. A., Miller, P. E., Minelli, M., Rockway, T. W., and Enemark, J. H. (1981). *Inorg. Chim. Acta* **56**, L27.

Clark, R. J. H., Davies, A. G., Puddephatt, R. J., and McFarlane, W. (1969). *J. Am. Chem. Soc.* **91**, 1334.

Clawley, S., and Danyluk, S. S. (1964). *J. Phys. Chem.* **68**, 1240.

Cooper, M. J., Holliday, A. K., Makin, P. H., Puddephatt, R. J., and Smith, P. J. (1974). *J. Organomet. Chem.* **65**, 377.

Cox, R. H. (1979). *J. Magn. Reson.* **33**, 61.

Crocker, C., and Goodfellow, R. J. (1981). *J. Chem. Res. Synop.* 38.

Davisson, A., McCleverty, J. A., and Wilkinson, G. (1963), *J. Chem. Soc.* 1133.

Dehmelt, H. G. (1953). *J. Chem. Phys.* **21**, 380.

Delpuech, J. J., Peguy, A., Rubini, P., and Steinmetz, J. (1977). *Nouv. J. Chim.* **1**, 133.

Deubzer, B., and Kaesz, H. D. (1968). *J. Am. Chem. Soc.* **90**, 3276.

De Vos, D. (1976). *J. Organomet. Chem.* **104**, 193.

De Vos, D., and Wolters, J. (1978). *J. R. Neth. Chem. Soc.* **97**, 219.

De Vos, D., Van Beelen, D. C., and Wolters, J. (1979a). *J. Organomet. Chem.* **172**, 303.

De Vos, D., Van Barneveld, W. A. A., Van Beelen, D. C., and Wolters, J. (1979b). *Rec. Trav. Chim. Pays-Bas* **98**, 202.

De Vos, D., Van Beelen, D. C., and Wolters, J. (1980). *Bull. Soc. Chim. Belg.* **89**, 791.

Dharmatti, S. S., and Weaver, H. E. (1951). *Phys. Rev.* **83**, 845.

Doddrell, D. *et al.*, (1974). *Aust. J. Chem.* **27**, 417.

Drew, D. A., and Morgan, G. L. (1977). *Inorg. Chem.* **16**, 1704.

Dupree, R., and Gardner, J. A. (1980). *J. Phys. Colloq.* (*Orsay Fr.*) **C8**, 20.

Dykstra, R. W., and Harrison, A. M. (1981). *J. Magn. Reson.* **45**, 108.

Dykstra, R. W., and Harrison, A. M. (1982). *J. Magn. Reson.* **46**, 338.

Dysart, S., Georgü, I., and Mann, B. E. (1981). *J. Organomet. Chem.* **213**, C10.

Egozy, Y., and Loewenstein, A. (1969). *J. Magn. Reson.* **1**, 494.

Ellis, P. D., Walsh, H. C., and Peters, C. S. (1973). *J. Magn. Reson.* **11**, 426.

Emsley, J. W., Feeney, J., and Sutcliffe, L. H. (1966). "High Resolution NMR Spectroscopy", Vol. 2, p. 1082. Pergamon, Oxford.

Epperlein, B. W., Krüger, H., Lutz, O., and Nolle, A. (1975). *Z. Naturforsch.* **30A**, 1237.

Faure, R., Vincent, E. J., Ruiz, J. M., and Lena, L. (1981). *Org. Magn. Reson.* **15**, 401.

Finke, R. G., Droege, M., Hutchison, J. R., and Gansow, O. (1981). *J. Am. Chem. Soc.* **103**, 1587.

Fischer, E. O., Knauss, L., Keiter, R. L., and Verkade, J. G. (1972). *J. Organomet. Chem.* **37**, C7.

Fischer, P., Stadelhofer, J., and Werden, J. (1976). *J. Organomet. Chem.* **116**, 65.

Flitcroft, N., and Kaesz, H. D. (1963). *J. Am. Chem. Soc.* **85**, 1377.

Flynn, C. M., Pope, M. T., and O'Donnell, S. E. (1974). *Inorg. Chem.* **13**, 831.

Freeman, M. A., Schultz, F. A., and Reilly, C. N. (1982). *Inorg. Chem.* **21**, 567.

Freeman, W., Pregosin, P. S., Sze, S. N., and Venanzi, L. M. (1976). *J. Magn. Reson.* **22**, 473.

Fritz, H. P., and Schwarzhaus, K. E. (1964a). *Ber.* **97**, 1390.

Fritz, H. P., and Schwarzhaus, K. E. (1964b). *J. Organ. Chem.* **1**, 297.

Fukushima, J. E. (1971). *J. Chem. Phys.* **55**, 2463.

Fukushima, J. E., and Mastin, S. H. (1969). *J. Magn. Reson.* **1**, 648.

Gaines, D. F., Coleson, K. M., and Hillenbrand, D. F. (1981). *J. Magn. Reson.* **44**, 84.

Gansow, O. A., Ho, R. K. C., and Klemperer, W. G. (1980). *J. Organomet. Chem.* **187**, C27.

Gheller, S. F. *et al.* (1981). *Inorg. Chim. Acta* **54**, L131.

Gillen, K. T., and Noggle, J. H. (1970). *J. Chem. Phys.* **53**, 801.

Gornostansky, S. D., and Stazer, C. V. (1967). *J. Chem. Phys.* **46**, 4959.

Gossard, A. C. (1966). *Phys. Rev.* **149**, 246.

Green, P. J., and Brown, T. H. (1971). *Inorg. Chem.* **10**, 206.

Grigoreo, A. I., Sipacheo, V. A., and Novoselova, A. V. (1971). *Dokl. Akad. Nauk. SSSR* **199**, 603.

Grim, S. O., and Wheatland, D. A. (1968). *Inorg. Nucl. Chem. Letters* **4**, 187.

Grim, S. O., and Wheatland, D. A. (1969). *Inorg. Chem.* **8**, 1716.

Grim, S. O., Wheatland, D. A., and McFarlane, W. (1967). *J. Am. Chem. Soc.* **89**, 5573.

Grim, S. O., McAllister, P. R., and Singer, R. M. (1969). *J. Chem. Soc. Chem. Commun.* 38.

Groning, O., Drakenberg, T., and Elding, L. I. (1982). *Inorg. Chem.* **21**, 1820.

Grüning, R., Krommes, P., and Lorberth, J. (1977). *J. Organomet. Chem.* **127**, 167.

Habayeb, M. A., and Hileman, O. E. (1980a). *Can. J. Chem.* **58**, 2115.

Habayeb, M. A., and Hileman, O. E. (1980b). *Can. J. Chem.* **58**, 2255.

Harris, R. K., and Mann, B. E. (1978). "NMR and the Periodic Table." Academic Press, New York.

Haslinger, E., Robien, W., Schlögl, K., and Weissensteiner, W. (1981). *J. Organomet. Chem.* **218**, C11.

Hatton, J. V., Saito, Y., and Schneider, W. G. (1965). *Can. J. Chem.* **43**, 47.

Hawk, R. M., and Sharp, R. R. (1973). *J. Magn. Reson.* **10**, 385.

Hawk, R. M., and Sharp, R. R. (1974a). *J. Chem. Phys.* **60**, 1009.

Hawk, R. M., and Sharp, R. R. (1974b). *J. Chem. Phys.* **60**, 1522.

Hays, G. R., Gillies, D. G., Blaauw, L. P., and Clague, A. D. H. (1981). *J. Magn. Reson.* **45**, 102.

Heath, E., and Howarth, O. W. (1981). *J. Chem. Soc. Dalton Trans.* 1105.

Herrmann, W. A. *et al.* (1981). *J. Am. Chem. Soc.* **103**, 1692.

Hildenbrand, K., and Dreeskamp, H. (1970). *Z. Phys. Chem.* **69**, 171.

Hirayama, M., and Sasaki, Y. (1982). *Chem. Lett.* 195.

Howarth, O. W., and Hunt, J. R. (1979). *J. Chem. Soc. Dalton Trans.* 1388.

Howarth, O. W., and Jarrold, M. (1978). *J. Chem. Soc. Dalton Trans.* 503.

Howarth, O. W., and Richards, R. E. (1964). *Proc. Phys. Soc. London* **84**, 326.

Howarth, O. W., and Richards, R. E. (1965). *J. Chem. Soc.* 864.

Howell, J. A. S., and Moss, K. C. (1971a). *J. Chem. Soc. A* 270.

Howell, J. A. S., and Moss, K. C. (1971b). *J. Chem. Soc. A* 2481.

Inglefield, P. T., and Reeves, W. T. (1964). *J. Chem. Phys.* **40**, 2425.

Ismail, I. M., Kerrison, S. J. S., and Sadler, P. J. (1980). *J. Chem. Soc. Chem. Commun.* 1175.

Jeannin, Y., and Martin-Frère, J. (1981). *J. Am. Chem. Soc.* **103**, 1664.

Jenny, T., Von Philipsborn, W., Kronenbitter, J., and Schwenk, A. (1981). *J. Organomet. Chem.* **205**, 211.

Jesson, J. P. (1971). "Transition Metal Hydrides" (E. L. Muetterties, ed.), p. 75. Dekker, New York.

Karr, C., and Schultz, H. D. (1968). *Spectrosc. Lett.* **1**, 205.

Kaufmann, J. (1964). *Z. Phys.* **182**, 217.

Kaufmann, J., Kronenbitter, J., and Schwenk, A. (1975). *Z. Phys.* **274A**, 87.

Kautt, W. D., Kruger, H., Lutz, O., Maier, H., and Nolle, A. (1976). *Z. Naturforsch.* **31A**, 351.

Kawasaki, Y., and Aritoni, M. (1976). *J. Organomet. Chem.* **104**, 39.

Kazanskii, L. P., and Spitsyn, V. I. (1975). *Dokl. Akad. Nauk. SSSR* **223**, 381 (English translation), *Proc. Acad. Sci. USSR Phys. Chem. Sec.* **223**, 721.

Kazanskii, L. P., Fedotov, M. A., Ptushkina, M. M., and Spitsyn, V. I. (1975). *Proc. Acad. Sci. USSR Phys. Chem. Sec.* **224**, 1029.

Keiter, R. L., and Verkade, J. G. (1969). *Inorg. Chem.* **8**, 2115.

Kennedy, J. D., and McFarlane, W. (1973). *Rev. Silicon, Germanium, Tin Lead Compd.* **1**, 235.

Kennedy, J. D., and McFarlane, W. (1974). *J. Organomet. Chem.* **80**, C47.

Kennedy, J. D., McFarlane, W., Pyne, G. C., and Wrackmeyer, B. (1975). *J. Chem. Soc. Dalton. Trans.* 386.

Kennedy, J. D., McFarlane, W., and Wrackmeyer, B. (1976). *Inorg. Chem.* **15**, 1299.

Kennedy, J. D., McFarlane, W., and Pyne, G. S. (1977). *J. Chem. Soc. Dalton Trans.* 2332.

Kennedy, J. D., McFarlane, W., Pyne, G. S., and Wrackmeyer, B. (1980). *J. Organomet. Chem.* **195**, 285.

Kerrison, S. J. S., and Sadler, P. J. (1978). *J. Magn. Reson.* **31**, 321.

Kidd, R. G. (1980). "Annual Report on NMR Spectroscopy," Vol. 10A, p. 2.

Kidd, R. G., and Spinney, H. G. (1973). *Inorg. Chem.* **12**, 1967.

Kidd, R. G., and Spinney, H. G. (1981). *J. Am. Chem. Soc.* **103**, 4759.

Kim, K. S., and Bray, P. J. (1974). *J. Magn. Reson.* **16**, 334.

Kitching, W., Das, V. G. K., and Wells, P. R. (1967). *J. Chem. Soc. Chem. Commun.* 356.

Kitching, W., Marriott, M., Adcock, W., and Doddrell, D. (1976). *J. Org. Chem.* **41**, 1671.

Köhler, F. H., Kalder, H. J., and Fischer, E. O. (1976). *J. Organomet. Chem.* **113**, 11.

Koridze, A. A., Petrovskii, P. V., Gubin, S. P., and Fedin, E. I. (1975a). *J. Organomet. Chem.* **93**, C26.

Koridze, A. A. *et al.* (1975b). *Izv. Akad. Nauk. SSSR* **239,** 1675.

Koridze, A. A., Astachova, H. M., Petrovskii, P. W., and Luzenko, A. (1978). *Izv. Akad. Nauk. SSSR* **242,** 117.

Kotz, J., Schaeffer, R., and Clouse, A. (1967). *Inorg. Chem.* **6,** 620.

Kovar, R. A., and Morgan, G. L. (1970). *J. Am. Chem. Soc.* **92,** 5067.

Krebs, P., and Dreeskamp, H. (1969). *Spectrochim. Acta* **25A,** 1399.

Kroneck, P., Lutz, O., Nolle, A., and Oehler, H. (1980a). *Z. Naturforsch.* **35A,** 221.

Kroneck, P., Lutz, O., and Nolle, A. (1980b). *Z. Naturforsch* **35A,** 226.

Kroneck, P., Kodweiss, J., Lutz, O., Nolle, A., and Zepf, D. (1982). *Z. Naturforsch.* **37A,** 186.

Kronenbitter, J., and Schwenk, A. (1977). *J. Magn. Reson.* **25,** 147.

Krueger, K., Lutz, O., Nolle, A., and Schwenk, A. (1973). *Z. Naturforsch.* **28A,** 119.

Kusaba, O., Hinoishi, T., and Kawasaki, Y. (1982). *J. Organomet. Chem.* **228,** 223.

Lallemand, J. Y., Soulie, J., and Chottard, J. C. (1980). *J. Chem. Soc. Chem. Commun.* 436.

Lassigne, C. R., and Wells, E. J. (1977). *Can. J. Chem.* **55,** 927; *J. Magn. Reson.* **26,** 55.

Lassigne, C. R., and Wells, E. J. (1978). *J. Magn. Reson.* **31,** 195.

Lauterbur, P. C., and Burke, J. J. (1965). *J. Chem. Phys.* **42,** 439.

Lauterbur, P. C., and King, R. B. (1966). *J. Am. Chem. Soc.* **87,** 3266.

Lee, K. (1968). *Phys. Rev.* **172,** 284.

Lefebvre, J., Chauveau, F., Doppelt, P., and Brevard, C. (1981). *J. Am. Chem. Soc.* **103,** 4589.

Le Gall, J. Y., Kubicki, M. M., and Petillon, F. Y. (1981). *J. Organomet. Chem.* **221,** 287.

Lehnig, M. (1981). *Chem. Phys.* **54,** 323.

Lehnig, M., Newmann, W. P., and Seifert, P. (1978). *J. Organ. Chem.* **162,** 145.

Lindman, B., Forsén, S., and Lilja, H. (1977). *Chem. Scripta* **11,** 91.

Lucchini, V., and Wells, R. P. (1980). *J. Organ. Chem.* **199,** 217.

Lutz, O., and Nolle, A. (1980). *Z. Phys.* **36B,** 323.

Lutz, O., and Stricker, G. (1971). *Phys. Letters* **35A,** 397.

Lutz, O., Nolle, A., and Schwenk, A. (1973). *Z. Naturforsh.* **28A,** 1370.

Lutz, O., Nolle, A., and Kroneck, P. (1976a). *Z. Naturforsh.* **31A,** 454.

Lutz, O., Nepple, W., and Nolle, A. (1976b). *Z. Naturforsch.* **31A,** 978.

Lutz, O., Nepple, W., and Nolle, A. (1976c). *Z. Naturforsch.* **31A,** 1046.

Lutz, O., Nolle, A., and Kroneck, P. (1977a). *Z. Phys.* **282A,** 157.

Lutz, O., Nolle, A., and Kroneck, P. (1977b). *Z. Naturforsh.* **32A,** 505.

Lutz, O., Oehler, H., and Kroneck, P. (1978a). *Z. Naturforsch.* **33A,** 1021.

Lutz, O., Oehler, H., and Kroneck, P. (1978b). *Z. Phys.* **288A,** 17.

Lutz, O., Messner, W., Mohn, K. R., and Kroneck, P. (1981). *Z. Phys.* **300A,** 111.

Lyčka, A., Šnobl, D., Handlick, K., Holeček, J., and Nádvornick, M. (1981). *Collect. Czech. Chem. Commun.* **46,** 1383.

Lynch, G. F., and Segel, S. L. (1972). *Can. J. Phys.* **50,** 567.

McConnell, H. M., and Berger, S. B. (1957). *J. Chem. Phys.* **27,** 230.

McConnell, H. M., and Weaver, H. E. (1956). *J. Chem. Phys.* **25,** 307.

McFarlane, H. C. E., McFarlane, W., and Rycroft, D. S. (1976). *J. Chem. Soc. Dalton Trans.* 1616.

McFarlane, H. C. E., McFarlane, W., and Turner, C. J. (1979). *Mol. Phys.* **37,** 1639.

McFarlane, W. (1967). *Mol. Phys.* **13,** 587.

McFarlane, W. (1976). *J. Organ. Chem.* **116,** 315.

McFarlane, W., and Rycroft, D. S. (1973). *J. Chem. Soc. Chem. Commun.* 336.

McFarlane, W., and Rycroft, D. S. (1976). *J. Magn. Reson.* **24,** 95.

McFarlane, W., Noble, A. M., and Winfield, J. M. (1970). *Chem. Phys. Lett.* **6,** 547.

McFarlane, W., Noble, A. M., and Winfield, J. M. (1971). *J. Chem. Soc. A,* 948.

Maciel, G. E., and Dallas, J. L. (1973). *J. Am. Chem. Soc.* **95,** 3039.

Majima, T., and Kawasaki, Y. (1979). *Bull. Chem. Soc. Japan* **52,** 73.

Malish, W., Maisch, R., Colquhoun, I. J., and McFarlane, W. (1981). *J. Organomet. Chem.* **220,** C1.

Mann, B. E. (1971). *J. Chem. Soc. Chem. Commun.* 1173.

Mann, B. E. (1973). *J. Chem Soc. Dalton. Trans.* 2012.

Mann, B. E. (1974). *Adv. Organ. Chem.* **12,** 135.

Marker, A., and Gunter, M. J. (1982). *J. Magn. Reson.* **47,** 118.

Masters, A. F., Brownlee, R. T. C., O'Connor, M. J., Wedd, A. G., and Cotton, J. D. (1980). *J. Organomet. Chem.* **195,** C17.

Masters, A. F., Brownlee, R. T. C., O'Connor, M. J., and Wedd, A. G. (1981). *Inorg. Chem.* **20,** 4183.

Mathiasch, B., and Mitchell, T. N. (1980). *J. Organomet. Chem.* **185,** 351.

Meadows, M. D. *et al.* (1982). *Proc. Natl. Acad. Sci. USA* **79,** 1351.

Milbrath, D. S., Verkade, J. G., and Clark, R. J. (1976). *Inorg. Nucl. Chem. Lett.* **12,** 921.

Minelli, M., Rockway, T. W., Enemark, J. H., Brunner, H., and Muschiol, M. (1981). *J. Organomet. Chem.* **217,** C34.

Mitchell, T. N. (1976). *Org. Magn. Reson.* **8,** 34.

Mitchell, T. N., Gmehling, J., and Huber, F. (1978). *J. Chem. Soc. Dalton Trans.* 960.

Mitchell, T. N., and Marsmann, H. C. (1981). *Org. Magn. Reson.* **15,** 263.

Morgan, G. L., and McVicker, G. B. (1968). *J. Am. Chem. Soc.* **90,** 2789.

Morishima, I., Inubishi, T., and Sato, M. (1978). *J. Chem. Soc. Chem. Commun.* 106.

Moriyama, H., Aoki, T., Shinoda, S., and Saito, Y. (1981). *J. Chem. Soc. Dalton Trans.* 639.

Muetterties, E. L., and Phillips, W. D. (1959). *J. Am. Chem. Soc.* **81,** 1084.

Mühlbach, G., Rausch, B., and Rehder, D. (1981). *J. Organomet. Chem.* **205,** 343.

Müller, I., and Rehder, D. (1977). *J. Organomet. Chem.* **139,** 293.

Nakano, T. (1977). *Bull. Chem. Soc. Jpn* **50,** 661.

Narasimhan, P. T., and Rogers, M. T. (1961). *J. Chem. Phys.* **34,** 1049.

Narita, K., Umeda, J., and Kusumoto, H. (1966). *J. Chem. Phys.* **44,** 2719.

Näumann, F., and Rehder, D. (1981). *J. Organomet. Chem.* **204,** 411.

Nielsen, P. S., Hansen, R. S., and Jakobsen, H. J. (1976). *J. Organ. Chem.* **114,** 145.

Nixon, J. F., and Pidcock, A. (1969). "Annual Review of NMR Spectroscopy" (E. Mooney, ed.). Academic Press, New York.

Nolle, A. (1977). *Z. Naturforsh.* **32A,** 964.

Ochsenbein, U., and Schläpfer, C. W. (1980). *Helv. Chim. Acta* **63,** 1926.

O'Donnell, S. E., and Pope, M. T. (1976). *J. Chem. Soc. Dalton Trans.* 2290.

Ostoja-Starzewski, K. H. A., and Pregosin, P. S. (1980). *Angew. Chem. Int. Engl. Ed.* **19,** 316.

Otera, J., Kusaba, A., Hinoishi, T., and Kawasaki, Y. (1982). *J. Organomet. Chem.* **228,** 223.

Packer, K. J., and Muetterties, E. L. (1963). *J. Am. Chem. Soc.* **85,** 3035.

Paulsen, K., and Rehder, D. (1982). *Z. Naturforsh.* **37A,** 139.

Paulsen, K., Rehder, D., and Thoennes, D. (1978). *Z. Naturforsch.* **33A,** 834.

Pereyre, M., Quintard, J. P., and Rahm, A. (1982). *Pure Appl. Chim.* **54,** 29.

Pesek, J. J., and Mason, W. R. (1977). *J. Magn. Reson.* **25,** 519.

Petrosyan, V. S. (1977). "Progress in NMR Sepctroscopy" (J. M. Emsley, J. Feeney, and L. H. Sutcliffe, eds.), Vol. 11, p. 115. Pergamon, Oxford.

Pregosin, P. S. (1981). "Annual Reports on NMR Spectroscopy" (G. A. Webb, ed.), Vol. 11A, p. 227. Academic Press, New York.

Pregosin, P. S. (1982). *Coord. Chem. Rev.* **44,** 247.

Pregosin, P. S., and Kunz, R. W. (1979). "NMR Basic Principles and Progress" (P. Diehl, E. Fluck, and R. Kosfeld, eds.), Vol. 16. Springer Verlag, Berlin and New York.

Proctor, W. G., and Yu, F. C. (1950). *Phys. Rev.* **78,** 471.

Proctor, W. G., and Yu, F. C. (1951). *Phys. Rev.* **81,** 20.

Puttfarcken, U., and Rehder, D. (1980). *J. Organ. Chem.* **185,** 219.

Rehder, D. (1972). *J. Organomet. Chem.* **37,** 303.

Rehder, D. (1976). *Z. Naturforsh.* **31B,** 273.

Rehder, D. (1977a). *J. Organomet. Chem.* **137,** C25.

Rehder, D. (1977b). *J. Magn. Reson.* **25,** 177.

Rehder, D. (1977c). *Z. Naturforsh.* **32B,** 771.

Rehder, D. (1980). *J. Magn. Reson.* **38,** 419.

Rehder, D. (1982). *Bull. Magn. Reson.* **4,** 33.

Rehder, D., and Dorn, W. L. (1976). *Trans. Metal. Chem.* **1,** 233.

Rehder, D., and Schmidt, J. (1974). *J. Inorg. Nucl. Chem.* **36,** 333.

Rehder, D., and Schmidt, J. (1977). *Trans. Metal. Chem.* **2,** 41.

Rehder, D., and Wieghardt, K. (1981). *Z. Naturforsh.* **36B,** 1251.

Rehder, D., Dahlenburg, L., and Muller, I. (1976a). *J. Organomet. Chem.* **122,** 53.

Rehder, D., Dorn, W. L., and Schmidt, J. (1976b). *Trans. Metal. Chem.* **1,** 74.

Rehder, D., Paulsen, K., and Lechert, H. (1978). *Z. Naturforsh.* **33A,** 1597.

Rehder, D., Bechthold, H. C., and Paulsen, K. (1980). *J. Magn. Reson.* **40,** 305.

Rehder, D., Bechthold, H. C., Keçeci, A., Schmidt, H., and Siewing, M. (1982). *Z. Naturforsh.* **37B,** 631.

Retcofsky, H. L., and Friedel, R. A. (1970). *Appl. Spectrosc.* **24,** 379.

Retcofsky, H. L., and Friedel, R. A. (1972). *J. Am. Chem. Soc.* **94,** 6579.

Rudolph, R. W., Wilson, W. L., Parker, F., Taylor, R. C., and Young, D. C. (1978). *J. Am. Chem. Soc.* **100,** 4629.

Sahm, W., and Schwenk, A. (1974). *Z. Naturforsh.* **29A,** 1763.

Schmidt, H., and Rehder, D. (1980). *Trans. Metal. Chem.* **5,** 214.

Schneider, W. G., and Buckingham, A. D. (1962). *Disscus. Faraday Soc.* **34,** 147.

Schumann, C., and Dreeskamp, H. (1970). *J. Magn. Reson.* **3,** 204.

Schutz, H. D., Karr, C., and Vickers, G. D. (1971). *Appl. Spectrosc.* **25,** 363.

Schwenk, A. (1968). *Z. Physik* **213,** 482.

Schwenk, A. (1970). *Phys. Letters* **31A,** 513.

Schwenk, A. (1971). *J. Magn. Reson.* **5,** 376.

Shier, G. D., and Drago, R. S. (1966). *J. Organomet. Chem.* **6,** 359.

Shortland, A., and Wilkinson, G. (1972). *J. Chem. Soc. Chem. Comm.* 318.

Simonnin, M. P., Lequan, M., and Lecourt, M. J. (1970). *Org. Magn. Reson.* **2,** 369.

Singh, G. (1968). *J. Organ. Chem.* **11,** 133.

Singh, G. (1975). *J. Organ. Chem.* **99,** 251.

Smith, P. J., and Smith, L. (1973). *Inorg. Chim. Acta Rev.* **7,** 11.

Smith, P. J., and Tupcianskas, A. P. (1978). "Annual Report on NMR Spectroscopy" (G. A. Webb, ed.), Vol. 8, p. 292. Academic Press, New York.

Talay, R., and Rehder, D. (1978). *Chem. Ber* **111,** 1978.

Talay, R., and Rehder, D. (1981). *Z. Naturforsh.* **36B,** 451.

Tarasov, V. P., Privalov, V. I., and Buslaev, Y. A. (1978a). *Mol. Phys.* **35,** 1047.

Tarasov, V. P., Privalov, V. I., and Buslaev, Y. A. (1978b). *Magn. Reson. Relat. Phenom. Proc. Congr. Ampere 20th,* 508.

Tarasov, V. P., Sinitsyna, S. M., Kopanev, V. D., Khlebodarov, V. G., and Buslaev, Y. A. (1980). *Koord. Khim.* **6,** 1568.

Tarasov, V. P., Privalov, V. I., Gorbik, A. A., and Buslaev, Y. A. (1981). *Dokl. Akad. Nauk. SSSR* **257,** 678.

Umeda, J., Kusumoto, H., Narita, K., and Yamada, E. (1965). *J. Chem. Phys.* **42,** 1458.

Vaickus, M. J., and Anderson, D. G. (1980). *Org. Magn. Reson.* **14,** 278.

Van Beelen, D. C., and Wolters, J. (1980). *J. Organomet. Chem.* **195,** 185.

Van Beelen, D. C., De Vos, D., Bots, G. J. M., Van Doorn, L. J., and Wolters, J. (1976). *Inorg. Nucl. Chem. Lett.* **12,** 581.

Van Beelen, D. C., Van der Kool, H. O., and Wolters (1979). *J. Organomet. Chem.* **179,** 37.

Van Beelen, D. C., Van Kampen, A. E. J., and Wolters, J. (1980). *J. Organomet. Chem.* **187,** 43.

Verkade, J. G. (1972). *Coord. Chem. Rev.* **9,** 1.

Vold, R. R., and Vold, R. L. (1974). *J. Chem. Phys.* **61,** 4360.

Vold, R. R., and Vold, R. L. (1975). *J. Magn. Reson.* **19,** 365.

Vold, R. R., Sparks, S. W., and Vold, R. L. (1978). *J. Magn. Reson.* **30,** 497.

Wehrli, F. W. (1976). *J. Magn. Reson.* **23,** 181.

Wehrli, F. W. (1978). *J. Magn. Reson.* **30,** 193.

Wehrli, F. W., and Wehrli, S. L. (1982). *J. Magn. Reson.* **47,** 151.

Whitesides, G. M., and Mitchell, H. L. (1969). *J. Am. Chem. Soc.* **91,** 2245.

Wrackmeyer, B. (1979). *J. Organomet. Chem.* **166,** 353.

Wrackmeyer, B. (1981). *J. Magn. Reson.* **42,** 287.

Yamamoto, Y., Haraguchi, H., and Fujiwara, S. (1970). *J. Phys. Chem.* **74,** 4369.

Zharkova, G. I., Igumenov, I. K., Tkachev, S. V., and Zemskov, S. V. (1982). *Koord. Khim.* **8,** 74.

Index for Volumes 1 and 2

Boldface numerals refer to volume numbers.